Coronal Mass Ejections

Astrophysics and Space Science Library

VOLUME 376

For other titles published in this series, go to
http://www.springer.com/series/5664

Timothy Howard

Coronal Mass Ejections

An Introduction

Springer

Dr. Timothy Howard
Space Studies
Southwest Research Institute
1050 Walnut Street, Suite 300
Boulder, CO 80302, USA
howard@boulder.swri.edu

Composite image of a coronal mass ejection observed on 28 October 2003 (the so-called Halloween event) combining a solar disk image with a coronagraph image. Image available courtesy of NASA.

ISSN 0067-0057
ISBN 978-1-4614-2879-4 ISBN 978-1-4419-8789-1 (eBook)
DOI 10.1007/978-1-4419-8789-1
Springer New York Dordrecht Heidelberg London

Printed on acid-free paper

Springer is part of Springer Science+Business Media (www.springer.com)

To my strength and inspiration.
My father, Gary Howard.

Preface

Coronal mass ejections (CMEs) are large eruptions of plasma, magnetic field and energy from the Sun. They are the largest individual solar eruptions, releasing more than an order of magnitude of energy than the more popularly known solar flare. Much has been learnt about CMEs over the years regarding their composition, structure, onset and evolution. There is, however, still much we do not know. This book presents a brief review of CMEs, from the history of their observation to methods by which we detect them, to the status quo regarding related phenomena and models describing their onset and evolution. This is not a book written for an expert but rather it is aimed at those who are just starting out in the field of CMEs and space weather or would simply like to know little more about them: for example, graduate students or those from other fields who would like some background on the topic.

The motivation for writing this book arises from my desires when I was starting out in this field. The main problem was that, while I had ideas for areas in which I would like to do research, I did not know what work in these areas had already been accomplished, or what questions had yet to be answered. All of the texts written about coronal mass ejections have been written for experts by experts and so there was not much a beginner to the field could understand without spending a great deal of time trawling through the literature. I wished that there was a single source from which I could gain a brief overview of the status quo regarding coronal mass ejections and a number of references with expert details that I could investigate further. This is the purpose I hope this book will serve.

The book does contain a large amount of mathematics. This is difficult to avoid in a book involving physics as it is the language with which we communicate. The reader may disregard the mathematics involved if they so wish, but it does contribute greatly to our understanding of these topics and, in some places, the concepts cannot really be explained adequately in any other way.

It is important to highlight the structure of this book as it relates to the expertise of the author. There are many related areas on the topics of CMEs and space weather that I am not an expert on. While a world expert may not gain a great deal of insight from the following text (although it is hoped that it will be useful resource for those areas in which they do not specialise), this book is not intended for an expert audience. This is an introductory book primarily aimed at those who are just entering the field and would like to gain a brief oversight of coronal mass ejections – what they are, what their importance is, and the status quo regarding data and theory developments. The book is also full of references from people far more qualified than I that discuss the many topics covered in more detail. I gladly yield to these experts and the references to elaborate on these topics.

The reader may also notice that some topics are covered in more detail than others. This is a reflection of the knowledge and expertise of the author. For example, Chaps. 4, 5 and 6 cover the theory of CME detection and data analysis techniques with great detail while Chaps. 8 and 9 only briefly review the models describing their onset and evolution. You may also notice that my history chapter (Chap. 2) almost exclusively discusses the history of CME observation and not modelling. This is not to imply that there is not a rich history of theoretical and modelling developments in describing CMEs, but rather because my expertise lie in the areas of observation and data analysis. I am not a mathematical modeler and so can only provide a brief review of these topics. The reader is encouraged, as with all of these topics, to learn more from the many references I have provided and to read further from other sources cited in those texts.

Finally, the reader will notice throughout the book certain opinions that are offered by the author. Chapter 11, in particular, has many of these. As with much of science, opinions are divided on many topics and some, including some experts, probably disagree with what has been discussed in places throughout the book. The reader themselves may come to different conclusions than those expressed by the author (or by the experts). If anything stated here can be proven incorrect then I will gratefully alter my views. It is important to realise that science is an evolutionary process, and what is "known" today may be "known to be incorrect" tomorrow. Our role as scientists is to express our viewpoints in as objective a manner as possible, and where necessary to yield those viewpoints when they are proven incorrect.

If there is a single message the reader should take away from this book it is this: The most severe space weather effects, (geo)magnetic storms, are not caused by solar flares. They are caused by coronal mass ejections. Therefore if one is interested in studying the causes of severe space weather it is the CME, and not the flare, that we should be investing our resources into. As this book shows, there is much that is not yet understood about this important and fascinating phenomenom.

Boulder, Colorado *Tim Howard*
May 2011

Acknowledgments

Firstly, I need to emphasise that this text often refers to the work Russell (Russ) Howard of the Naval Research Laboratory (NRL). Russ is a pioneer of CME research and has played a major role in many hardware missions, including *Solwind*, *SOHO* and *STEREO*. He is the Principal Investigator of the LASCO and SECCHI instrument suites, and continues to perform crucial work on the future of CME and ICME research. While he and I have worked together on occasion and share a common sir name, we are not in any way related to one another. The reference list in each chapter refers to a number of works with a Howard as a lead or co-author. I ask the readers not to confuse the works of T.A. Howard (myself) and R.A. Howard (Russ).

I would like to especially thank my mentors who have helped to develop my knowledge and skills over the years. Without their support I would have given up this field many years ago. Thanks to my PhD supervisor, Fred Menk and my first postdoctoral advisor George Simnett, who opened the door for me into the scientific world. Thanks also to my current boss, Craig DeForest, who has continued to enhance my integration into the scientific community.

My gratitude to my Air Force (AFRL) colleagues who have provided me with so much support over the years and continue to do so. Thanks especially to Janet Johnston, Dave Webb and Rich Radick.

Special thanks are extended to my long-time colleague and friend, James Tappin. His knowledge on space weather, CMEs and data analysis seem to be endless and much of the information provided in this book has originated from my many discussions with him over the years.

My gratitude is especially extended to my many colleagues who have kindly reviewed some of the chapters of this text. Especial thanks to Jack Gosling for reviewing Chaps. 1, 2, 7, 8, 9 and 10, and for providing invaluable insight and guidance. His assistance has greatly improved the quality of these chapters and corrected some things I had said that were downright wrong. Thanks also to James Tappin for reviewing Chap. 6 and for providing much of the substance of Chap. 4, as he co-wrote the *Space Science Reviews* series of papers with me from which much of this material arose. Last, but not least, my gratitude is extended to Craig DeForest for reviewing Chap. 8 and Dave Webb for Chaps. 7 and 9.

To my colleagues in the space physics community I send my regards and thanks for their support and for welcoming me into the community. To name a few, thanks to my colleagues and friends at NOAA/SWPC, SwRI, NSO and CU.

I would like to acknowledge and thank my family and friends and extend my supreme gratitude to my father and inspiration, Mr. Gary Howard. His understanding, compassion and support are endless and I do not have the words to express how truly grateful I am.

Contents

Acronyms

3-D	Three dimensions, or three-dimensional.
ACE	*Advanced Composition Explorer*: Spacecraft launched in 1997 into L1 orbit, designed to monitor the interplanetary medium with a suite of in-situ instruments.
ACRIM	Active Cavity Radiometer Irradiance Monitor: Radiometer on board *SMM*.
AIA	Atmospheric Imaging Assembly: Visible light, UV and EUV imaging suite on board *SDO*.
A_p	An index describing geomagnetic activity. It is a daily index derived from the eight three-hourly a indices. A_p may be regarded as a daily average of the three-hourly K_p index.
APL	Applied Physics Laboratory.
ASCII	American Standard Code for Information Interchange: The standard code for storing text.
ATM	Apollo Telescope Mount: The mount on *Skylab* on which scientific instruments were situated, including the solar observing instruments.
AU	Astronomical Unit: The average distance of the Earth from the Sun ($\sim 1.5 \times 10^8$ km).
B_{IMF}	The interplanetary magnetic field vector: Can also represent the magnitude of the field.
BCS[1]	Bent Crystal Spectrometer: X-ray spectrometer on board *SMM*.
BCS[2]	Bragg Crystal Spectrometer: Gamma-ray spectrometer on board *Yohkoh*.
Caltech	California Institute of Technology.
CASS	Center for Astrophysics and Space Sciences (University of California, San Diego).
CCD	Charge-Coupled Device: May be regarded as a digital version of photographic film.
CCMC	Community Coordinated Modeling Center: NASA-based site for space weather related models (ccmc.gsfc.nasa.gov).
CDS	Coronal Diagnostic Spectrometer: EUV spectrometric imager on board *SOHO*.

CELIAS	Charge, Element, and Isotope Analysis System: Particle detector on board *SOHO*.
CGS	Centimetre, Gramme, Second: Unit system favoured by astronomers.
CIR	Corotating Interaction Region: Interaction in the heliosphere between fast and slow solar wind streams.
CME	Coronal Mass Ejection: An eruption of plasma and magnetic field from the Sun.
COR	Coronagraphs on board *STEREO*: There are two, COR1 (inner) and COR2 (outer).
COSPIN	Cosmic and Solar Particle Investigation: Cosmic ray detector on board *Ulysses*.
COSTEP	Comprehensive SupraThermal and Energetic Particle analyser: Particle detector on board *SOHO*.
C/P	Coronagraph/Polarimeter: White light coronagraph/polarimeter on board *SMM*.
CRIS	Cosmic Ray Isotope Spectrometer: Energetic particle instrument on board *ACE*.
CRS	Cosmic Ray Subsystem: Cosmic ray instrument on board *Voyager*.
Dst	An index describing geomagnetic activity. Dst is derived from measurements of the equatorial region of the geomagnetic field, and thus monitors the Earth's ring current. When the ring current is enhanced, the equatorial geomagnetic field is reduced, indicated by a reduction in the Dst index. A sudden large reduction in the Dst index is an indicator of a geomagnetic storm.
DUST	Dust analyser on board *Ulysses*.
E1	Plasma experiment on board *Helios*.
E2	Fluxgate magnetometer on board *Helios*.
E3	Fluxgate magnetometer on board *Helios*.
E4	Search coil magnetometer on board *Helios*.
E5	Plasma wave experiment on board *Helios*.
E6	X-ray experiment on board *Helios*.
E7	Cosmic ray experiment on board *Helios*.
E8	Electron and ion spectrometer on board *Helios*.
E9	Zodiacal light experiment on board *Helios*.
E10	Micrometeoroid analyser on board *Helios*.
EIS	Extreme-ultra-violet Imaging Spectrometer: On board *Hinode*.
EIT	Extreme-ultra-violet Imaging Telescope: EUV imager on board *SOHO*.
EMS	Erupting Magnetic Structures: A term describing ICMEs observed at large distances from the Sun without a coronagraph CME counterpart.
EPAC	Energetic Particles Composition: Energetic ion detector on board *Ulysses*.

EPACT	Energetic Particle Acceleration, Composition, and Transport: Energetic particle detector on board *WIND*.
EPAM	Electron, Proton, and Alpha Monitor: Energetic particle instrument on board *ACE*.
EPD	Energetic Particle Detector: Dust instrument to be part of Solar Orbiter.
EPI	Energetic Particle Investigation: Currently proposed for the Solar Sentinels.
ERNE	Energetic and Relativistic Nuclei and Electron experiment: Particle detector on board *SOHO*.
ESA	European Space Agency: Main institution of European space research and exploration.
EUI	Extreme Ultra-violet Imager: To be part of Solar Orbiter.
EUV	Extreme Ultra-Violet: The highest frequency band of the ultra-violet spectrum (prior to the x-ray band).
EUVI	Extreme-Ultra-Violet Imager: On board *STEREO*.
EVE	Extreme-ultra-violet Variability Experiment: EUV irradiance instrument on board *SDO*.
FCS	Flat Crystal Spectrometer: X-ray spectrometer on board *SMM*.
FGM	FluxGate Magnetometer: On board *Ulysses*.
FIELDS	A magnetic and electric fields instrument to be part of Solar Probe Plus.
FITS	Flexible Image Transport System: A digital file format, typically used in the storage of scientific data. Fits files typically contain an ASCII header followed by the data, often in the form of a digital image.
FOV	Field of view: The region of sky observed by an imager.
GLE	Ground Level Enhancement: An increase of cosmic ray intensity at the Earth, as a result of the arrival of solar energetic particles.
GOLF	Global Oscillations at Low Frequencies: Helioseismology instrument on board *SOHO*.
GRB	Gamma-Ray Burst: X-ray and gamma-ray detector on board *Ulysses*.
GRS	Gamma-Ray Spectrometer : On board *SMM* and currently proposed for the Solar Sentinels.
GSFC	Goddard Space Flight Center: The main administration wing of NASA.
Hα	The hydrogen emission line in the visible light spectrum denoting the α transition: A common wavelength at which visible light telescopes observe the Sun.
HAE	Helium Abundance Enhancement: A sudden increase in the abundance of helium in the interplanetary medium, typically following an interplanetary shock.
HAF	Hakamada-Akasofu-Fry model (Also known as HAFv2): Model describing the propagation of an interplanetary shock.

HAO	High Altitude Observatory: Based in Boulder, CO.
HEOS-2	*Highly Eccentric Orbit Satellite 2*: A European spacecraft launched in 1972 designed to study the high altitude magnetosphere and near-Earth interplanetary medium.
HESSI	The original name of the *RHESSI* spacecraft.
HI	Heliospheric Imager: White light heliospheric imagers on board *STEREO*. There are two, HI-1 (inner) and HI-2 (outer).
HICA	High energy Ion Composition Analyzer: Energetic particle instrument currently proposed for the Solar Sentinels.
HISCALE	Heliosphere Instrument for Spectra, Composition and Anisotropy at Low Energies: Energetic ion detector on board *Ulysses*.
HMI	Helioseismic and Magnetic Imager: Helioseismology instrument on board *SDO*.
HXIS	Hard X-ray Imaging Spectrometer: On board *SMM*.
HXRBS	Hard X-Ray Burst Spectrometer: On board *SMM*.
HXT	Hard X-ray Telescope: Hard x-ray imager on board *Yohkoh*.
ICE	*International Cometary Explorer*: Originally the *ISEE-3* spacecraft but left its L1 orbit in 1982 to investigate Comet P/Giacobini-Zinner.
ICME	Interplanetary Coronal Mass Ejection: The heliospheric counterpart of a CME.
IMF	Interplanetary Magnetic Field: The magnetic field within the interplanetary medium.
IMP-8	*Interplanetary Monitoring Platform 8*: A spacecraft launched into High Earth Orbit in 1973, designed to monitor the interplanetary medium.
IMPACT	In-situ Measurements of Particles And CME Transients: Solar wind particle detector suite on board *STEREO*.
IPS	InterPlanetary Scintillation: A short-duration distortion of the signal from a distant radio source as a result of a dense structure passing between it and the observer. This is a technique that has been used for ICME detection and tracking.
IR	Infra-Red: The region of the electromagnetic spectrum in the frequency band immediately smaller than visible light.
IRIS	Infra-Red Interferometric Spectrometer: IR spectrometer on board *Voyager*.
ISEE-3	*International Sun–Earth Explorer 3*: A spacecraft launched in 1978 into the L1 Lagrange point, where it monitored the interplanetary medium before later becoming ICE and heading off to explore a comet.
ISIS	Integrated Science Assembly: Two particle instruments to be part of Solar Probe Plus.
ISPM	Interplanetary Shock Propagation Model: Model describing the propagation of an interplanetary shock.

ISS	Imaging Science Subsystem: Wide and narrow angle imagers on board *Voyager*.
JAXA	Japanese Aerospace eXploration Agency: Main scientific and space hardware institution in Japan.
JPL	Jet Propulsion Laboratory: NASA scientific and hardware group based at Caltech.
K_P	An index describing geomagnetic activity. K_P is determined every 3 h and ranges from 0 (quiet) to 9 (active) in increments of 1/3. They are derived using a network of ground-based magnetometers and measure the most disturbed horizontal component.
L1	The first Lagrange point: A location on the Sun–Earth line where the gravitational pull of the Sun is exactly canceled out by that of the Earth. This is around 1.5×10^6 km from the Earth, or 1% of the distance from the Earth to the Sun.
LASCO	Large Angle Spectroscopic Coronagraph: Coronagraphs on board *SOHO*. There were three, C1 (inner), C2 (middle) and C3 (outer) but C1 ceased to operate following the *SOHO* incident in 1998.
LECP	Low-Energy Charged Particles: Energetic particle instrument on board *Voyager*.
LET	Low Energy Telescope: Energetic particle instrument on board *STEREO*.
LICA	Low energy Ion Composition Analyzer: Energetic particle instrument currently proposed for the Solar Sentinels.
LOS	Line Of Sight: The vector from the observer through the point of interest and out to infinity.
LWS	Living With a Star: A NASA program focused on the Sun, Earth and space weather.
MAG	General acronym for a magnetometer on board a spacecraft. The same acronym is used for the magnetometers on board *Voyager*, *ACE* and *STEREO*, and proposed for the Solar Sentinels.
MDI	Michelson Doppler Imager: Helioseismology instrument on board *SOHO*.
METIS/COR	Coronagraph package to be part of Solar Orbiter.
MFI	Magnetic Field Investigation: Magnetometer on board *WIND*.
MHD	Magnetohydrodynamics.
MIR	Merged Interaction Region: Regions of compressed plasma and dense magnetic field at very large distances from the Sun, often caused as a result of a CME interacting with the surrounding medium.
MSFC	Marshall Space Flight Center: NASA scientific and hardware group based in Alabama.
NASA	National Aeronautic and Space Administration: The leading authority in US space research and travel.
NOAA	National Oceanic and Atmospheric Administration.
NRL	Naval Research Laboratory.

NS Neutron Spectrometer: Currently proposed for the Solar Sentinels.

NSSDC National Space Science Data Center: A collection of space science data managed by NASA originating in book form but now available online.

OSO-7 *Orbiting Solar Observatory* 7: A solar observing spacecraft launched in 1971.

P78-1 A spacecraft launched in 1979 which carried the *Solwind* coronagraph.

PA Position Angle: The projected angle from the northward vector in a solar or heliospheric image. It begins at zero degrees (north) and moves counterclockwise toward the east. So due north, east, south and west are at PA = 0°, 90°, 180° and 270° respectively.

PEA Post-Eruptive Arcade: A highly structured structure often observed in EUV in the low corona following the launch of a CME. EUVs take the form of a group of magnetic loops forming an arcade, and are believed to be aligned with the underlying CME structure.

PHI Visible imager and magnetograph to be part of Solar Orbiter.

PLASTIC PLAsma and SupraThermal Ion Composition: Solar wind particle experiment on board *STEREO*.

PLS PLasma Science: Solar wind instrument on board *Voyager*.

PR Public Relations. The effort behind the publication and public outreach of a mission or programme.

PWS Plasma Wave Subsystem: Plasma wave instrument on board *Voyager*.

$R_\odot$ Unit of distance: 1 $R_\odot$ = 1 Solar radius ~695,500 km.

R_E Unit of distance: 1 R_E = 1 Earth radius ~6,360 km.

R_H Unit of distance: 1 R_H = 1 Hermian (Mercury) radius ~2,440 km.

R_J Unit of distance: 1 R_J = 1 Jovian (Jupiter) radius ~71,500 km.

R_S Unit of distance: 1 R_S = 1 Saturn radius ~60,270 km.

R_U Unit of distance: 1 R_U = 1 Uranus radius ~25,560 km.

R_N Unit of distance: 1 R_N = 1 Neptune radius ~24,770 km.

RHESSI *Ramaty High Energy Solar Spectroscopic Imager*: Spacecraft launched in 2002 designed to monitor the Sun with an x-ray imager and spectrometer.

RMS Root Mean Square: A statistical measure of a varying parameter. It is obtained by averaging the square of each sample and then taking the square root.

RPW Radio and Plasma Waves: Radio and wave instrument to be part of Solar Orbiter.

SC Sudden (storm) Commencement: The onset of a geomagnetic storm following an abrupt increase in the strength of the horizontal component of the geomagnetic field (known as a sudden impulse).

SCM Search Coil Magnetometer: Currently proposed for the Solar Sentinels.

SDO	*Solar Dynamics Observatory*: Spacecraft launched in February 2010 designed to monitor the Sun with a suite of imagers.
SECCHI	Sun Earth Connection Coronal and Heliospheric Investigation: Imaging instrument suite on board *STEREO*.
SEP	Solar Energetic Particle: High-energy particles originating from the Sun and observed in the heliosphere.
SEPICA	Solar Energetic Particle Ionic Charge Analyser: Energetic particle instrument on board *ACE*.
SEPQ	Solar Energetic Particle Q-charge state and composition analyzer: Energetic particle instrument currently proposed for the Solar Sentinels.
SEPT	Solar Electron Proton Telescope: Solar wind particle detector on board *STEREO*.
SET	SupraThermal Electron telescope: Suprathermal electron analyser on board *STEREO*.
SETI	Search for Extra-Terrestrial Intelligence: Institute based in California dedicated to the search for extra-terrestrial life.
SI[1]	le Systéme International: The most commonly accepted base of units. SI units include the metre, kilogram, second and kelvin.
SI[2]	Sudden Impulse: Increase in the strength of the horizontal component of the geomagnetic field. Usually caused by the arrival of a high pressure packet of solar wind at the Earth such as a CME.
SID	Sudden Ionospheric Disturbance: A sudden increase in density of ionized particles in the ionosphere caused by electromagnetic radiation from a solar flare arriving at the Earth.
SIS	Solar Isotope Spectrometer: Energetic particle instrument on board *ACE*.
SIT	Suprathermal Ion Telescope: Suprathermal ion analyser on board *STEREO*.
SMEI	Solar Mass Ejection Imager: White light heliospheric imager on board *Coriolis*.
SMEX	SMall EXplorer: A NASA program involving the design, building and launch of small spacecraft (180–250 kg).
SMM	*Solar Maximum Mission*: A spacecraft launched in 1980 designed to measure the Sun with a suite of instruments and spectrometers.
SOHO	*SOlar and Heliospheric Observatory*: A spacecraft launched in 1995 into the L1 point with a suite of solar observing instruments on board.
SoloHI	Heliospheric imager to be part of Solar Orbiter.
SOT	Solar Optical Telescope: Optical imager on board *Hinode*.
SPICE	Spectral Imaging of the Coronal Environment: EUV spectrometer to be part of Solar Orbiter.
STE	SupraThermal Electrons: Energetic particle instrument currently proposed for the Solar Sentinels.

STEREO	*Solar TErrestrial RElations Observatory*: Pair of solar-observing spacecraft, launched in 2006, sharing an orbit about the Sun with the Earth.
STEREO-A	The *STEREO* spacecraft that is leading the Earth: All instruments on board this spacecraft also have the suffix "-A" (e.g. COR2-A, HI-1A).
STEREO-B	The *STEREO* spacecraft that is following the Earth: All instruments on board this spacecraft also have the suffix "-B" (e.g. COR2-B, HI-1B).
STIX	Spectrometer/Telescope for Imaging X-rays: X-ray spectrometer to be part of Solar Orbiter.
STOA	Shock Time Of Arrival: Model describing the propagation of an interplanetary shock.
SUMER	Solar Ultra-violet Measurements of Emitted Radiation: UV spectrometer on board *SOHO*.
SWA	Solar Wind Analyser: Solar wind instrument to be part of Solar Orbiter.
SWAN	Solar Wind Anisotropies: Neutral particle detector on board *SOHO*.
SWAVES	*STEREO*-Waves: Radio burst instrument on board *STEREO*.
SWComp	Solar Wind Composition: Solar wind particle instrument currently proposed for the Solar Sentinels.
SWE^1	Solar Wind Experiment: Solar wind particle detector on board *WIND*.
SWE^2	Solar Wind Electrons: Solar wind particle instrument currently proposed for the Solar Sentinels.
SWEA	Solar Wind Electron Analyser: Solar wind particle detector on board *STEREO*.
SWEAP	Solar wind Electrons, Alphas and Protons Investigation: Solar wind particle detector to be on Solar Probe Plus.
SWEPAM	Solar Wind Electron, Proton, and Alpha Monitor: Solar wind particle instrument on board *ACE*.
SWI	Solar Wind Ions: Solar wind particle instrument currently proposed for the Solar Sentinels.
SWICS^1	Solar Wind and Ion Composition Studies: Solar wind particle detector on board *Ulysses*.
SWICS^2	Solar Wind Ion Composition Spectrometer: Suprathermal ion detector on board *WIND* and solar wind ion instrument on board *ACE*.
SWIMS	Solar Wind Ions Mass Spectrometer: Solar wind ion instrument on board *ACE*.
SWOOPS	Solar Wind Observations Over the Poles of the Sun: Solar plasma detector on board *Ulysses*.
SWPC	Space Weather Prediction Center: NOAA-based prediction center for space weather.
SXT	Soft X-ray Telescope: On board *Yohkoh*.

TGRS	Transient Gamma-Ray Spectrometer: Gamma-ray spectrometer on board *WIND*.
TR*ACE*	Transition Region And Coronal Explorer: Spacecraft launched in 1998 designed to monitor the Sun with a UV/EUV imager.
ULEIS	Ultra Low Energy Isotope Spectrometer: Energetic particle instrument on board *ACE*.
URAP	Unified Radio And Plasma: Radio and plasma wave instrument on board *Ulysses*.
USAF	US Air Force.
UV	Ultra-Violet: The region of electromagnetic spectrum in the frequency band immediately larger than visible light.
UVCS	Ultra-Violet Coronagraph Spectrometer: UV spectrometer on board *SOHO*.
UVS	Ultra-Violet Spectrometer: UV spectrometer on board *Voyager*.
UVSP	Ultra-Violet Spectrometer and Polarimeter: UV spectroscopic imager on board *SMM*.
VHM	Vector Helium Magnetometer: Magnetometer on board *Ulysses*.
VIRGO	Variability of Solar Irradiance and Gravity Oscillations: Helioseismology instrument on board *SOHO*.
WAVES	Radio and plasma wave experiment on board *WIND*.
WBS	Wide Band Spectrometer: X-ray spectrometer on board *Yohkoh*.
WINDSAT	Microwave polarimetry oceanography instrument on board *Coriolis*.
WISPR	Wide-field Imager for Solar Probe Plus.
WSA	Wang-Sheeley-Arge model: Photospheric and coronal field and plasma model based on photospheric measurements.
XRI	X-Ray Imager: X-ray spectrometer currently proposed for the Solar Sentinels.
XRP	X-Ray Polychromator: Soft x-ray spectroscopic imager on board *SMM*.
XRT	X-ray telescope on board Hinode.

Chapter 1
Introduction

We live in an era of nanoelectronics and space technology. In electronic storage alone, data that occupied an entire building 30 years ago can now be stored on a device the size of a thumbnail. While the need for smaller and faster electronics provides technical obstacles, the natural environment imposes its own difficulties. The Earth's magnetic field responds to variations in the heliosphere, and as such is constantly changing in dynamics and configuration. The changes are relatively small, and so in the past technology has for the most part only been affected by large disturbances. We also spend more time in aircraft and continue to launch people into space. Smaller technology and radiation dosage to people are affected by smaller variations in the space environment, and so a greater understanding of how the Earth's magnetosphere and ionosphere is affected by the Sun is needed.

The last three decades have seen a significant amount of research into the phenomenon now known as space weather. It is almost impossible to see a funding proposal for solar or magnetospheric research today without space weather being addressed to some extent. The reason is that researchers have identified it as a large source of funding and so even those who are not interested in space weather may use it as a way to increase their chances of gaining funding. This is because space weather has implications that extend across many areas, from defence to telecommunications to commercial, and space flight to power station maintenance. The effects are increased at higher altitudes (and latitudes), and so spacecraft and high-flying aircraft are particularly at risk.

There are a number of space weather forecasting programs around the world, among them the Space Weather Prediction Center at NOAA in the US, and the Ionospheric Prediction Service in Australia. Groups working in this area attempt to predict the arrival of space weather related phenomena through observation and modeling of solar, magnetospheric and ionospheric processes, and issue a forecast of space weather at the Earth. The goal is to predict with days in advance the onset time, the intensity and duration of so-called (geo)magnetic storms at Earth. A geomagnetic storm is characterised by a sudden injection of particles into the magnetosphere, causing a large disturbance in the Earth's magnetic field.

The major contributor to intense geomagnetic storms are not solar flares, as is commonly believed, but coronal mass ejections. These are large ejections of matter and magnetic field from the Sun, moving with several hundred (or thousand)

T. Howard, *Coronal Mass Ejections: An Introduction*, Astrophysics and Space Science Library 376, DOI 10.1007/978-1-4419-8789-1_1,

kilometres per second with a regularity dependent on the 11 year solar cycle. Upon their occasional impact with the Earth they may, if their magnetic field is oriented in a certain way, enable the injection of a large amount of their material into the Earth's upper atmosphere, thereby initiating a geomagnetic storm.

As coronal mass ejections, or CMEs are the prime source of major space weather effects at the Earth, an understanding of these phenomena is now essential to any who work in the technological field, particularly those working at high-altitudes or with space technology. However, understanding CMEs and their effects on space weather requires not only an understanding of solar physics, but also the physics of the interplanetary medium, and the Earth's magnetosphere and ionosphere. Each is a specialised topic and is studied by specialists in each field, and rarely is the gap between disciplines bridged. Consequently, texts on this topic tend to focus heavily on a single part of the space weather picture, and hence tend to be highly specialised and with a great deal of depth and detail.

In short, the nature of the literature tends to force researchers to specialise in only a small aspect of space weather, thereby removing the necessary view of the "big picture". As a result, the space physics community today is populated mostly by solar physicists, magnetospheric physicists and ionospheric physicists – even solar physicists are divided into upper and lower solar atmosphere, and heliospheric physicists. Magnetospheric physics also tend to be divided into categories of latitude and altitude. There are, to be perfectly honest, very few true space weather physicists in the community today.

1.1 Overview

It is the intention of this book to introduce the concept of space weather in a non-specialised way, focusing primarily on the phenomenon responsible for major space weather effects at the Earth: the coronal mass ejection (CME). We consider the CME from the solar perspective, from its launch from the corona and through the interplanetary medium to its arrival at the Earth and creation of a geomagnetic storm and to its eventual fate deep in the heliosphere. We look at the history of the study of CMEs and also at the theory behind their detection, launch and evolution. Spacecraft and ground-based instruments for CME and storm monitoring are discussed, along with associated phenomena. We discuss popular models describing CME onset and evolution, many of which are used for space weather forecasting today. We conclude with a summary of our current understanding of CMEs and the direction of future research, along with a story of the life of a typical CME. This book is aimed mostly at the (post)graduate student level or those who would simply like to know more about CMEs. My main intention is to encourage students and scientists alike to look at the "big picture" when it comes to CMEs and space weather, and to realise the benefits in considering the entire Sun–Earth system as a collective. It is also my intention to reinforce the view that it is the CME, not the solar flare, that is usually responsible for major space weather events. This view is widely accepted in the space physics community, but not by the general public. Finally, it is my hope that this book will provide a resource that I wish was available when I was starting out in this field, providing a brief overview of the status quo on CMEs with valuable references from which more specific and specialised information can be sought.

1.2 A Brief Review of Fundamental Questions

1.2.1 What is a CME?

A CME, or coronal mass ejection, is a large eruption of plasma and magnetic field from the Sun. It can contain a mass larger than 10^{13} kg [41] and may achieve a speed of several thousand kilometres per second [35]. A typical CME has a mass of around 10^{11}–10^{12} kg and has a speed between 400 and 1,000 km/s. It also typically spans several tens of degrees of heliographic latitude (and probably longitude). By comparison, the Earth has a mass of around 6×10^{24} kg and is around $(5 \times 10^{-3})^{\circ}$ in heliographic latitude. An image of a CME is shown in Fig. 1.1, as observed by the LASCO coronagraph on board *SOHO*. This is a "classic" three-part CME, with the leading edge (probably a flux rope) followed by a dark cavity region, followed by a bright filament within.

CMEs may erupt from any region of the corona but are more often associated with lower latitude regions, particularly near solar minimum. They are also often associated with the heliospheric current sheet (the surface where the polarity of the Sun's magnetic field changes) [33], but this by no means explains the location of every CME. They erupt from the Sun with an occurrence of around once a day during typical solar minimum, and around four or five times per day during solar maximum [57]. Only a small percentage of CMEs are directed toward the Earth.

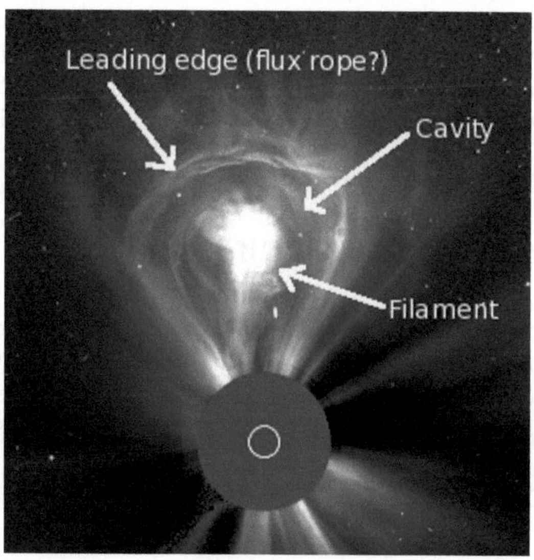

Fig. 1.1 Image of a "classic" three-part CME as observed by the LASCO coronagraph on board *SOHO*. The *white circle* towards the bottom of the image represents the surface of the Sun while the *grey disk* is the occulter of the coronagraph. Labeled are the three parts of this CME: The leading edge, the following cavity and the filament within (This image was obtained from the SOHO webpage and is provided courtesy of NASA/ESA)

1.2.2 What is a CME Made Of?

There has been some debate over the years about whether a CME is a "coronal-mass ejection", that is an ejection of coronal mass, or a "coronal mass-ejection", that is an ejection of mass that is observed in the corona.[1] When they are at a few solar radii from the Sun we can currently only observe CMEs with white light imagers, so here we can only observe light scattered from the free electrons in the CME. We can therefore only conclude that a component (probably large component) of the CME is comprised of plasma.

The composition of the elements and ions comprising the plasma in a CME remains uncertain. The answer may lie in a complete understanding of the mechanism responsible for CME launch or from assuming that the CME flux rope cannot interact with the environment during its evolution. Assuming the CME launch near the Sun is magnetically dominated, then it seems reasonable that the material dominant in the launch region would comprise the bulk of the mass of the CME. So, for example, if the CME was initiated in the corona, then one would expect the CME to be comprised of coronal material. Spectrographic studies of the solar corona have revealed it to contain heavy ions that are formed at very high temperatures [42]. If, on the other hand, the CME was initiated in the lower solar atmosphere such as the photosphere, then we may expect the CME abundance to be similar to that of the photosphere, that is mostly hydrogen, helium and other lighter elements [22]. It has been shown that emerging regions in the corona, such as from new active regions, are composed of photospheric material [44,45].

In reality, the CME is probably a combination of material from many regions on the Sun, and some CMEs may have different amounts of different solar components. There has been a suggestion that the associated erupting filament/prominence may be the lower solar atmospheric component erupting along with the coronal component of the CME [25,34]. It has been argued (and some would say convincingly so) that because there is no interaction between the magnetic flux rope comprising the CME and the surrounding flux, then the composition within the flux rope observed at 1 AU must be the same as that originating at the Sun. In this case, the composition of the ICME (Sect. 1.2.4) is the same as the CME itself.

1.2.3 What is an ICME?

An ICME, or interplanetary coronal mass ejection, is generally regarded as the heliospheric counterpart of the CME. That is, a CME when it is at much larger distances from the Sun (>50 $R_\odot$). ICMEs have been detected with a larger range of instruments than have CMEs, and so much more is known about them. Like CMEs

[1] For this reason, many workers refer to a CME as simply a Solar Mass Ejection.

they have large masses and contain a magnetic field, but generally are not as fast. This is almost certainly due to the large deceleration imposed on fast CMEs by the surrounding solar wind.

While they may not be as fast overall as their CME counterparts, many ICMEs are still supersonic (i.e. they have speeds faster than the speed of sound in the surrounding solar wind). This means that they often cause shocks in the interplanetary medium which are subject to other secondary effects, such as energetic particle acceleration and electromagnetic radio bursts. It is the ICME and its shock that impact with the Earth and cause large space weather effects.

The magnetic structure inside an ICME is varied, but is typically greater in magnitude than that of the surrounding interplanetary magnetic field.[2] Often, a highly structured helical magnetic field is observed within ICMEs, and these are called magnetic clouds [9, 10]. The spiral nature of the cloud can be regarded as a continuation of the twisting of the CME magnetic field that is often observed in prominence eruptions [6,7] and sometimes observed in CMEs [23], although this is by no means certain. The twist in the magnetic structure is called the helicity and can also be regarded as a measure of magnetic complexity.

1.2.4 What is an ICME Made Of?

Thanks to in-situ observations, we have a very good idea of what an ICME is made of. This is because ICMEs pass through interplanetary spacecraft regularly, some of which contain instruments to measure particle composition. Direct measurements of ICMEs from the early 1970s revealed a helium abundance enhancement following interplanetary shocks, and high ionisation states of oxygen and iron [3–5, 17]. ICMEs are also known to contain heavy elements in high ionisation states, such as Fe^{10+} and even Fe^{16+} [18,40]. They also contain cooler ions as well, such as singly-charged helium, magnesium and neon [5]. The high-temperature ions are generally regarded to originate low in the solar corona or from heating during the launch of the CME, while the low temperatures are probably associated with the filament material that erupted behind the CME [11, 36]. While different ICMEs have different compositions there do appear to be some repeatable patterns, like He^{++}/H^{+} and Fe^{16+} enhancements, that are common to many ICMEs.

It is important to remember that just because we know in detail the composition of the ICME does not necessarily mean we can make deductions of the composition of the CME. The interplanetary medium, through which the CME has often spent several days travelling before arriving at the spacecraft, also contains material which may interact with the CME en-route. Hence the ICME near 1 AU may have a different composition to that which was launched from the Sun. The combined ICME

---

[2] It should be noted that the sheath region ahead of the ICME, not generally regarded as being part of the ICME itself, sometimes has a larger field than the ICME itself.

structure may therefore be composed of a combination of its initial material and the solar wind material, meaning that by removing the solar wind material we could deduce what it was initially made of. At worst, all of the initial material has been removed from the ICME, perhaps through re-entry into the Sun, or has been rendered insignificant by the amount of solar wind material accumulated in the regions neighbouring the ICME flux rope. A great number of workers maintain that there is little or no difference between the composition of the ICME within the flux rope and its CME counterpart. After all how, if the CME is a closed magnetic structure, can its material interact with the environment? (Except via magnetic reconnection.)

1.2.5 What is the Difference Between CMEs and ICMEs?

The main differences between CMEs and ICMEs are the regions in space in which they are observed and the methods by which they are detected. CMEs are detected directly mostly through white light observations, (although closer to the Sun they can also be observed in other ways, e.g. radio, UV), but are often related to solar surface eruptions in the lower solar atmosphere. These eruptions, while not directly related to the eruption of the CME, are probably a consequence of the same eruptive mechanism, and so can provide information on CME location and structure. They cannot, unfortunately, provide us with information on CME composition or energy. ICMEs, on the other hand, are observed in white light, radio scintillation and directly, but not until they are some considerable distance from the Sun or in a particular region where the *STEREO* instruments can observe them. We can track their appearance across larger distances and directly measure their magnetic properties and composition when a spacecraft gets hit by one. While we can accumulate detailed information on the ICME, we cannot obtain a great amount of accurate information about its origins.

Because we do not have sufficient information on the transition from CME to ICME, we can only make general associations between them. For example, we can usually associate an ICME with a CME by investigating the timing of each. For example, an ICME observed near the Earth with an in-situ spacecraft can be matched with an Earth-directed CME from a reasonable extrapolation based on the speeds of each [12, 30], and ICME images can be matched for location and timing with their CME counterparts [29, 32]. We also now have the *STEREO* imaging suite, which enables the monitoring of CMEs in white light from coronagraphs through to beyond 1 AU. While limited to observing near the ecliptic plane, this provides the continuous monitoring of CMEs in white light across this entire region. Unfortunately, to date little scientific information has yet been extracted from these white light datasets.

The relationship between a CME and an ICME, i.e. the nature of the transition from one to the other, therefore remains uncertain. The problem is that the solar environment undergoes a change in regime somewhere between where the CME and ICME are observed. Close to the Sun the magnetic forces and gravity dominate, so the physics describing the behaviour of phenomena in this region is dominated by those describing these forces. Further from the Sun these forces become less important, and thermal and pressure forces play a more significant role. Hence in this regime, it is the physics of thermodynamics and hydrodynamics that probably describe the ICME evolution. ICMEs themselves, however, are low-β objects and are thus magnetically dominated, although the majority of the driving energy for the ICME lies in the bulk plasma motion, which is hydrodynamically dominated (low-β). At large distances from the Sun it is unknown which of these dominate the structure and kinematic evolution of the ICME, and no single model exists that accurately describes CME evolution through both of these regimes.

To summarise, an ICME is simply a CME that is a large distance from the Sun. Differences in each are therefore due to the interaction of the CME with its environment as it evolves, and to the change in regime from magnetic to fluid dominated. Now that we can more-or-less observe some of them continuously in white light across the entire inner heliosphere (in some regions), we are approaching the time when it will make little sense in distinguishing between the two.

1.2.6 How do We Detect CMEs?

CMEs are primarily detected by coronagraphs that block out the majority of light from the Sun leaving the relatively faint surrounding corona. The most successful coronagraph to date for CME detection has been the Large Angle Spectroscopic Coronagraph (LASCO) [8] on board *SOHO*, which has detected well over 10^4 CMEs since its launch in 1995. LASCO detects the CME by observing the white light scattered off the electrons within the plasma of the CME. More recently, other spacecraft-based coronagraphs have joined that of LASCO. These are the COR coronagraphs [27] on board the *STEREO* spacecraft and work on a similar principle to that of LASCO.

Coronagraph images are two-dimensional, and so CME images are in fact projections into the sky plane. Hence, the direction of propagation plays a large role in the appearance of a CME. For example, a CME moving very close to the plane of the sky would appear narrower and faster than the same CME with a component along the Sun-observer line. The effects of projection of CMEs into the sky plane is demonstrated in Fig. 1.2. Here the same CME is observed from four different instruments, with two (LASCO C2 and C3) on the Sun–Earth line and the other two (COR2-A and COR2-B) co-planar with the first (in the ecliptic), but separated by 20°. The narrowest and furthest from the Sun is the image in Fig. 1.2c, indicating that this CME is closest to the plane of the sky of COR2-A.

CMEs suffering extreme projection effects are of course those with a very large component along the Sun-observer line, particularly those heading directly toward or away from the observer. Such CMEs appear to completely encircle the Sun and are hence termed halo CMEs [26]. Those with a major component along the

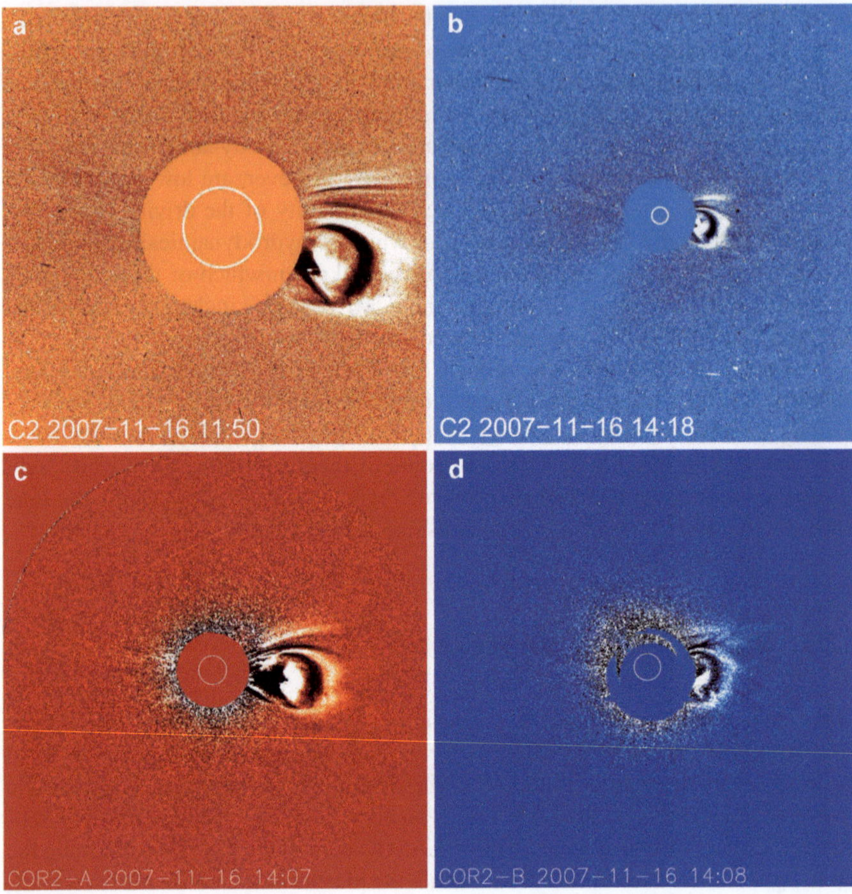

Fig. 1.2 Four images of the same CME viewed from three different viewpoints [31]. (**a**) LASCO C2, (**b**) LASCO C3, (**c**) *STEREO*/COR2-A and (**d**) *STEREO*/COR2-B images of an event observed on 16 November 2007. The two LASCO images were obtained at 11:50 and 14:18 UT and both *STEREO* images are at the same time, at 14:07 UT. The effects of projection are most apparent in panels c and d, where it is clear that the COR2-A image is closest to the plane of the sky (Reproduced with kind permission of Springer Science and Business Media)

Sun-observer line are seen to partially surround the Sun and are called partial halo CMEs. A partial halo may be regarded as a CME with an apparent angular width of more than 120°. Of course, one observer's halo CME is another observer's limb event. For example, at the time of writing the *STEREO* spacecraft were each close to the plane of the Sun (around 75° from the Sun–Earth line). So an Earth-directed CME will appear as a halo to *SOHO*/LASCO, but as a limb CME to the *STEREO*/CORs. Early space weather detection today can be summarised as the search for halo or partial halo CMEs from the perspective of the Earth.

Due to the problems posed by projection, it is helpful to look at solar surface eruptions as indicators of direction of CME propagation. These take the form of solar flares (typically observed in Hα [21, 48, 49], EUV [14, 53, 54] and x-ray [46, 56]), erupting prominences (observed in Hα [1] and EUV [19]), disappearing filaments (Hα [38]), post-eruptive arcades (EUV [52]), coronal dimming (x-ray [47–49] and EUV [24,28,51]) and a variety of other phenomena. These eruptions are not the cause of CMEs, but are associated by either a common cause, or by a secondary reaction to the CME launch itself. Hence, while they do not indicate the exact structure of the CME on the solar disk, they can give an indication of the source region of the CME. A halo CME, for example, can be identified as Earth-directed (or not) by looking for an associated eruption. A visible eruption is an indicator of an observer-directed CME, since if the eruption is there then it must be on the Earthward side of the Sun, meaning that so is the CME. Figure 1.3 shows an image of a halo CME (observed with LASCO C2 and C3) and its associated surface eruption, in this case a solar flare observed with EIT. The sunspots on the

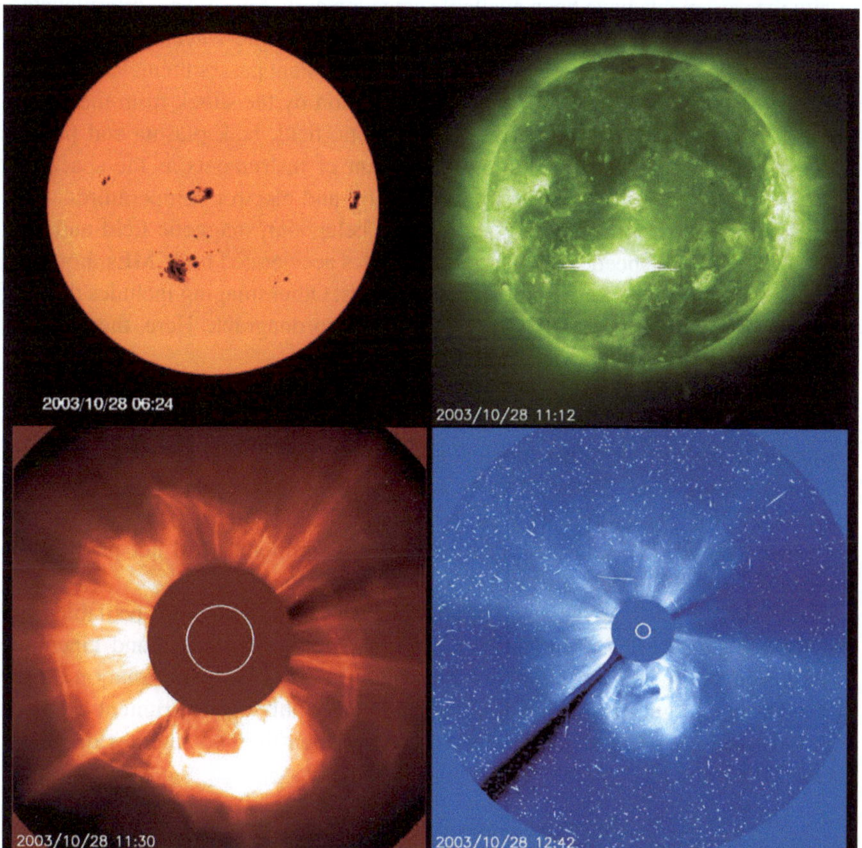

Fig. 1.3 Images associated with the so-called "Halloween" CME. *Top left*: *SOHO*/MDI magnetogram image; *Top right*: *SOHO*/EIT 195Å image with a flare south of centre; *Bottom left*: *SOHO*/LASCO C2 coronagraph image; *Bottom right*: *SOHO*/LASCO C3 coronagraph image (Courtesy of the NASA/ESA SOHO website at http://sohowww.nascom.nasa.gov/)

Sun at this time are also shown. This type of CME would be identified by a space weather forecaster as one to watch and when this particular CME impacted the Earth in October 2003 it caused the now famous "Halloween storm": a large geomagnetic disturbance which caused a blackout in Sweden and disabled two Japanese spacecraft [37, 39].

1.2.7 How do We Detect ICMEs?

As mentioned earlier, ICMEs are detected by a variety of instruments and techniques. This is because they pass across a large region of the sky and sometimes pass by the Earth, making direct detection much more accessible. The oldest methods of ICME detection are through interplanetary scintillation, radio burst observations and in-situ measurements have been available since the early Space Age. Interplanetary scintillation involves the study of radio sources at the metre-wavelength level, and monitoring the change to their signals as a dense transient passes them. Radio bursts vary in frequency as a result of particle acceleration by the shock from the ICME. In-situ detection involves measuring the magnetic field, bulk plasma and particle signatures directly in detail using an assortment of instruments as the spacecraft moves through the ICME. These include proton and electron temperatures, ionic and elemental composition, energetic particle behaviour, magnetic field and bulk particle density and speed. The energetic particles accelerated by ICMEs, however, have been studied since long before these methods. Other imaging includes those of scattered white light, along the same lines as the coronagraph. Here, the ICME is imaged directly as it moves through the heliosphere.

One disadvantage of the in-situ spacecraft is that they are only able to monitor a single track through the ICME as it moves past the spacecraft. This leads to confusion to the complete structure and composition of ICMEs. For example, many ICMEs have not been observed to be associated with magnetic clouds (perhaps one-third [20] to one-half [13]). Could this be because the ICME does not contain a magnetic cloud or because the magnetic cloud component did not pass by the spacecraft? Figure 1.4 demonstrates this confusion.

Even with the latest advances in ICME detection, only images are available from the Sun out to around 1 AU. Thus we can only monitor structural and kinematic evolution of the ICME as it evolves, but we still do not have information on the internal structure until it reaches 1 AU. Also to date, only one instrument has been able to observe ICMEs in the crucial region between the two regimes described above. This is the HI-1 instrument on board *STEREO*.

1.2.8 Why do CMEs Erupt?

There is still speculation about the reason for the eruptions of CMEs. One explanation is because the Sun is trying to do what all things in nature try to do: reduce

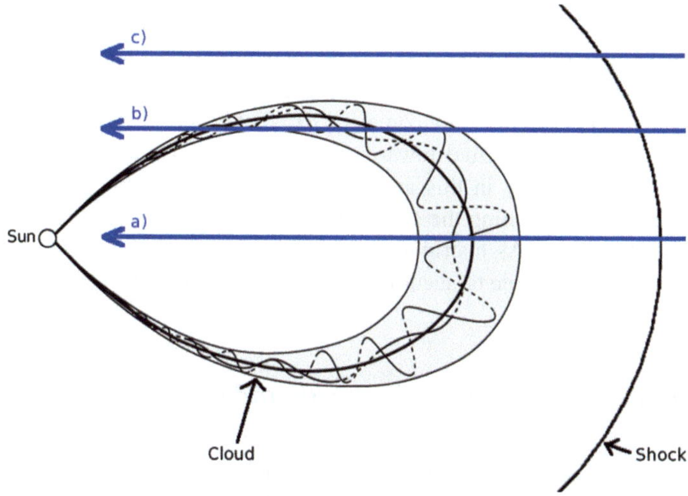

Fig. 1.4 Diagram demonstrating confusion in using in-situ data to measure ICMEs (Modified from Russell and Mulligan [43]). Three spacecraft tracks are shown through a single ICME overall structure, including the shock and sheath regions: (**a**) Passing directly through the centre, (**b**) passing through a flank of the magnetic structure, and (**c**) passing through the sheath region but missing the cloud entirely. It is impossible to identify with track (**c**) (and difficult with track (**b**)) whether the ICME contained the magnetic structure or not

its energy. As the Sun evolves through its cycle its coronal magnetic field becomes twisted and entangled and is continuously being joined by new fields emerging from the solar photosphere. It requires energy to sustain these complex structures. When the level of complexity reaches a certain threshold, it becomes more beneficial to the Sun (from an energy point of view) to remove the magnetic field rather than sustain it. The result is an eruption of a component of the field. As the magnetic structure evolves the energy contained within is converted to gravitational potential and kinetic energy (i.e. the CME gains altitude and speed). CMEs have been known to achieve over 10^{39} J in kinetic energy alone [41].

Also, the natural state of the solar corona is one of expansion. This is how the solar wind is formed. CMEs originate as closed coronal magnetic field structures which act to inhibit this expansion in those regions. Hence, a CME launch may be initiated by re-configuring the closed structure.

Chapter 8 reviews some of the theories describing CME onset. If either of these theories is correct, then the CME likely originates from the magnetic field in the low corona, and its energy is provided from the magnetic energy stored in the complex structure and from the tendency for the coronal to move toward an expanding state. The launch mechanism itself is currently unknown, but it seems clear that energy is transmitted both away and towards the Sun. The energy away is converted mostly to kinetic and gravitational potential energy (in the form of an erupting CME), while that transmitted towards is converted to emission (in the form of a flare), kinetic (moving solar plasma around and accelerating solar energetic particles), and thermal energy. It is known, for example, that the energy within a solar flare is less than 10% of that of the associated CME [16,55], and that combining the known energies of all other eruptive events still does not come close to that of the CME itself [16]. Maybe there is less energy transmitted towards the Sun or maybe we cannot accurately measure all of the energy in this area. A solar flare, for example, is typically a broadband eruption, spanning the electromagnetic spectrum from (at least) visible light through to x-rays. As instruments measuring solar flares are narrowband by comparison, it is impossible to accurately measure their total energies.

1.2.9 How do CMEs Affect the Earth and its Inhabitants?

The Earth is enclosed within the solar wind, which is a continually-flowing "ocean" of plasma and magnetic field moving outward from the Sun. For the most part, the solar wind is deflected around the Earth by the geomagnetic field, which itself is distorted by this interaction. The dynamics of this distortion and the interaction with the Earth's atmosphere result in a buildup of plasma in various regions around the Earth. This resulting combination of plasma and field is called the magnetosphere, and its behaviour is strongly influenced by that of the surrounding solar wind.

The aurora, for example, is caused by particle precipitation from the solar wind and accumulated plasma in the magnetotail (Table 10.1), from particles which are able to directly enter the Earth's atmosphere via the divergence of magnetic field lines near the poles (the so-called cusp region), and the open field lines in the tail. These energetic particles provide energy to the particles of the atmosphere, resulting in an emission of light. Poleward of the cusp region, geomagnetic field lines are no longer connected to each other, but rather to the interplanetary magnetic field. Hence, from the perspective of the magnetospheric system, these field lines are open. When a strong southward-directed magnetic field arrives at the magnetosphere, its field can temporarily connect with that of the Earth's – a process known a magnetic reconnection (Sect. 10.3). This causes more field lines on the dayside of the magnetosphere to open, resulting in a movement of the cusp toward the equator. Ram pressure from the solar wind also plays a role on the structure of the magnetosphere. While changes in pressure do not open or close field lines, they do change the size of those field lines, and high ram pressure can further enhance the equatorial movement of the auroral oval. Hence, the latitude on the Earth at which an aurora occurs depends on the magnetic configuration of the medium in which the Earth is submerged and the ram pressure on the dayside. Figure 1.5 demonstrates this phenomenon.

This simple example demonstrates the dependence of not only the structure and dynamics of the magnetosphere on the behaviour of the solar wind, but also a method by which energy and particles may enter the Earth's atmosphere. Now consider the arrival of an ICME, which from the Earth's perspective may be regarded

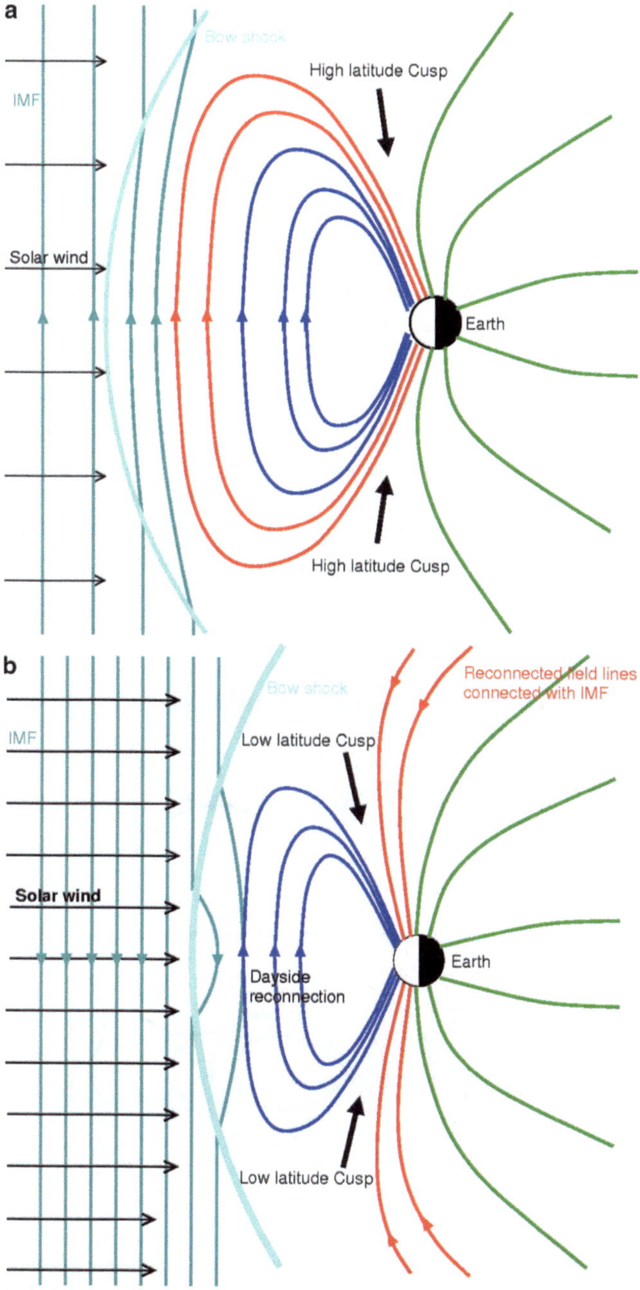

Fig. 1.5 Diagram representing a simplified version of the behaviour of high-latitude geomagnetic field lines. (**a**) Under low geomagnetic activity conditions, where the cusp region is at high latitudes. (**b**) When a southward-directed magnetic field arrives at the Sun, magnetic reconnection enables more dayside field lines to open, moving the cusp region towards the equator resulting in an aurora being observed at lower latitudes. Increased pressure also reduces the size of the magnetosphere, further enhancing this effect. See also Fig. 10.4

as a dense version of the solar wind (both in magnetic field and in plasma). When this occurs, two major processes may occur:

1. If the magnetic field of the ICME is directed southward (relative to the Earth), magnetic reconnection exposes the Earth to the plasma contained within the CME, which is injected directly into the geomagnetic field. Reconnection causes closed field lines to open, accessing them to the solar wind and allowing a larger proportion of the Earth's atmosphere to be exposed to its plasma.
2. The increased pressure impacting the magnetosphere causes it to compress and closed magnetic field lines to be reduced in size. This results in a further expansion of the auroral ovals, where the effects of direct impact of solar wind particles with the atmosphere are exposed to more dense populations of people on Earth.

The combination of these two effects results in an increase in geomagnetic activity called a (geo)magnetic storm. The first of these effects is more significant for geomagnetic activity than the second. One popular index for measuring geomagnetic activity is called the Dst index [50], but many other indices are used to monitor activity (e.g. K_P). Figure 1.6a provides a simple classic diagram showing the reconnection process for both a southward (a) and northward (b) interplanetary field.

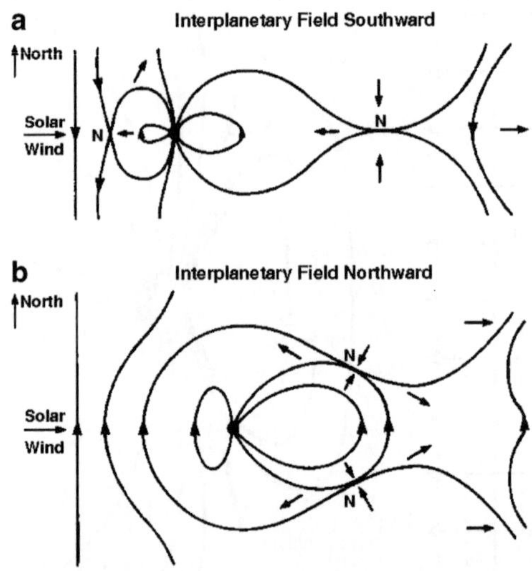

Fig. 1.6 Classic diagram showing the effects of magnetic reconnection at the Earth. (**a**) A CME with a southward magnetic field arrives at the Earth, allowing magnetic reconnection to occur between it and the geomagnetic field. This allows an injection of particles from the CME directly into the magnetosphere, and a further reconnection in the tail. The result is a large disturbance to the geomagnetic field. (**b**) The arrival of a northward magnetic field arriving at the Earth. No magnetic reconnection occurs on the dayside of the Earth but it does still occur on the nightside. Such reconnection, however, does not lead to magnetic storm conditions [15]

Some of the ways in which the Earth and its inhabitants are affected by geomagnetic storms are listed below [2].

1. Interference of telecommunication through phone lines and satellites.
2. Increase in radiation exposure to high-altitude and/or high-latitude aircraft fliers and astronauts.
3. Increase in atmospheric drag on orbiting spacecraft, thereby reducing orbit speed (potential crash landing).
4. Interference in spacecraft circuitry.
5. Damage to spacecraft hardware (e.g. solar cells).
6. Interference/damage to ground-based micro – and nanocircuitry.
7. Unexpected current generation in power lines, resulting in power station damage.

It should be noted that if the CME has a magnetic field directed northwards, then a large magnetic storm will likely not occur (Fig. 1.6b). This is because the geomagnetic field is from (geographic) south to north (i.e. the geographic south pole corresponds to the magnetic north pole), rendering magnetic reconnection at the dayside impossible. Some less significant results may still occur due to the increased pressure, e.g. an expansion of the auroral oval or a reduction in the size of the magnetosphere. It should be noted that under a northward or zero north–south field, reconnection does still occur on the nightside of the magnetosphere, but this does not lead to storm conditions.

1.3 Summary

A coronal mass ejection, or CME is a large eruption of magnetic field and plasma from the Sun. CMEs may contain masses in excess of 10^{13} kg, achieve speeds greater than 4,000 km/s and may span several tens of degrees heliospheric latitude and/or longitude. They are probably ejected in order for the Sun to reduce the energy required to maintain large complex magnetic fields in the low corona and because of the natural tendency of the solar corona toward an expanding state and the overcoming of temporary impedances to that state. As they evolve through the interplanetary medium they become interplanetary coronal mass ejections (ICMEs), which, although the ICME flux rope itself likely consists of originating solar wind material, the entire ICME structure may be composed of both solar and solar wind material. CMEs and ICMEs are detected using white light cameras, which detect light scattered from the electrons in the CME/ICME and CMEs close to the Sun may be observed with radio and UV instruments. ICMEs are also detected using radio techniques and directly when they impact with in-situ spacecraft containing a large variety of particle and magnetic field instruments.

When they occasionally impact the Earth, CMEs alter the behaviour of the Earth's magnetosphere. If the magnetic field of the arriving CME has a strong southward component, magnetic reconnection between it and the dayside geomagnetic field may occur, resulting in an opening of field lines and a large injection of particles into the magnetosphere. The increased pressure and/or shock from the CME may cause a compression of the magnetosphere. These result in a large disturbance to the Earth's magnetic field known as a (geo)magnetic storm. Such storms are known to cause a variety of potentially serious deleterious effects.

Thus, the study of CMEs is important not only from a scientific basis, but also for technical interests. Scientifically CMEs provide information on the evolution of the Sun through a crucial process by which it removes built-up energy stored in the complexity of its magnetic field. Technically CMEs are an obstacle to the continuing development of electrical and space technology, so their understanding is crucial not only to assist in the design of "storm resistant" hardware, but also in the prediction of their arrival and the consequences to space weather at the Earth when they do.

References

1. Athay, R.G., Illing, R.M.E.: J. Geophys. Res. **91**, 10961–10973 (1986).
2. Baker, D.N., Balstad, R., Bodeau, J.M., Cameron, E., Fennel, J.F., Fisher, G.M., Forbes, K.F., Kintner, P.M., Leffler, L.G., Lewis, W.S., Reagan, J.B., Small, A.A., III, Stansell, T.A., Strachan, L., Jr.: Severe Space Weather Events - Understanding Societal and Economic Impacts, NRC Workshop Rep. (2009).
3. Bame, S.J., Asbridge, J.R., Feldman, W.C., Fenimore, E.E., Gosling, J.T.: Solar Phys. **62**, 179–201 (1979).
4. Borrini, G., Gosling, J.T., Bame, S.J., Feldman, W.C.: J. Geophys. Res. **87**, 7370–7378 (1982).
5. Borrini, G., Gosling, J.T., Bame, S.J., Feldman, W.C.: Solar Phys. **83**, 367–378 (1983).
6. Bothmer, V., Rust, D.M.: In Crooker, N.U., Joselyn, J.-A., Feynmann, J. (eds.), Geophys. Monog. Ser. **99**, p.139, AGU, Washington DC (1997).
7. Bothmer, V., Schwenn, R.: Ann. Geophys. **16**, 1–24 (1998).
8. Brueckner, G.E., Howard, R.A., Koomen, M.J., Korendyke, C.M., Michels, D.J., Moses, J.D., Socker, D.G., Dere, K.P., Lamy, P.L., Llebaria, A., Bout, M.V., Schwenn, R., Simnett, G.M., Bedford, D.K., Eyles, C.J.: Solar Phys. **162**, 357–402 (1995).
9. Burlaga, L.F.: In Schwenn, R., Marsch, E. (eds.), Physics of the Inner Heliosphere, p.1, Springer-Verlag, New York (1991).
10. Burlaga, L.F., Sittler, E.C., Mariani, F., Schwenn, R.: J. Geophys. Res. **86**, 6673–6684 (1981).
11. Cane, H.V., Kahler, S.W., Sheeley, N.R., Jr.: J. Geophys. Res. **91**, 13321–13329 (1986).
12. Cane, H.V., Richardson, I.G.: J. Geophys. Res. **108**, 1156–1168 (2003).
13. Cane, H.V., Richardson, I.G., Wibberenz, G.: J. Geophys. Res. **102**, 7075–7086 (1997).
14. Cheng, C.-C.: Solar Phys. **65**, 283–298 (1980).
15. Dungey, J.W.: In De Witt, C., Hieblot, J., Lebeau, A. (eds.), Geophysics: The Earth's Environment, Gordon Breach, New York (1963).
16. Emslie, A.G., Kucharek, H., Dennis, B.R., Gopalswamy, N., Holman, G.D., Share, G.H., Vourlidas, A., Forbes, T.G., Gallagher, P.T., Mason, G.M., Metcalfe, T.R., Mewaldt, R.A., Murphy, R.J., Schwartz, R.A., Zurbuchen, T.H.: J. Geophys. Res. **109**, doi:10.1029/2004JA 010571 (2004).
17. Fenimore, E.E.: Astrophys. J. **235**, 245–257 (1980).
18. Gloeckler, G., Fisk, L.A., Hefti, S., Schwadron, N.A., Zurbuchen, T.H., Ipavich, F.M., Geiss, J., Bochsler, P., Wimmer-Schweingruber, R.F.: Geophys. Res. Lett. **26**, L157–L160 (1999).
19. Gopalswamy, N., Hanaoka, Y., Hudson, H.S.: Adv. Space Res. **25**, 1851–1854 (2000).
20. Gosling, J.T.: In Russell, C.T., Priest, E.R., Lee, L.C. (eds.), Geophys. Monog. Ser. **58**, p.9, AGU, Washington DC (1990).
21. Gosling, J.T., Hildner, E., MacQueen, R.M., Munro, R.H., Poland, A.I., Ross, C.L.: J. Geophys. Res. **79**, 4581–4587 (1974).

22. Grevesse, N., Sauval, A.J.: Space Sci. Rev. **85**, 161–174 (1998).
23. Gurman, J.B.: In: The Best of SOHO gallery (1999), available via SOHO. http://sohowww.nascom.nasa.gov/gallery/images/c2helix.html. Cited 2 June 1998.
24. Harrison R.A., Bryans, P., Simnett, G.M., Lyons, M.: Astron. Astrophys. **400**, 1071–1083 (2003).
25. House, L.L., Wagner, W.J., Hildner, E., Sawyer, C., Schmidt, H.U.: Astrophys. J. **244**, L117–L121 (1981).
26. Howard, R.A., Michels, D.J., Sheeley, N.R., Jr., Koomen, M.J.: Astrophys. J. **263**, L101–L104 (1982).
27. Howard, R.A., Moses, J.D., Vourlidas, A., Newmark, J.S., Socker, D.G., Plunkett, S.P., Korendyke, C.M., Cook, J.W., Hurley, A., Davila, J.M., Thompson, W.T., St. Cyr, O.C., Mentzell, E., Mehalick, K., Lemen, J. R., Wuelser, J.P., Duncan, D.W., Tarbell, T.D., Wolfson, C.J., Moore, A., Harrison, R.A., Waltham, N.R., Lang, J., Davis, C.J., Eyles, C.J., Mapson-Menard, H., Simnett, G.M., Halain, J.P., Defise, J.M., Mazy, E., Rochus, P., Mercier, R., Ravet, M.F., Delmotte, F., Auchere, F., Delaboudinière, J.-P., Bothmer, V., Deutsch, W., Wang, D., Rich, N., Cooper, S., Stephens, V., Maahs, G., Baugh, R., Mcmullin, D.: Space Sci. Rev. **136**, 67–115 (2008).
28. Howard, T.A., Harrison, R.A.: Solar Phys. **219**, 315–342 (2004).
29. Howard, T.A., Simnett, G.M.: J. Geophys. Res. **113**, doi:10.1029/2007JA0129209 (2008).
30. Howard, T.A., Tappin, S.J.: Astron. Astrophys. **440**, 373–383 (2005).
31. Howard, T.A., Tappin, S.J.: Solar Phys. **252**, 373–383 (2008).
32. Howard, T.A., Webb, D.F., Tappin, S.J., Mizuno, D.R., Johnston, J.C.: J. Geophys. Res. **111**, doi:10.1029/2005JA011349 (2006).
33. Hundhausen, A.J.: J. Geophys. Res. **98**, 13177–13200 (1993).
34. Hundhausen, A.J.: In Strong, K.T., Saba, J.L.R., Haisch, B.M. (eds.), Many Faces of the Sun, p.143, Springer, New York (1999).
35. Hundhausen, A.J., Burkepile, J.T., St. Cyr, O.C.: J. Geophys. Res. **99**, 6543–6552 (1994).
36. Illing, R.M.E., Hundhausen, A.J.: J. Geophys. Res. **90**, 275–282 (1985).
37. Jacob, B.: In: Killer Electrons, NASA press release, available via NASA/GSFC. http://www.nasa.gov/vision/universe/solarsystem/killer_electrons.html. Cited 15 April 2004.
38. Joselyn, J.A., McIntosh, P.S.: J. Geophys. Res. **86**, 4555–4564 (1981).
39. Lundstedt, H.: Proc. 35th COSPAR Sci. Assem., Paris, p.3791 (2004).
40. Lepri, S.T., Zurbuchen, T.H., Fisk, L.A., Richardson, I.G., Cane, H.V., Gloeckler, G.: J. Geophys. Res. **106**, 29231–29238 (2001).
41. MacQueen, R.M.: Phil. Trans. R. Soc. Lond. A. **297**, 605–620 (1980).
42. Meyer, J.-P.: Astrophys. J. Supp. **57**, 173–204 (1985).
43. Russell, C.T., Mulligan, T.: Adv. Space Res. **29**, 301–306 (2002).
44. Sheeley, N. R., Jr.: Astrophys. J. **440**, 884–887 (1995).
45. Sheeley, N. R., Jr.: Astrophys. J. **469**, 423–428 (1996).
46. Sheeley, N.R., Jr., Bohlin, J.D., Brueckner, G.E., Purcell, J.D., Scherrer, V.E., Tousey, R., Smith, J.B., Jr., Speich, D.M., Tandberg-Hanssen, E., Wilson, R.M.: Solar Phys. **45**, 377–392 (1975).
47. Sterling A.C., Hudson, H.S.: Astrophys. J. **491**, L55–L58 (1997).
48. Stewart, R.T., Howard, R.A., Hansen, F., Gergely, T., Kundu, M.: Solar Phys. **36**, 219–231 (1974a).
49. Stewart, R.T., McCabe, M.K., Koomen, M.J., Hansen, R.T., Dulk, G.A.: Solar Phys. **36**, 203–217 (1974b).
50. Sugiura, M.: Ann. Int. Geophys. Year **35**, 945–948 (1964).
51. Thompson, B.J., Plunkett, S.P., Gurman, J.B., Newmark, J.S., St. Cyr, O.C., Michels, D.J.: Geophys. Res. Lett. **25**, 2465–2468 (1998).
52. Tripathi, D., Bothmer, V., Cremades, H.: Astron. Astrophys. **422**, 337–349 (2004).
53. Wood, A.T., Jr., Noyes, R.W.: Solar Phys. **24**, 180–196 (1972a).
54. Wood, A.T., Jr., Noyes, R.W., Dupree, A.K., Huber, M.C.E., Parkinson, W.H., Reeves, E.M., Withbroe, G.L.: Solar Phys. **24**, 169–179 (1972b).

55. Webb, D.F., Cheng, C.-C., Dulk, G.A., Edberg, S.J., Martin, S.F., McKenna-Lawlor, S., McLean, D.J.: In Sturrock, P.A (ed.), Solar Flares: A Monograph From Skylab Workshop II, Colo. Assoc. Uni. Press, Boulder, p.471 (1980).
56. Webb, D.F., Hundhausen, A.J.: Solar Phys. **108**, 383–401 (1987).
57. Yashiro, S., Gopalswamy, N., Michalek, G., St. Cyr, O.C., Plunkett, S.P., Rich, N.B., Howard, R.A.: J. Geophys. Res. **109**, doi:10.1029/2003JA010282 (2004).

Chapter 2
History

In this chapter the history of coronal mass ejection study is reviewed. The emphasis is on CME and ICME observation, and the scientific contributions made to our understanding of these phenomena from those observations. Less emphasis has been placed on what I consider to be secondary effects of the CME (e.g. radio bursts, solar surface activity, solar energetic particles). The study of these secondary effects has made significant contributions to our understanding of the CMEs and space physics in general, much of which predates the discovery of the CME. A large number of texts has been written on these phenomena, and discussing how they are all related can be confusing or even misleading in a brief review. Hence, this chapter will address these secondary phenomena only in their early historical context and mostly before the CME was directly observed. Chapter 7 addresses these secondary phenomena in more detail.

It is also important to note that significant contributions were made to our understanding of CMEs by way of modelling, work on which has continued throughout the observational history of CMEs. In this chapter, we do not consider the contributions of modelling, but rather saving these for complete chapters describing their onset and evolution (Chaps. 8 and 9).

2.1 The Early Years

The history of the observation of CMEs probably dates back to a very fortunate catch in the nineteenth century, when the solar corona was beginning to be studied in great detail for the first time. The solar corona can only be observed naturally during a solar eclipse, which only achieves totality for a few minutes, and so acquiring information on the corona proved difficult. This is most likely why early detailed descriptions are lacking, even though solar eclipses have been observed and documented for centuries (the first identification of the solar corona was probably in 968 AD [102], but observations date back to the first eclipse recording in 1223 BC [283]). Given that on average a CME occurs only a few times a day and that we only have a window of a few minutes to observe the corona during eclipse totality, one can estimate that the probability of actually observing a CME during a solar eclipse is low, even during solar maximum.

T. Howard, *Coronal Mass Ejections: An Introduction*, Astrophysics and Space Science Library 376, DOI 10.1007/978-1-4419-8789-1_2,

As with observations of the corona, the effects of space weather has its roots in antiquity as well. In 34 AD, for example, the Roman Emperor Tiberius mistook the red glow of the aurora for fires at Ostia (the port of Rome), and dispatched troops to investigate. There must have been a major geomagnetic storm for the aurora to be above Ostia, which lies at a latitude just south of 42°N. By the eleventh century the concept of magnetism was known by the Chinese and the magnet was in wide use by the Europeans by the twelfth century [192], although the magnetised Earth theory did not emerge until the turn of the seventeenth century (in Gilbert's *De Magnete*, published in 1600 [65]). Thanks to the wide use of magnets for navigation, a vast database of geomagnetic measurements was built up in the century which followed and in 1724, two workers (George Graham in London and Anders Celsius in Sweden) independently found a simultaneous deviation by a small angle in the compass needle that lasted around a day [86]. These were later named "magnetic storms" by von Humboldt in 1805.[1]

Early observations of the Sun include those of sunspots by the Chinese dating back as far as the fourth century BC, and in the west in the eighth century AD. Galileo is often (incorrectly) accredited with the discovery of sunspots (in his letters to Mark Welser in 1612) but he is the first to have observed them with a telescope [250]. The nineteenth century brought a wave of solar discoveries, including solar spectroscopy in 1817 [61], the sunspot cycle in 1843 [220], and solar flares, differential rotation and chemical composition in 1859 [30, 31, 99]. The era of photography helped here, with the first solar photograph obtained in 1845 [47].[2] Figure 2.1 shows a drawing of the eclipse observed on 18 July 1860 in Torreblanca (Spain). Toward the southwest (lower-right) of the image appears to be a bubble-shaped structure that is disconnected from the Sun and remaining corona. Drawings of the same eclipse by other workers also reveal an extended structure in this region of the Sun. This is believed to be the first direct observation of a coronal mass ejection, although none realised what it was at the time.

In 1852, the sunspot cycle was "absolutely" connected with geomagnetic activity by Edward Sabine, based on an accumulation of data since the 1830s [216]. This relationship was confirmed by two other researchers, working independently, at around the same time [219]. Later that decade in 1859 the now famous Carrington Event (or Carrington Storm) occurred. Here a powerful flare erupted from a large active region on the Sun (recorded by Richard Carrington [30]) and 18 h later the most intense magnetic storm in recorded history occurred at Earth. As a result, telegraph systems failed across Europe and North America and aurora were observed at latitudes as low as the Caribbean. Contemporary estimates of the Dst index for the Carrington Event range from −1,600 nT [248] to −850 nT [228]. In the following years, associations between flares and geomagnetic storms continued, although the relationship was not one-to-one. For example, Maunder [184] and Greaves and Newton [87, 88] showed that the great geomagnetic storms were usually

[1] For a review of geomagnetism, refer to Stern [231].

[2] An excellent summary on the history of the study of the Sun may be found at the High Altitude Observatory (HAO) webpage, at http://www.hao.ucar.edu/Public/education/spTimeline.html.

Fig. 2.1 Drawing of the 1860 eclipse recorded by Tempel [205] and identified later by Jack Eddy. This is believed to be the first observation of a coronal mass ejection

accompanied on the Sun by groups of large-area sunspots. In 1931, Chapman and Ferraro [33–35] proposed that this correlation could be explained if there was a sporadic ejection of ionised material from the Sun. In the same year, the coronagraph was invented, allowing the continuous monitoring of the corona without the need to wait for a solar eclipse. This was achieved by permanently blocking the brighter light from the photosphere using a disk, known as an occulting disk [178].

2.2 Coronal Transients

Although the coronagraph was invented by Bernard Lyot in 1931, it was not until later that the sensitivity of the instrument was reduced to a level where faint coronal eruptions could be observed. For the most part, this required the utilisation of space-based coronagraphs, and it is Richard Tousey, using *OSO-7* coronagraph observations who is accredited with the discovery of the CME. In a review published in the Proceedings of the Fifteenth Plenary Meeting of COSPAR (*Space Research XIII*) in 1973, Tousey referred to transients in the K-corona moving with speeds of 400–1,000 km/s [244]. He described the observation of the first CME thusly:

> The prominence erupted at 1701 UT and at 1938 UT the OSO image showed a cloud just emerging from the occulter shadow... The next image frame was at 2111 UT. The extreme right edge of this frame recorded a portion of a bright cloud whose leading edge extended barely into the inner polarizing ring. In the following frame at 2123 UT the cloud had moved outward, and its recorded portion suggested a cloud of circular shape, with diameter about equal to the sun's radius, located at 35° N radially above the prominence. The next two images taken at $11\frac{1}{2}$ minute intervals, clearly show the motion of the plasma cloud through the corona. (pp. 724–725 [244])

At around the same time, coronal disturbances were being monitored using the ground coronagraph at Sacramento Peak in New Mexico. These were reported by Howard DeMastus, Bill Wagner and Rich Robinson in the *Solar Physics* journal in 1973. They refer to a number of "fast green line events" or "coronal transients" observed on the solar limb from 1956 to 1972, and they attempted to associate them with other forms of solar limb activity [45]. By this time coronal transients had also been recognised by workers using the Mauna Loa coronagraph, with observations published the following year [64, 163]. It seems highly likely that all groups had observed manifestations of the same phenomenon.

The coronagraph on board *OSO-7* continued to observe CMEs and a total of 20 were confirmed before it re-entered the Earth's atmosphere in 1974 [120]. The previous year, in 1973 the US space station *Skylab* was launched. Around 77 transients were observed by the *Skylab* coronagraph from May 1973 to February 1974 [196], and they were immediately identified as mass ejections [81]. The first appearance of the term "coronal mass ejection" appears to be in Gosling et al. [83], although the term "mass ejection coronal transient" appears in Hildner [110]. Initially, workers preferred to adhere to the more conservative "coronal transient", and the coronal mass ejection term was initially reserved for a particular type of eruption observed, but over time this term began to dominate. By 1990, virtually all workers were referring to all large ejecta observed with a coronagraph as a coronal mass ejection or CME.

Observations of CMEs continued into the 1980s with the launch of the US Department of Defense Test Program satellite *P78-1* in February 1979, and of NASA's *Solar Maximum Mission* (*SMM*) in February 1980. On board each, amongst an assortment of other solar instruments, was the Naval Research Laboratory's coronagraph, *Solwind* [188] and NASA's coronagraph/polarimeter C/P [182] respectively. Among the discoveries of this next generation of space-based coronagraphs was the first Earth-directed CME by Russ Howard and co-workers. This transient was observed in November 1979 and was associated with an interplanetary shock detected near the Earth. The results were published in the *Astrophysical Journal* in 1982 [121]. The term "halo CME" arises from this publication. The "classic" three-part CME structure shown in Fig. 1.1 was also first identified by the *SMM* C/P in this era [138]. Figure 2.2 shows images of CMEs obtained by these early instruments.

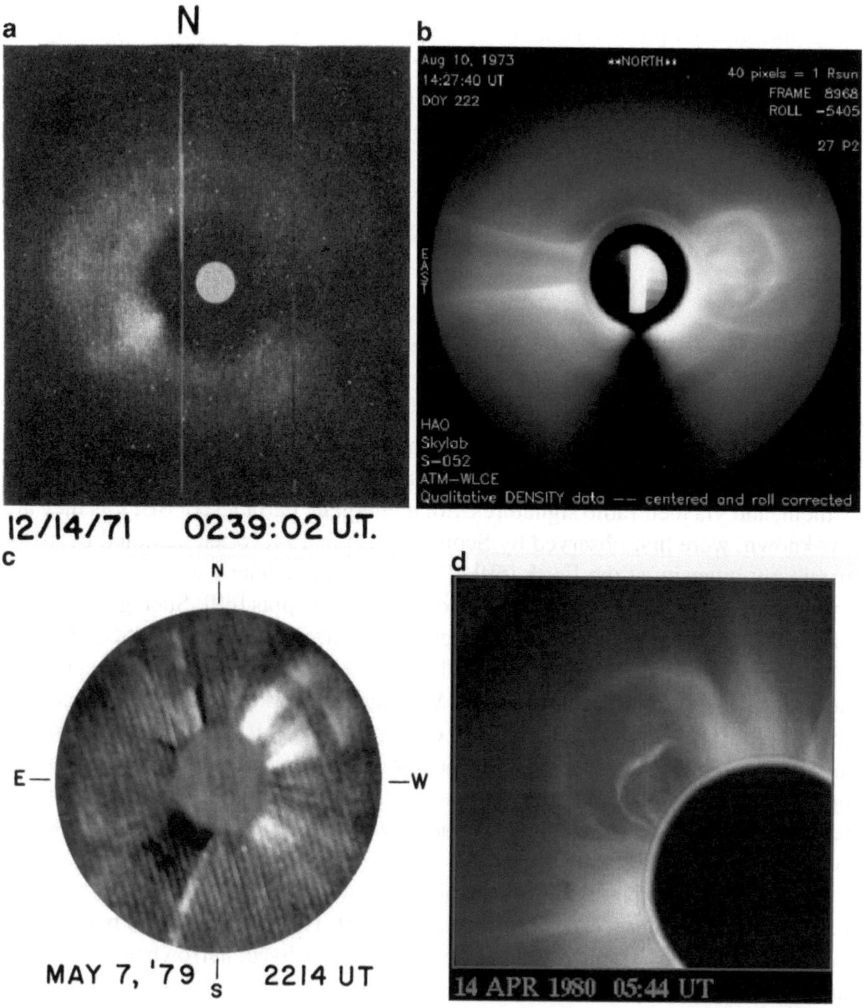

Fig. 2.2 Images of some of the early CMEs observed by space-based coronagraphs. (**a**) One of the first CMEs observed with *OSO-7* by Tousey [244]. This image was obtained on 14 December 1971 (Copyright Wiley-VCH Verlag GmbH & Co. KGaA. Reproduced with permission). (**b**) The coronagraph on board *Skylab* (available courtesy of the High Altitude Observatory (HAO)), obtained on 10 August 1973. Images from (**c**) *Solwind* on 7 May 1979 [151] (Reproduced with kind permission of Springer Science and Business Media), and (**d**) C/P on 14 April 1980 [230] (courtesy of HAO) follow

The combination of the two coronagraphs from *Solwind* and C/P resulted in the observation of over 2,000 CMEs, thereby enabling the detailed statistical analysis of their properties for the first time. Hundhausen et al. [137] using C/P reported that the location of the CME was more evenly distributed around the Sun than the events observed by *Skylab*, which were localised around the equator. Howard et al. [122] surveyed almost a thousand CMEs over 3 years (March 1979 to December 1981) using *Solwind* and provided extensive statistical results on structure, mass, angular span, location and kinetic energy. Both reported a "major" CME occurrence of around one per day. Similar statistical results using the complete C/P CME dataset were reported by Hundhausen et al. [136]. The fastest CME by that time was reported in this paper and determined to be 2,101 km/s. Hence, by 1995 solar physicists had a good picture of CME occurrence, structure, speed, mass and energy via an investigation of case studies as well as statistical surveys.

2.3 Interplanetary Coronal Mass Ejections (ICMEs)

Long before the discovery of the CME, investigations of the interplanetary counterparts of solar eruptions were being investigated, in studies pre-dating the space age. Here, interplanetary shocks were being studied via particles that were accelerated by them, and via their radio signatures. Solar energetic particles, or SEPs as they are now known, were first observed by Scott Forbush in 1946 when he noted bursts of cosmic ray intensity at the Earth [59]. They were immediately associated with solar flares and with variable magnetic fields around sunspots [60]. Such ground level enhancements, or GLEs, were later detected by neutron monitors in 1956 [187] and with riometers in 1959 [206]. Although always associated with flares, solar energetic particles (SEPs) were described by John Wild and co-workers in 1963 to be accelerated by two stages: Flare acceleration of electrons up to $\sim$100 keV, and acceleration caused by an outward-moving fast magnetohydrodynamic shock. The second phase appeared to be necessary for substantial acceleration of protons and higher-energy electrons [271]. This two-stage acceleration was confirmed using in-situ observations through the 1980s and 1990s [68].

The study of radio bursts arose from observations with the first radiospectrograph at Penrith in New South Wales (Australia) in 1950. The observed radio bursts were classified into three "Types": Type I bursts were short-lived, narrowband bursts occurring during storm periods, Type II bursts were longer in duration, accompanied solar flares and drifted gradually in frequency, while Type III bursts were short-lived broadband bursts where the frequency of maximum intensity drifted rapidly [267]. Later in 1957, using an interferometer at the Nançay observatory in France, a fourth type of burst event, designated Type IV, was discovered [14]. Type IV bursts were long-duration, associated with solar flares, and often followed a Type II burst. Finally, a new burst which often followed Type III was identified in 1959 and classified as a Type V [270]. According to McLean and Labrum [186]:

> The observations of Type II and Type III bursts contributed significantly to the developing subject of solar flare 'anatomy' [269]. It was found repeatedly that groups of Type III bursts occurred at the very start of flares, coincident with the arrival of X-rays as signified by the onset of sudden ionospheric disturbances. The Type II burst, if one occurred, began some minutes later. (pp. 12–13 [186])

The first height-time plots of CMEs (or rather, their shocks) were plotted indirectly using analysis of Type II bursts. Figure 2.3 provides an example of such a plot

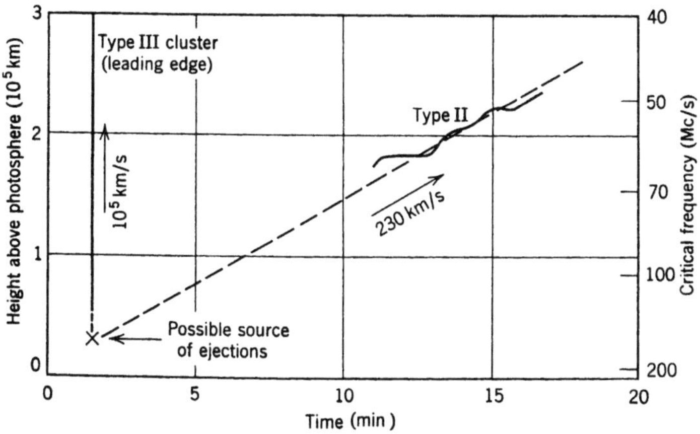

Fig. 2.3 An early example of a height-time plot, derived from Type II and Type III bursts [268]. This plot results in a Type II speed of 230 km/s

from Wild et al. [268]. While it was not entirely understood at the time what was being observed, relatively precise measurements of the kinematic evolution of solar eruptions were being made well before the discovery of the CME, and even before the space age.

2.3.1 *In-Situ Observations*

Into the space age, and still before the detection of CMEs by coronagraphs, interplanetary shocks were being observed by in-situ spacecraft as they were impacted by, and passed through these shocks in space. It was suggested by Thomas Gold (1955) that high-speed plasma ejected from the Sun would produce a collisionless shock in the interplanetary medium [70]. So, while the shock is not the ICME itself, it is a convenient associated signature. Interplanetary shocks were first directly observed in 1962 by the *Mariner 2* spacecraft [229] and a further two were reported in 1968 by Jack Gosling and co-workers using the *Vela 3* spacecraft pair [78]. Hundhausen et al. [135] used solar wind observations of shock disturbances to estimate that a large shock was associated with an ejection of 10^{13} kg and 10^{32} ergs from the Sun. By 1973, several publications had emerged reporting interplanetary shock observations, many of which were connected with geomagnetic activity [50,132,133,139,140,166,199,242]. Hence by the discovery of the CME, the theory of the formation and propagation of interplanetary shocks was firmly established, and had been confirmed with direct observation using in-situ spacecraft. They were associated with eruptions from the Sun (then mostly believed to be solar flares), and were known to cause increases in geomagnetic activity, particularly in the form of a sudden-(storm)-commencement, or S(S)C.

The first direct association between interplanetary shocks and CMEs was made by Gosling et al. [82] by comparing a CME observed by the coronagraph on *Skylab* with an interplanetary shock detected by *Pioneer 9* [82]. Other early works include Dryer [48], Burlaga et al. [21] and Michels et al. [189]. The term "ICME" appears to have been coined by Xuepu Zhao in a paper presented at the Solar-Terrestrial Predictions Workshop in Ottawa in 1991, and a paper published in the *Journal of Geophysical Research* in 1992 [282]. It was brought into mainstream use following the publication of a review by Murray Dryer [49].

The next question that was addressed with the usage of in-situ data was the elemental/ionic composition and thermal behaviour of ICMEs. The first observations of the composition of the plasma behind interplanetary shocks revealed a helium abundance enhancement, and pre-dates the discovery of the CME [111]. The association of such enhancements with solar flares was made even earlier, dating back to the late 1960s [9, 165]. Following a statistical study in which 73 cases of helium abundance enhancements (HAEs) were measured, it was suggested by Borrini et al. [15, 16] that HAEs were the interplanetary signatures of CMEs. In 1979, high ionisation states of oxygen and iron were detected following interplanetary shocks [8, 56]. This also provided information on the thermal state of ICMEs, indicating that they were hotter than the surrounding solar wind. It is now believed that the cooler, singly charged helium ions may be associated with filament material known to be associated with CMEs [27, 79, 138, 221]. Other ions and temperature measurements followed, including magnesium and neon, further suggesting the presence of filament material or dense plasma from the low corona or chromosphere [16]. Detailed measurements of ion composition, however, would need to await the next generation of in-situ explorers in the 1990s.

2.3.1.1 Magnetic Clouds

The quest for the identification of a magnetic structure within the ICME was fulfilled in 1981, when Len Burlaga and co-workers identified a smoothly rotating magnetic field vector following an interplanetary shock for an ICME observed with five spacecraft (*Voyager 1* and *2*, *Helios 1* and *2* and *IMP-8*) [25]. They called it a "magnetic cloud" citing early theoretical work dating back to the 1950s [193]. Figure 2.4 shows their sketch of this event, including the structure that later became synonymous with ICMEs: a shock, followed by a sheath, followed by the magnetic cloud.[3] While this paper did not make the connection with solar transients, an accompanying paper [161] did. This paper presented a statistical survey of 45 magnetic clouds, and directly associated many of them with CMEs. It is this paper that first identified the combination of characteristics of magnetic clouds that are still

[3] This is not to be confused with the classic three part CME structure observed in coronagraphs, where the shock and sheath are not involved. Relative to that structure the shock and sheath will form ahead of the leading structure (flux rope).

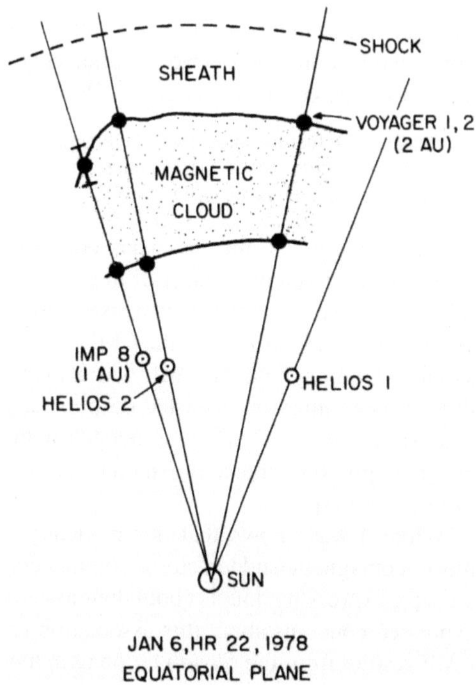

Fig. 2.4 Sketch of the geometry of the magnetic cloud observed by Burlaga et al. [25] from spacecraft observations on 6 January 1978 (their Fig. 5). The dots show where the observed boundaries of the cloud would be at 2200 UT on that day, assuming they moved at constant speed [25]

used today: low temperatures, high magnetic field strength, and a smoothly rotating magnetic field vector. Magnetic clouds also have a long duration, typically about 10–48 h with an average of around 27 h [170].

By 1990, it was accepted that magnetic clouds were probably manifestations of coronal mass ejections [20, 273] and/or filaments [274], that they were regarded as a strong source of increased geomagnetic activity [281] and that they were often drivers of interplanetary shocks [161]. It was also known that only a subset of in-situ ICMEs (30–50%) showed a clear magnetic cloud signature [29, 74]. The rest were identified by other signatures in the solar wind, such as the presence of an interplanetary shock, expansion signatures in the speed and density profiles, energetic particle and temperature decreases, and chemical composition signatures such as HAEs.

2.4 The Solar Flare Myth

In the same year that Chapman & Ferraro made their suggestion of the ionised material ejection as possibly being responsible for geomagnetic storms, Hale [89] suggested that this material came from large solar flares [89]. Dellinger [44] associated flares with the geomagnetic disturbance known as a sudden ionospheric disturbance (or SID), and Newton [197, 198] found a statistical correlation between large flares and magnetic storms. Later in 1950, Chapman [32], who had not mentioned flares in his and Ferraro's initial suggestion of the cause of magnetic storms, then cited flares as the likely cause. Quoting Kahler [156]:

> Thus, we see that by about 1960 there appeared little reason to doubt that all three solar-terrestrial disturbances — large geomagnetic storms, SIDs and SEP events — were directly caused by the flare itself. (p. 115 [156])

This idea continued into the space era. For example, when interplanetary shocks were first observed by spacecraft in the 1960s, they were assumed to be caused by solar flares, even though effective associations were made with only mixed success [78, 132, 133, 229]. So, when the CME was discovered in the 1970s, it was naturally assumed by many that the CME was also the result of a shock wave from the solar flare. This assumption persisted despite early revelations that CMEs and geomagnetic storms were often not associated with flares [45, 81, 155, 180], and that the energy required to launch the mass ejection was much greater than that of the flare itself [180, 263].

While it was known that interplanetary shock waves were the likely cause of most geomagnetic sudden storm commencements, by the early 1970s some workers were expressing doubts about their association with flares. In 1972, Hundhausen expressed concerns about this association [132, 133], and workers using the early CME results from the *Skylab* coronagraph noted the inconsistency between CME and flare occurrence [81]. Joselyn and McIntosh [155] expressed surprise at the small percentage of flare-related geomagnetic storms, and proceeded to question the validity of previous work that found a large percentage of such storms. Sime et al. [223] questioned the validity of describing a CME as a shock front with the observation that the flanks of the CME did not move laterally as the loop top moved outward through the corona. Further evidence, including the movement of surrounding plasma ahead of the CME (implying that the CME cannot be a shock because the shock should be the leading feature) were presented by Sime and Hundhausen [224]. At the same time, Simnett and Harrison [97, 226, 227] found that the flare associated with the CME was confined to a loop at only one footpoint of the CME, while Harrison and co-workers [94, 96, 98] back-projected CMEs to determine their onset time, and found that none of them were coincident with a solar flare onset. They found that typically the flare onset occurred some time later than that of the CME. Figure 2.5 shows two diagrams produced by Harrison [92] demonstrating the relationship between the CME and its associated flare.

By 1992, evidence of a CME-centred concept had been accumulated from virtually every area of space physics research. In his excellent review in the *Annual Reviews of Astronomy and Astrophysics*, Steve Kahler addressed the questions of

> [H]ow did we form such a fundamentally incorrect view of the effects of flares after so much observational and theoretical work... [and] what is the... evidence to support a primary role for CMEs? (p. 114 [156])

The review presented evidence from CME and flare observations themselves, to metric radio bursts, interplanetary shocks and magnetic fields, solar-energetic

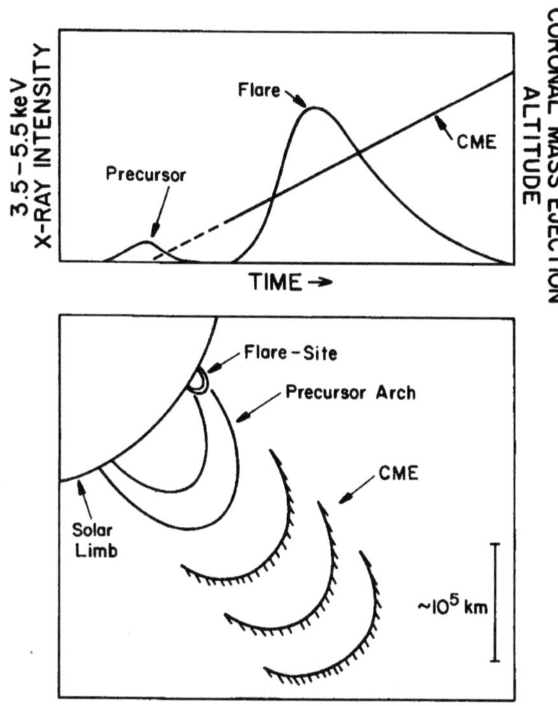

Fig. 2.5 Diagram representing the relationship between the CME and its associated flare (originally from Harrison [92] (his Fig. 6), and reproduced in Hundhausen [134] and Gosling [75]). Reproduced here with permission © ESO. The *top panel* shows the temporal relationship, showing the flare onset time occurring later than that of the CME, while the *bottom panel* shows the structural relationship, with the flare associated with one footpoint of the CME

particles and their geomagnetic consequences. Kahler demonstrated that it is the CME, not the flare, which played the central role in major heliospheric and geomagnetic phenomena [156].

Despite the solidity of evidence, most of the solar physics community continued to advocate the flare as the primary source of space weather. A review by Hudson in 1987 listed 42 great discoveries in solar physics and did not even mention CMEs [129], while a Lockheed Martin x-ray flare poster distributed at the AGU Fall Meeting in 1992 explicitly cited flares as the source for major geomagnetic storms. Finally, following a presentation of a soon-to-be travelling AGU exhibit addressing the Sun–Earth connection but not mentioning CMEs, Jack Gosling decided to write his now famous paper "The Solar Flare Myth", which was published in the *Journal of Geophysical Research* in late 1993 [75]. This paper reviewed and consolidated previous work with the express intention of removing the operation of the flare from "centre stage", at which it was still firmly placed in the eyes of much of the solar-terrestrial community. He confirmed that the source of interplanetary shocks and of most geomagnetic storms was the CME and not the flare, and that the

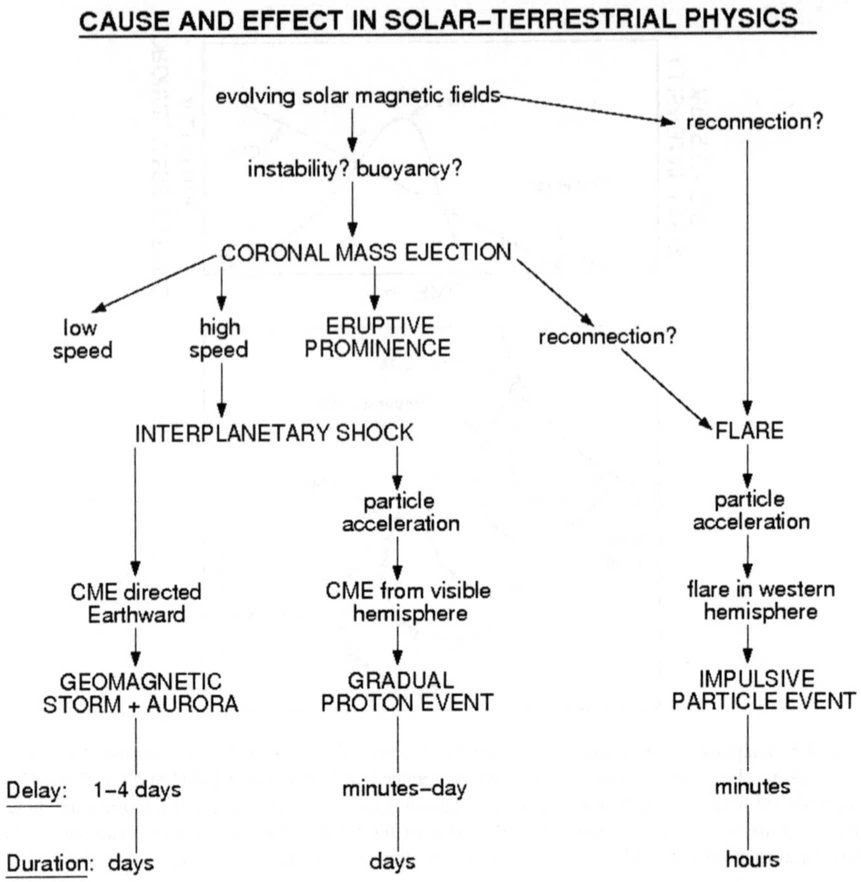

Fig. 2.6 The Gosling "modern paradigm" of cause and effect in solar-terrestrial physics. The events labeled in all caps refer to observational phenomena while lowercase letters indicate physical processes or descriptive characteristics. Reproduced from Gosling [75]

relationship between the flare and CME was secondary at best. He proposed a so called "modern paradigm", shown in Fig. 2.6 (his Fig. 16), describing the relationship between flares, CMEs and geomagnetic activity. Note that two possibilities are suggested for the occurrence of the flare. Either they are connected as secondary to a common physical process (labeled as simply "evolving solar magnetic fields"), or they are a secondary process to that of the CME launch.

This publication caused outrage among the solar physics (particularly the flare) community and the debate intensified. A special session of the AGU Meeting in Baltimore in May 1995 entitled 'Is "The Solar Flare Myth" Really a Myth?' was convened (a session to which Gosling himself was not invited). A challenge paper by Švestka [234] referred to Gosling's conclusions as "faulty and dangerous" and the response by Gosling and Hundhausen (p. 57) accused Švestka and others of attempting to re-classify the definition of a solar flare. A further response by Harrison [93] referred to the attempted reclassification to encompass virtually all eruptions from the Sun as "very misleading". Other challenges (e.g. Hudson et al. [131] and Pudovkin [203]) and responses were issued throughout 1995 and it seemed likely that this debate would remain unresolved for years to come.

Towards the end of 1995, however, the intensity of the Solar Flare Myth debate suddenly appeared to die away. Not coincidentally, the *SOHO* spacecraft was launched in December 1995. Perhaps the clarity of CME data from LASCO proved more conclusive, or perhaps the community was in awe at the quality of the data delivered by *SOHO*, but when the dust settled it appeared that the CME community had prevailed.[4] While the CME and flare "camps" remain largely divided in the solar physics community, it is generally accepted today that CMEs, and consequently large transient solar wind disturbances and geomagnetic storms, are not caused by solar flares. The Gosling paradigm remains the commonly accepted "big picture" of the relationship between CMEs, flares and geomagnetic activity, although it is worth noting that despite this almost unanimous acceptance the remnants of the flare confusion remain. To this day prominent solar physicists still carelessly refer to a flare associated with a CME as its "source". This is often intended to imply the source region of the CME as projected onto the solar surface, but even this association is likely to be inaccurate. Using flares to assist in CME identification is a fundamentally flawed process for two main reasons. Firstly, as often as not the CME is not associated with a significant flare, and secondly, at best, the flare is located at a single footpoint of large CMEs only.

2.5 Interplanetary Scintillation

Also before the detection of CMEs by coronagraphs, the possibility for ICME detection was being investigated from a completely different direction. Sometime in the 1960s (probably 1964 [106]), Tony Hewish and co-workers at Cambridge University discovered that radio signals from distant sources vary as a result of variations in the interplanetary medium This is known as interplanetary scintillation (IPS) and the distortions were observed at radio sources around the metre wavelength level (frequencies around 100 MHz). Using IPS one can monitor the solar wind.[5] and so can monitor density perturbations in the medium. Hence one could track ICME density using IPS. By the discovery of the CME in 1973 several papers on this detection had appeared [107, 114–116]. It was not known at the time whether the transients observed were the same ones observed in the low corona, but it was clear that these were dense structures moving through the interplanetary medium between the Sun and the Earth.

---

[4] The settling of the debate may have occurred at a meeting on CMEs in Bozeman Montana in 1996 where a large number of those from the flare camp were present. According to Jack Gosling (private communication (2009)), Loren Acton, a main player in solar flares, was instrumental in getting the solar community to take notice.

[5] For a review of early work, refer to W. A. Coles in *Space Science Reviews* [37].

By 1978 a number of interplanetary transients had been detected using the IPS technique. Houminer and Hewish [116] and Houminer [115] investigated density enhancements in the interplanetary medium that were at low solar latitudes and appeared to co-rotate with the Sun. Watanabe and co-workers [258, 260] reported on disturbances in the interplanetary medium which they attributed to flare shock waves. Interplanetary scintillation and proton density observed by the *Pioneer 6* and *7* spacecraft from January–April 1971 were found to be strongly correlated by Houminer and Hewish [117], and a relationship with the geomagnetic A_P index was also confirmed [117]. Three transients were identified using IPS by Rickett [211] and these were correlated with *Pioneer 9* and *HEOS 2* at the Earth [211]. A further relationship between the A_P index and scintillation parameters was found by Vlasov et al. [252] and later confirmed using data later than 1978 [253, 254]. The work was performed using four separate ground-based radio arrays, two early arrays at Lords Bridge near Cambridge (the $4\frac{1}{2}$ Acre Array [46][6]) and Pushchino near Moscow (BSA Large-Phased Array [251]), and two later ones in San Diego County [4] and Toyokawa [259].

It should be noted that many of the events mentioned above were more likely the result of enhanced density regions of the Sun brought about by the merger of fast and slow streams, phenomena now known as corotating interaction regions (CIRs). Vlasov et al. (1981) identified two types of large-scale perturbations moving away from the Sun from 0.3 to 1.2 AU away from the Sun, those which vary over times of the order of 24 h (ICMEs), and those that existed for several days (CIRs) [252].

Despite these efforts it remained unclear whether some of the transients observed using IPS were related to the CMEs being observed in the coronagraphs. The CME review paper by MacQueen [180] includes the comment that:

> Radio scintillation measurements have, for the most part, proved to be a disappointment [at associating IPS interplanetary transients with coronagraph mass transients], due principally to the limited temporal and angular coverage brought about by the paucity of suitable radio sources, and also as a result of the low [signal-to-noise ratio] present in the observations of a single event. (p. 618 [180])

It would require an improvement of the IPS technique along with a large statistical database of CMEs produced by the next generation of coronagraphs in order for a firm connection between the two phenomena to be made.

2.5.1 Connecting CME and ICME Images Using IPS Observations

In 1978, the $4\frac{1}{2}$ Acre IPS radio array was upgraded. The collecting area was doubled to create the 3.6 Hectare Array, and the receivers were upgraded. This enabled more radio sources to be monitored, thereby increasing the spatial resolution of the maps

[6] The $4\frac{1}{2}$ Acre Array was the telescope used by Hewish and co-workers to identify the first pulsar, a discovery for which he received the Nobel Prize in Physics in 1974.

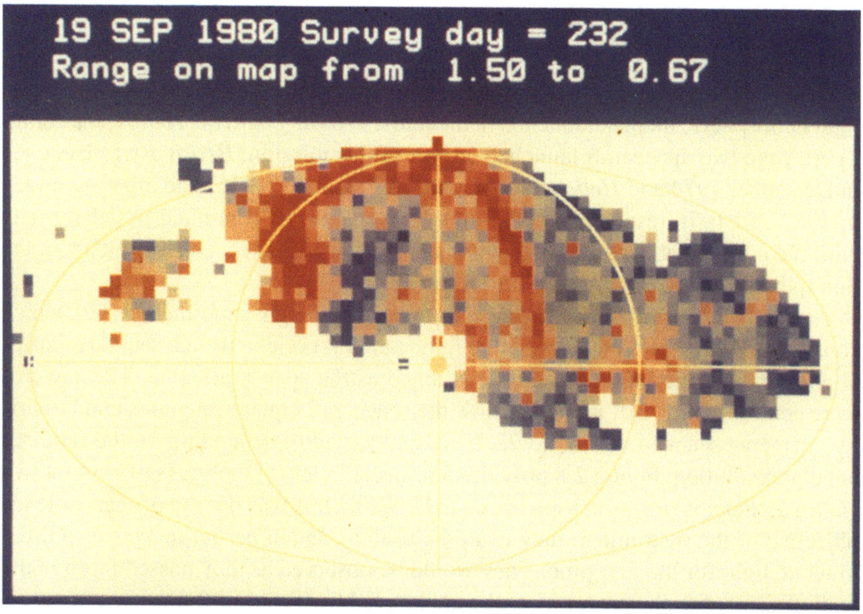

Fig. 2.7 IPS map of an ICME from 19 September 1980. This is a Mollweide projection with the Sun at the centre and the 90° contour shown. Each *square* represents a radio source detected with the 3.6 Hectare Array and a *red square* indicates an increase in density (i.e. part of a possible ICME) [237]

produced by IPS. This, along with the coronagraph dataset provided by *Solwind* and C/P allowed the first comparison of IPS transients with coronagraph CME images. Thus, for the first time ICMEs observed by IPS (the first effective ICME images) were connected with coronagraph CMEs. Figure 2.7 shows an IPS map with a CME from 19 September 1980 from Tappin [237]. Each square on this figure represents a radio source and the red squares are those from which an increase in density has been identified.

In his PhD research work, James Tappin [237] analysed results from a survey of IPS observations from February 1980 to March 1981 using the 3.6 Hectare Array. He identified nine transients with a likely association between CME and IPS transients, three of which were also associated with disappearing filaments. Other papers by Hewish and co-workers later emerged connecting IPS ICMEs with a solar surface feature, but most of these were associated with a low latitude coronal holes [104, 105, 118, 240]. Also through 1981–1985, Woo and co-workers reported on using IPS to study interplanetary shocks, which they connected with blast waves from solar flares [276–278]. It appears that apart from the work of Tappin, a direct association between coronagraph CMEs and IPS ICMEs was not made again until the end of the 1980s (e.g. [222]), or at least not in the literature. Into the 1990s, however, the association was made more readily [144, 275].

2.5.2 White Light ICME Images

Also of vital importance in the connection of CMEs and ICMEs was the first white light heliospheric imager, launched in the mid-1970s on board the *Helios* spacecraft. There were two spacecraft launched as part of this mission, *Helios 1* was launched in December 1974 and *Helios 2* was launched in January 1976. The missions ended in 1982 and 1976 respectively, although both spacecraft continued to deliver data until the mid 1980s. To this day they remain in their highly eccentric orbit about the Sun (perihelion ~0.3 AU, aphelion ~1.0 AU).

Each contained a white light imager as the zodiacal light experiment [169], which consisted of three photometers (white light cameras) oriented such that large strips at constant ecliptic latitude could be scanned as the spacecraft spun. The cameras were centred at 15°, 30° and 90° below the spacecraft equatorial plane, and the first two cameras scanned at 5.6° to 22.5° longitude width, depending on the required angular resolution. Figure 2.8 provides a projected view of the scans of the first two *Helios 2* cameras, from Jackson et al. [151]. While this was not the primary science objective of the instrument, they could be used to obtain partial images of ICMEs in white light for the first time: They would be observed as they passed through the field of view of each camera (note the arc labeled CME in Fig. 2.8).

The usage of *Helios* to detect ICMEs was first demonstrated by Richter et al. [210]. They noted high-latitude "plasma clouds" and measured speeds for a number of them at around 300 km/s. They even associated one event with a CME observed by *Solwind* on 5 June 1979. This CME had a measured speed of 500 km/s and the *Helios* plasma cloud, observed on 6 June 1979, had a speed of 260–330 km/s. Thus they noted that if this was the same event then it had experienced a deceleration en-route. Jackson and co-workers took the observations further, and attempted to

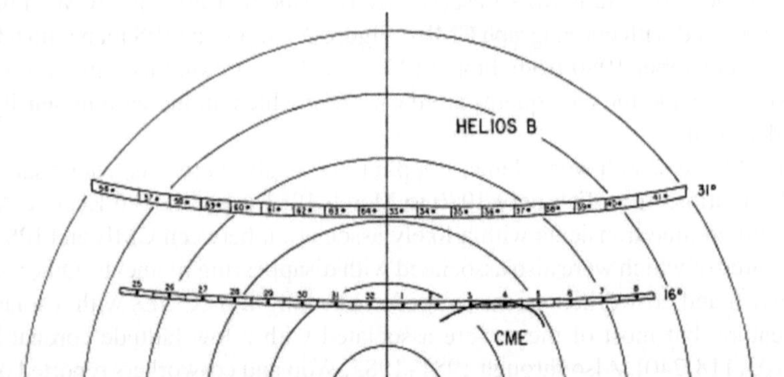

Fig. 2.8 Projection of the field of view of *Helios 2* with a CME leading edge included. This is from a study on the 7 May 1979 CME [151]. The strips shown are from the 15° and 30° photometers from *Helios 2*, with a longitudinal width of 5.6° (Reproduced with kind permission of Springer Science and Business Media)

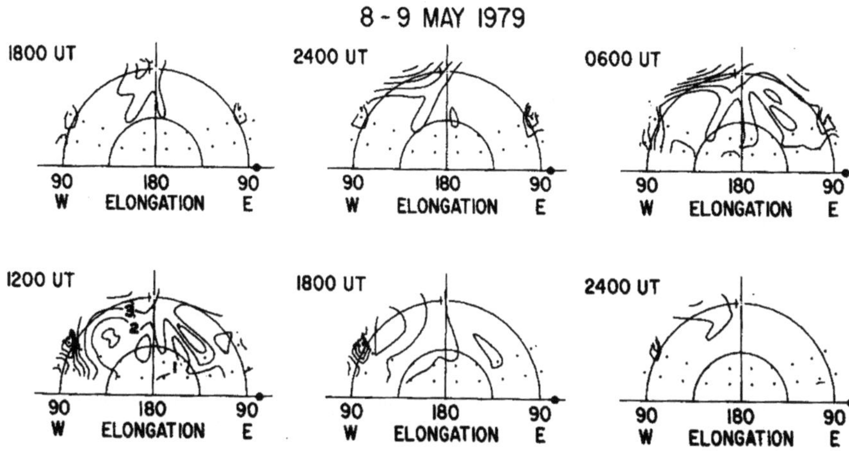

Fig. 2.9 Early attempts at ICME 3-D production for the event on 8–9 May 1979 CME [150]

produce low-resolution images of the density changes. Early works using this technique were published in 1985, 1988 and 1989 [141, 143, 150, 151] and Fig. 2.9 show the results from one of these early production attempts. The technique was developed further through the 1990s [109, 147] and eventually *Helios* white light data were used with IPS data [108]. This is now termed tomography [143] and is still in use for ICMEs mostly by the University of California, San Diego (CASS) today. Other contemporary work compared *Helios* transients with coronagraph CMEs and interplanetary shocks [142, 149, 265, 266], and by the end of the *Helios* era the association between coronagraph CMEs and white light and IPS ICMEs had been firmly established.

Although the early attempts at 3-D construction by *Helios* have been questioned by some, the zodiacal light experiment was a success as a proof of concept for a white light heliospheric imager. Without the success of this instrument, it is likely that the next generation of heliospheric imagers would not have been constructed, even though it took 20 years for the next one to emerge.

2.5.3 Contribution to the Solar Flare Myth Debate

As should be obvious, interplanetary imaging of ICMEs using white light imagers and IPS were contemporary with the Solar Flare Myth debate. So those working in IPS and with *Helios* made their contribution. Early IPS work attempted to connect interplanetary transients with flares [258] and Tappin tried to make this connection with little success [237, 240]. Studies which did make the flare connection include those by Woo and colleagues [276–278]. Hence, by 1985 the connection between IPS ICMEs and solar flares remained ambiguous.

From the white light ICME side, although the majority of the *Helios* photometer workers were in the solar flare "camp", studies attempting to link the white light

ICMEs were also met with mixed success. Richter et al. [210] in their first results paper tried, but failed to associate solar flares with their observed transients [210], yet others successfully made the connection [235]. Most of the studies, however, wisely focused mostly on the CME connection, and spent less time on the flares.

The latter half of the 1980s saw a series of studies accumulating evidence of a lack of flare association with IPS ICMEs. In a survey of 96 interplanetary transients using IPS, Hewish and Bravo, writing in *Solar Physics* in 1986, found that many events had no association with flares or even disappearing filaments [105]. Houminer and Hewish [118] showed that the large geomagnetic disturbance in August 1972 was not associated with a flare, and presented a case in favour of a coronal hole source. In the same year Hewish, in an article in *New Scientist* stated:

> The widely held textbook theory is that solar flares are responsible... [T]he energy launches blast waves upwards into the solar atmosphere, producing an interplanetary shock wave that could ultimately reach the Earth and produce magnetic storms. This has been the generally accepted theory for the past 50 years, but our new method of mapping interplanetary weather with a radio telescope does not agree with it. (p. 48 [103])

Note that here Hewish refers to "interplanetary weather", which has now been replaced with the more catchy term "space weather". While Hewish and colleagues clearly rejected the notion that magnetic storms were caused by solar flares, they mostly believed that coronal holes were the source of ICMEs and of major space weather at the Earth. Today it is generally accepted that major magnetic storms are probably caused by erupting closed magnetic field structures (CMEs), which do create the occasional coronal hole (see coronal dimming in Sect. 7.2.6). It is also quite clear that recurrent geomagnetic activity is caused by corotating fast streams from equatorial coronal holes or the equatorial extension to polar coronal holes, but the storms caused by these are generally (but not always) not particularly large.

2.6 The 1990s: The Next Generation of Imaging and In-Situ Spacecraft

A list of the spacecraft making significant contributions to our understanding of CMEs is provided in Chap. 3. Section 3.3 reveals that only one such spacecraft was launched in the 1980s (*SMM*), and had that been launched only 2 months earlier it would have been a 1970s launch. While many of the spacecraft launched in the 1970s continued to function well into the 1980s (and some into the 1990s and later), after *SMM* no new missions of significance to CME study were launched throughout the 1980s decade. Furthermore, once *SMM* re-entered Earth's atmosphere in 1989, no continuous surveillance of the outer corona occurred until 1996.[7]

[7] The low corona did continue to be observed by ground coronagraphs throughout, solar flare activity was monitored by the *GOES* spacecraft, and brief studies of the outer corona were provided by the SPARTAN-201 flights with the Shuttle in 1993, 1994, 1995 and 1998.

So, after such a productive decade of space-based solar observatories, why was there such a sudden and prolonged slump? There are two reasons. Firstly, the solar community was largely focused on solar flares, not CMEs. The second reason, I'm convinced, lies in the changes in priority for NASA launches during the 1980s. My understanding is that following the success of the Shuttle (first launched in April 1981), an edict was issued requiring scientific (particularly NASA-funded) spacecraft to be launched via the Shuttle. Unfortunately, launches fell behind in schedule from the very beginning, and thus did the launch of many missions. *Ulysses*, for example, was scheduled to be launched on board *Challenger* in its very next flight following STS-51L on 31 January 1986. History remembers the tragic events of that morning, and the loss of the spacecraft with all hands. The grounding of the Shuttle fleet following the *Challenger* disaster created further delays, and *Ulysses* was finally launched on board *Discovery* in October 1990. It seems likely that NASA had changed its priorities for launches by then, for the next generation of solar and interplanetary medium space observatories would not be launched from the Shuttle.[8]

2.6.1 In-Situ Probes: Ulysses, WIND and ACE

As mentioned in the previous section, the launch of *Ulysses* was delayed considerably through the 1980s and was launched in 1990. On board was an assortment of magnetic field, energetic particle and other experiments (Sect. 3.4) but what was unique about *Ulysses* was its orbit. This spacecraft was charged with the exceptionally difficult task of leaving the ecliptic plane and achieving a near polar orbit around the Sun. It achieved this by a gravitational assist around the planet Jupiter, and it needed to pass closer to the planet than any previous artificial object to do so. *Ulysses* passed within six Jovian radii in February 1992, and provided unprecedented information about the planet's magnetosphere in the process. It then moved into its (almost) polar orbit about the Sun, where it has performed over three complete orbits to date. Figure 2.10 shows the third orbit of *Ulysses* from ~2002–2008.

Given its unique orbit, the contributions to solar and interplanetary exploration made by *Ulysses* were mainly of observations of the polar regions of the Sun. It was already known that the polar regions lay on open magnetic field lines and at the surface were polar coronal holes, long known to be the source of fast-flowing solar wind. *Ulysses* found, for example, that there were two solar wind types, with fast solar wind emanating from the polar regions and slow solar wind at lower latitudes, and no intermediate speed solar wind in-between [185, 255]. Other discoveries include the behaviour of the solar dipole, which seemed to act more like two monopole fields due to the distortion by the solar wind [7], and the blockage of cosmic rays into the polar regions by high levels of Alfvén waves [113]. It was also found that polar magnetic fields were connected with the equatorial regions, if one moved far enough away from the Sun [58, 214].

[8] With the exception of a number of brief flights with the SPARTAN-201 coronagraph.

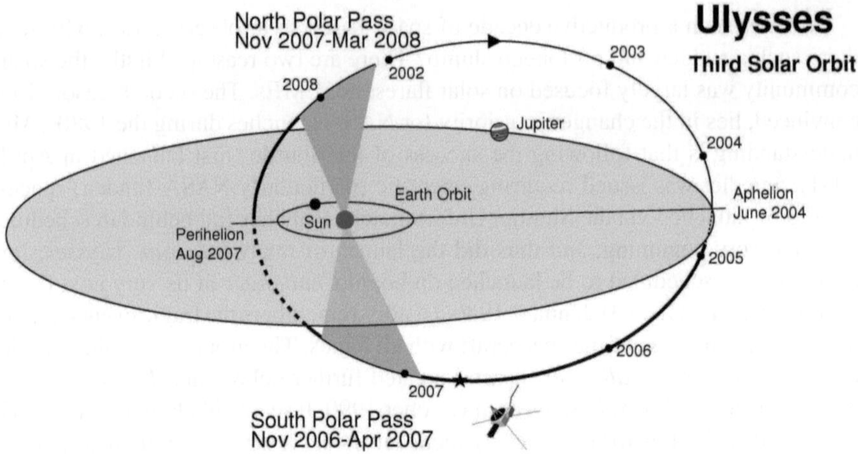

Fig. 2.10 Diagram of the third orbit of *Ulysses* (available courtesy of NASA)

Ulysses also contributed to our understanding of the nature of ICMEs. Gosling (1994) showed that CMEs can occur in the fast solar wind [77] and characteristics of ICMEs at distances from the Sun out to 5 AU were observed in large numbers [26, 80, 217, 282]. Later, ICMEs in *Ulysses* were compared with transients observed with IPS [153, 154] and with heliospheric white light images when they became available [238]. Charge state distributions of ICMEs were investigated using *Ulysses* data by Henke et al. (1998, 2001), who found that the charge state ratios of heavy solar wind ions (C^{6+}/C^{5+}, O^{7+}/O^{6+}, Si^{10+}/Si^{9+}, Fe^{12+}/Fe^{11+}) were related to the structure of the internal magnetic field [100, 101].

In February 2008, *Ulysses* lost its secondary X-band transmitter which, among other things, allowed the regulation of temperature of the spacecraft. As it was on its way out away from the Sun, operators predicted 6 weeks before it froze to a point beyond operation. Despite this, the spacecraft continued to function for a further 18 months, and was finally turned off on 30 June 2009, just 4 months shy of its 19th anniversary from launch.

The next two solar/interplanetary in-situ spacecraft (*WIND*, launched November 1994 and *ACE*, launched August 1997: refer to Sect. 3.4) contained sophisticated instrumentation and have since been used as scientific and monitoring probes. The instruments on board the *WIND* and *ACE* spacecraft were mostly improved or modified versions of those already tested on previous missions. Many of the instruments on board *ACE*, for example, are actually flight spares from *Ulysses* and *WIND*,[9] and *ACE*/SEPICA is an upgraded version of the *ISEE-3* solar wind particle analyser. Likewise, the orbits from each spacecraft are not unlike those that had been

[9] EPAM is the flight spare for *Ulysses*/HI-SCALE, SWEPAM is the flight spare for *Ulysses*/SWOOPS, SWICS is the flight spare for *Ulysses*/SWICS, and MAG is the flight spare from *WIND*/MFI.

seen before. The L1 location of *ACE* had been previously occupied by *ISEE-3*, and *WIND* was certainly not the first spacecraft to assume a high-Earth orbit that passed into the upstream solar wind region beyond the magnetosphere (the *Vela* spacecraft were in such orbits in the early 1960s (Sect. 3.1)), although its orbit did take on a slightly unique posture.

So, along with detailed observations of the solar wind and ICMEs, these spacecraft provided a continuous datastream of interplanetary medium and ICME properties that remain continuous to this day. Their enhanced instrumentation also provided more in-depth studies of phenomena in the interplanetary medium. This, coupled with the next generation of imaging instruments (discussed in the following sections), allowed for the first time a reliable continuous monitoring of solar, interplanetary and magnetospheric activity simultaneously from the perspective of instruments in long-term stable orbits. This greatly enhanced the space weather forecasting product that is still in use. *ACE* remains the most crucial early-detection system for space weather, as it is always in the sunward direction of the Earth and is capable of monitoring both ICME ram pressure and (crucially) magnetic field orientation. *WIND* provides a similar monitoring capacity as well, but its new orbit does not place it Sunward of the Earth very often, and it is too close to the Earth to provide sufficiently advanced forecasting. Also, *WIND* data are not available in real time.

Statistical studies of ICMEs using *WIND* and *ACE* include Cane and Richardson [28,208], Lynch et al. [176,177] and Howard and Tappin [124]. The main scientific contributions provided by these spacecraft seem to be advances in our understanding of the composition of ICMEs, and of magnetic clouds. Iron charge distribution of ICMEs were investigated by *ACE* by Gloeckler et al. [67] and Lepri et al. [173] who found typical charges of 9+–11+, but charges greater than 16+ were also identified [67, 173]. As with earlier studies, the higher charge states were attributed to hot plasma originating low in the solar corona or from initial heating during the launch of the CME. Later work includes the investigation of solar wind heating by ICME-driven shocks [164], and the relationship between composition and solar surface parameters [164,207].

As mentioned in Sect. 2.3.1.1, it was already known by 1990 that only a small fraction of ICMEs were observed to contain the recognised magnetic cloud structure. However, by that year a global picture of the structure of a magnetic cloud had already been formed [22]. This picture, shown in Fig. 2.11 is described at 1 AU by Lepping [170]:

> [M]agnetic clouds at 1 AU are approximately force-free structures. The magnetic cloud's geometry is that of a nested set of helical magnetic field (**B**) lines confined to a flux tube, which is curved on a scale of about 1 AU (or maybe a little smaller at its nose) ... when considered globally... When examined locally, the structure is approximately cylindrically symmetric, and the pitch angle of the helical field lines increases with increasing distance from the axis of the cloud, such that the field is aligned with the axis of symmetry at the position of the axis and perpendicular to it on the cloud's boundary, in most cases. (pp. 80–81 [170])

Consequently, empirical modelling of magnetic clouds using in-situ data were based on this global picture. A model developed by Burlaga [19] and refined by

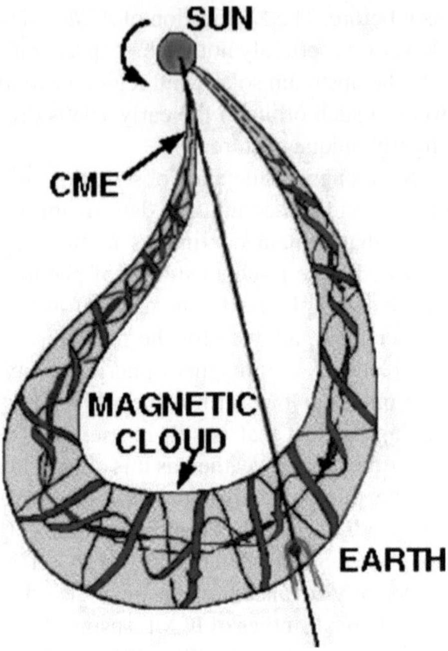

Fig. 2.11 Sketch of a global view of a magnetic cloud through the ecliptic plane, with solar rotation taken into account [183]

Lepping et al. [171] using *ISEE-3* and *IMP-8* (and described further in Sect. 7.3.2) set the scene for magnetic cloud reconstruction techniques for when the next generation of in-situ data became available. Several different methods for such reconstruction followed, and Riley et al. [212] provides a review of the more popular models.

Another important discovery using *ACE* data is magnetic reconnection within ICMEs, including magnetic clouds. Briefly, magnetic reconnection is the process where field lines from different magnetic regimes are connected, which violates the frozen-in field condition of magnetohydrodynamics (MHD). This enables the transfer of energetic particles across magnetic regimes, and the release of large amounts of stored energy in magnetised plasma environments. We discuss this in further detail with regard to the Earth's magnetic field in Sect. 10.3. The theory of reconnection as applying to solar flares dates back to the electromagnetic "neutral point" idea of Giovanelli [66], which was developed for MHD by Dungey and others through the 1950s [51, 52, 200, 236] and established by Petschek [201]. The exact role of reconnection in CME launch and evolution remains to this day unknown (we discuss some theories surrounding this role in Chap. 8), and although some evidence for reconnection has been observed in coronagraph CME data [225], it had not been directly observed by in-situ spacecraft until recently. Observations by Gosling and co-workers using *ACE* identified this signature [84, 85] and by the end of 2005 over 40 reconnection events had been observed by *ACE* with a large fraction within ICMEs [76].

Since the launch of *WIND* in 1994 several hundred publications have appeared dealing with magnetic clouds with *Ulysses*, *WIND* and *ACE*. Studies have involved comparing them with CMEs [28, 208], solar surface structures [40, 168] and solar flares [204], geoeffectiveness [55, 159], magnetic reconnection [84, 85] and even internal whistler wave propagation [194]. The general picture of magnetic clouds remains as was defined from their discovery, but they are now an integral and essential part of ICME study.

2.6.2 Imaging Observatories: Yohkoh, TRACE

The next generation of imaging observatories began with the launch of the Japanese spacecraft *Yohkoh* in August 1991 (Sect. 3.4). The spacecraft was abundant with imagers, including soft and hard x-ray, and x-ray and gamma-ray spectrometers. The most popularly used instrument appears to be the Soft X-ray Telescope (SXT [246]), which provided whole-disk images of the Sun in soft x-rays. New features associated with CMEs were observed with this instrument, including the "sigmoid" of active regions [1, 215] (Sect. 7.2.8) and a renewed inspiration to study coronal dimming [73, 232] (Sect. 7.2.6). Of great significance was the overlap in observing between *Yohkoh* and *SOHO*, which does not have an x-ray imager on board. Studies comparing CMEs with *Yohkoh*/SXT include Webb [261], Hudson et al. [130] and Sterling et al. [233].

Another milestone in solar coronal observations was provided by the *Transition Region And Coronal Explorer* (*TRACE*) spacecraft. This is a dedicated NASA Small Explorer (SMEX) mission providing high-resolution, multiwavelength EUV images of a selected region of the Sun [90]. Because of its limited field-of-view, it is a campaign-based instrument, meaning that researchers apply for observing time on the instrument to study a region of interest on the Sun. *TRACE* provided views of the corona in unprecedented detail, along with some spectacular movies of eruptions in the low corona. It has allowed an investigation of the structure of flux ropes and loops in the low corona, along with the nature of the helicity (twist) of magnetic structures during an eruption [3, 6]. Thus it has also assisted in model development of CME eruptions. CME work involving *TRACE* includes Zhang and Wang [280], Goff et al. [69], and Qiu et al. [204].

Yohkoh observed the Sun for 10 years (almost an entire solar cycle) until it ceased operations in December 2001. *TRACE* completed its final observing sequence in June 2010 after just over 12 years in operation.

2.6.3 The Solar and Heliospheric Observatory (SOHO)

There are many reasons why the cornerstone of solar observing to this day is the *Solar and Heliospheric Observatory* (*SOHO*), and not just because it was identified by the European Space Agency (ESA) as part of the "cornerstone" of its long-term

Table 2.1 The 12 *SOHO* instruments in order of field of view. The instruments are identified first by their acronym then their full names, their field of view (if applicable), primary purpose and a reference to the instrument paper is also provided. Each of these papers were published in a special edition of the *Solar Physics* journal in 1995

Acron.	Name	Field of view	Primary purpose	Ref.
SWAN	Solar Wind Anisotropies	Whole-sky	Lyman alpha radiation detector	[11]
LASCO	Large Angle Spectroscopic Coronagraph	1.1–30 $R_\odot$	White light and EUV coronagraph	[17]
EIT	Extreme-ultraviolet Imaging Telescope	Full solar disk	Multiwavelength EUV imager.	[43]
MDI	Michelson Doppler Imager	Full solar disk	Solar oscillations and magnetic field investigation	[218]
UVCS	UltraViolet Coronagraph Spectrometer	$\sim 40 \times 60'$	UV spectroscopy and visible polarimetry studies	[162]
CDS	Coronal Diagnostic Spectrometer	$\sim 240 \times 240''$	EUV imaging spectrometer	[95]
SUMER	Solar Ultraviolet Measurements of Emitted Radiation	Thin slits	EUV analysis	[272]
CELIAS	Charge, Element, and Isotope Analysis System	N/A	Solar wind and particle detector	[119]
COSTEP	Comprehensive Suprathermal and Energetic Particle Analyzer	N/A	Energetic particle detector	[195]
ERNE	Energetic and Relativistic Nuclei and Electron experiment	N/A	Energetic particle detector	[243]
GOLF	Global Oscillations at Low Frequencies	N/A	Helioseismology observer	[63]
VIRGO	Variability of the Solar Irradience and Gravity Oscillations	N/A	Helioseismology and radiometry	[62]

"Horizon 2000" science program. First and foremost was the quality and variety of the data provided by its 12 instruments (summarised in Table 2.1). While many of these types of instruments had been used in the past, on *SOHO* they were of higher resolution and quality, and were all available on board a single spacecraft.

With regard to CME study, the EUV imager EIT and spectrometer CDS provided invaluable information on solar eruptions associated with CMEs, but the major contributors to CME research were of course the coronagraphs. LASCO originally consisted of three coronagraphs, C1 with a field of view (FOV) of 1.1–3.0 $R_\odot$, C2 (FOV 1.5–6.0 $R_\odot$) and C3 (FOV 3.7–30 $R_\odot$). C2 and C3 are white light imagers, while C1 observed at variable EUV wavelengths.

For $2\frac{1}{2}$ years *SOHO* returned images of unprecedented detail on the Sun, including CMEs. The sensitivity of LASCO led to halo (Earth directed) CMEs to be easily detected for the first time, and a large statistical database of CME observations had begun. Then on 25 June 1998, the spacecraft suddenly went into an uncontrollable spin and was lost for around a month. It was located on 23 July by a radio telescope and was dead in space, but careful analysis of its spin and trajectory enabled a prediction for when solar panels would be pointing at the Sun, providing power to the spacecraft. The first signal was received on 3 August and it was fully recovered by 16 September. Some ingenious engineering and scientific analysis went into the recovery of *SOHO*, including a study of the images of the Sun as they moved in and out of the field of view. The incident was later attributed to a sequence of operational errors leading to both gyroscopes being left off [245] but it was actually caused by a combination of blunders. It is very rare for a spacecraft to be recovered once such a malfunction has occurred. This is the only critical malfunction to occur in *SOHO* during its 15 years lifetime, although since January 2003 its data transmission capabilities have been limited following a malfunction in the pointing system of its high-gain antenna, which is now unable to move. Since the *SDO* launch some of *SOHO*'s instruments have become redundant and so have been turned off. Most recently MDI in April 2011.

Almost all of the instruments on board *SOHO* returned to operation unscathed during the 1998 incident, with the exception of the LASCO/C1 camera. Nobody knows for certain what happened to C1, but it generally believed that one of the glass plates of the Fabry-Perot was misaligned when the instrument froze (in the early stages of the spin, the side of the spacecraft on which LASCO was located was pointing away from the Sun). Thus, C1 was disabled before most workers had really figured out how to work with it.[10] Unlike C2 and C3 which were mandated to continuously observe the Sun (following a decision by the late Guenther Brueckner, original Principal Investigator of LASCO), C1 remained essentially a campaign instrument, so joint studies with the other coronagraphs were often difficult. Hence, relatively few scientific investigations of CMEs have been performed using C1. Some examples include Plunkett et al. [202], Cook et al. [38], and Mierla et al. [191].

The other two instruments, C2 and C3 have gone on to great heights. The actual number of publications using LASCO is virtually impossible to identify, but it easily numbers in the thousands and probably tens of thousands. Along with the detailed study of CMEs, LASCO has assisted in research from solar wind origination to space weather to comet discovery. While many of the parameters CMEs documented had been measured with previous instruments (kinematics, mass, energy), LASCO provided them with a sensitivity not before seen, and has now for the first time provided a continuous dataset of observations for more than an entire solar cycle. Two popular CME catalogs have appeared, managed by NRL (http://lasco-www.nrl.navy.mil/cmelist.html) and Goddard (http://cdaw.gsfc.nasa.gov/CME_list/). From 1996 to the end of 2008, the latter provided details on just under 14,000 CMEs observed with LASCO C2 and C3. Milestones achieved by these instruments include:

1. The largest database of CMEs, by an order of magnitude.
2. The first statistical database of CME properties over an entire solar cycle.

---

[10] I like to joke that LASCO should now be called LACO, as the spectroscopic capabilities of the instrument were lost when C1 ceased to operate.

3. Unprecedented statistical details on CME properties [279].
4. The first large database of halo CMEs and their relationship with space weather and interplanetary shocks [124].
5. The first attempts at 3-D reconstruction of CMEs [123].
6. Studies of the onset region of CMEs, including evidence of magnetic reconnection [225].

Although some of the instruments on board *SOHO* have been eclipsed by their successors (e.g. *SDO*), LASCO remains crucial to the solar mission despite its succession by newer space coronagraphs. It is probably the main reason for why the *SOHO* mission remains active – had it not been for the unique dataset provided by LASCO, the spacecraft would likely have been retired in 2008.

2.6.3.1 Information Exchange

Another reason for the success of *SOHO* is the manner with which its information was disseminated to the scientific and general community. The importance of the internet, or specifically the World Wide Web regarding information exchange cannot be overstated. For previous space missions, data were available either via specific publications (such as the King NSSDC Interplanetary Medium Data Book [39, 160] or from scientific publications), or via personal request to those responsible for collecting and maintaining the data (i.e. those who owned the dataset). The latter particularly applied to early coronagraph data. Hence, scientific information related to the Sun and CMEs was typically only available to the scientific elite, and public exposure was extremely limited. The World Wide Web became widely used around the time of the launch of *SOHO*,[11] which enabled NASA to introduce its open-data policy. Hence through the Web, data from *SOHO* were universally available, thereby increasing its popularity, accessibility and public exposure. Scientists and the general public alike can to this day access *SOHO* data in whichever form is preferred, from the high-quality "FITS" files to complete images constructed by the *SOHO* analysis teams. NASA now encourages an open data policy on many of its scientific missions, and it is now the accepted standard for the generations of solar and space missions (and not only the NASA ones) that have followed. Table 2.2 provides a selection of popular websites at which space science data are accessible to the general public. There are many others and the list continues to grow.

UFOs? Or Misuse of Data?

Provision of scientific data to the general public has proven to be an outstanding success in terms of exposure, scrutiny and public participation in scientific work. For example, the University of California, Berkeley's SETI@Home project, launched in

[11] The World Wide Web Consortium (WC3), for example, was founded in September 1994 [152].

Table 2.2 A selection of popular websites at which space science data are available. These range from the general to the specific, such as those managed by a single spacecraft team

Name	Manged by	URL
General		
National Space Science Data Center	NASA/GSFC	http://nssdc.gsfc.nasa.gov/space/
Coordinated Data Analysis Workshop (CDAW)	NASA/GSFC	http://cdaw.gsfc.nasa.gov/
Solar-Geophysical Data	NOAA	http://sgd.ngdc.noaa.gov/sgd/jsp/solarindex.jsp
Solar Data Analysis Center	NASA/GSFC	http://umbra.nascom.nasa.gov/sdac.html
Community Coordinated Modeling Center	NASA/GSFC	http://ccmc.gsfc.nasa.gov/models/index.php
Solar Data	NOAA	ftp://ftp.ngdc.noaa.gov/STP/SOLAR_DATA/
Specific		
SOHO Data	NASA/GSFC	http://sohowww.nascom.nasa.gov/data/data.html
STEREO Science Center	NASA/GSFC	http://stereo-ssc.nascom.nasa.gov/data/
ACE Science Center	Caltech	http://www.srl.caltech.edu/ACE/ASC/
WIND MFI Page	NASA/GSFC	http://lepmfi.gsfc.nasa.gov/
Yohkoh SXT Data	MSU	http://www.lmsal.com/SXT/data.html

1999 [2], has taken public participation to a new level, by analysing data using the processing power of home computers. Comet hunters have used the *SOHO*/LASCO coronagraphs to identify hundreds of new comets, assisting LASCO in becoming the instrument with the greatest number of comet discoveries (a review on the comet work can be found in Biesecker et al. [12] and a "Sungrazing Comet" website has been established by NRL which can be found at http://sungrazer.nrl.navy.mil/index.php). Accessibility has also improved public awareness and interest in space science and has encouraged new generations of scientists (myself included) to join the community.

The downside to an open data policy is the increased risk of misuse or misinterpretation. For example, there are no shortage of conspiracy theories surrounding data from Mars and the other planets. Regarding solar instruments, my personal favourite (probably because I was working in the UK Midlands at the time) involves the "UFO superhighway" supposedly passing through the field of view of some of the solar imagers. It was cited by UFO hunters as the most conclusive evidence of UFOs yet obtained. Here's the story:

In January 2003, the British Newspaper *The Daily Mail* reported that *SOHO* was beaming back:

> hundreds of images of UFOs travelling along a kind of super-highway. [249]

They were provided with one such image by the owner of an electronics company in Manchester, who stated that the images were provided to him by a Spanish businessman using a large dish to directly collect *SOHO* data.

The objects observed were claimed to be on-edge flying saucers that were only a few hundred miles from *SOHO* itself, and that they failed to navigate a straight course, indicating a form of intelligence. Other newspapers reported the claims, including the *Perth Sunday Times* and *The Evening News of Scotland* and they were due for presentation at the National Space Centre in Leicester.

Unfortunately (and not surprisingly), these claims turned out to be untrue. Suspicions were immediately raised when the claimants did not release the original images, but only the digitally enhanced ones. It was also noted that the "UFOs" were always seen edge-on. NASA responded later that month with a "How to Make Your Own UFO" page, which can be found on the *SOHO* website at http://sohowww.nascom.nasa.gov/hotshots/2003_01_17/. They showed that bright objects, such as planets and (in that case) cosmic ray hits saturate the cameras on *SOHO*, resulting in a "bleeding" of the intensity of the object to the left and right of the instrument's CCD. These can be isolated and heavily processed using any number of graphics-editing software packages, to produce the "UFO" image. Figure 2.12 shows how this can be achieved.

Finally, the success of *SOHO* can also be attributed to the publicity, or "PR machine" behind the mission. Along with the open-data policy, the "PR" team was responsible for public outreach programs, classes for students including an online "Ask Dr. SOHO" email exchange, a screen saver, a cutout module of the spacecraft, and even souvenirs ranging from cards with moving images to small satchels of sunscreen with an image of the Sun as observed in EUV. The *SOHO* team appears to have been the first solar research team to seriously attempt to reach the general public in this way, and its success has set a precedent for the missions that followed.

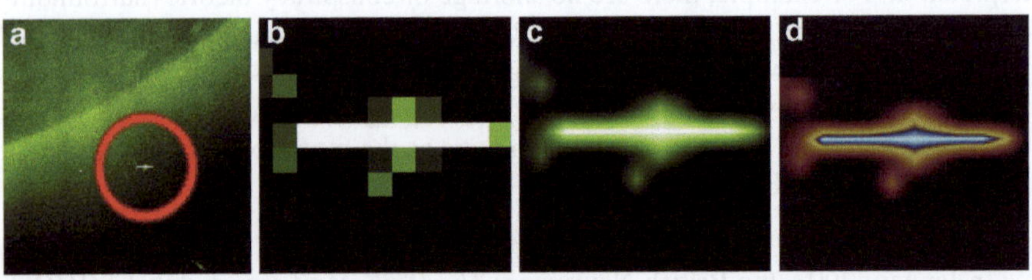

Fig. 2.12 How to make a UFO using *SOHO* images. (**a**) The original: a section of a *SOHO*/EIT image. The circle highlights a cosmic ray hit. (**b**) Step 1: Cut-out the region of interest and enlarge, this image shows the cosmic ray pixels highlighted on the image to the left, with a little different colour scaling. (**c**) Step 2: Interpolate the enlarged image (using any one of many methods) instead of simply re-sampling it. (**d**) Step 3: Change and "touch up" the colour table, and we have what may look like a nice UFO with a glow and exhaust fumes! (Courtesy of the NASA *SOHO* team)

2.7 The 2000s: Continuous Monitoring of CMEs, ICMEs and Space Weather

We now move towards the present day, and the contributions from missions launched since 2000. There are five missions of relevance in this decade, all of them containing imagers and one with in-situ instruments as well. Regarding their contribution to our understanding of CMEs, three of these missions are in a similar league with *Yohkoh* and *TRACE*, that is they provided detailed information on solar surface characteristics, but nothing beyond the low corona in the Sun. These are reviewed briefly and further details may be found in Sect. 3.5. The other two have contributed greatly to our understanding of CMEs and particularly ICMEs, and are discussed in greater detail.

2.7.1 Low Corona and Solar Observers: RHESSI and Hinode

The first of the twenty-first century dedicated solar observatories was the *High Energy Solar Spectroscopic Imager* (*HESSI*), launched in February 2002 and renamed *RHESSI* in honour of Reuven Ramaty who passed away in 2001 [112]. *RHESSI* is a hard and soft x-ray imager, and in such capacity cannot observe CMEs directly. However the solar surface phenomena associated with CMEs have been studied in great detail. Among the achievements of these instruments are microflare heating of the solar corona [10] and polarization measurements of solar gamma-ray flares [13], and while little work has been done with CMEs, some work includes flare/CME energy comparison [53] and CME-associated coronal waves [256].

In September 2006, the Japanese spacecraft *Hinode* was launched. As with *RHESSI*, the purpose of *Hinode* is to investigate the solar surface, not CMEs. On board are x-ray and visible light imagers (SXT [71] and SOT [247]) and an EUV spectrometer (EIS [41]). CME work with *Hinode* to date has been mainly for comparison purposes, to investigate the solar response to CME launches. Such work includes Harra et al. [91], Webb [262] and MacIntosh [179].

It is also important to note that although not dedicated solar observatories, *GOES* spacecraft also carry soft x-ray instruments for monitoring the Sun. They have been doing so since 1974 as part of the Sun Environment Monitor (SEM). *GOES-12* (launched in 2001) was the first to include an x-ray imager. The *GOES* instruments are used for cataloging x-ray flares – the A, B, C, M and X classes describe x-ray flux as observed by *GOES*.

2.7.2 2003: SMEI – The First Complete White Light ICME Images

In the same month in 2003 that the news of UFOs in the *SOHO* images were being addressed, the USAF/NRL spacecraft *Coriolis* was launched. On board was a polarimetric microwave radiometer designed for space-based monitoring of ocean wind speed and direction, called WINDSAT. Also on board was an almost full-sky camera payload, designed to monitor interplanetary transients in white light. The instrument, called the Solar Mass Ejection Imager or SMEI [54, 145] was the first instrument since *Helios* (20 years prior) to observe the outer corona/interplanetary medium in white light and the first instrument to provide complete white light images of ICMEs.

SMEI consists of three scanning cameras which build up an image of the sky throughout its 102 polar orbit about the Earth. It observes the sky starting from around 20° elongation, or around 0.35 AU (75 $R_{\odot}$). While parts of the images are often contaminated by aurora and particle noise from the passage of the spacecraft through various regions of the magnetosphere (i.e. the polar caps and South Atlantic Anomaly), SMEI has allowed for the first time direct measurement of complete ICMEs through white light. As with coronagraph images, the images are heavily projected, but unlike coronagraphs the projection effects may be reduced with the application of some sophisticated geometry (refer to Sect. 5.3). Hence, SMEI allows three-dimensional reconstructions of ICMEs.

Early work with SMEI involved mostly height-time comparisons with coronagraph CMEs, interplanetary shocks and geomagnetic storms [128, 238, 239, 264] but some three-dimensional work has been attempted from the start. For example, Jackson and co-workers have extended his tomographic reconstruction work (which originally used *Helios* and later IPS work) to include SMEI data [146] and myself and co-workers have performed more simplified reconstruction techniques based on leading edge measurements of SMEI ICMEs [123, 128]. Figure 2.13 shows results from one such event.

The most recent work with SMEI involves the utilisation of the projection effects in order to accurately reconstruct the three-dimensional structure and trajectory of ICMEs. The theory behind this reconstruction is discussed in Chap. 5, but briefly it involves applying the physics of CME appearance and geometry relative to an observer in order to reconstruct the ICME itself. This work is still in its infancy but discussion of the development and utility of this technique has appeared in a series of three papers by myself and James Tappin in *Space Science Reviews* [126, 127, 241].

2.7.3 2006: STEREO – A New Approach to Solar Observation

The successes of SMEI and *Helios* contributed to the launch of a white light heliospheric imager on board the next solar observatory. The *STEREO* spacecraft [158], launched a month after the launch of *Hinode* in 2006, assumed an orbit and suite of instruments never before seen on a solar mission. The purpose of *STEREO* was to provide multiple in-situ measurements and images of the Sun from different viewpoints from the traditional Sun–Earth line, and so each were placed in an orbit similar to that of the Earth about the Sun. The difference was that one spacecraft would orbit slightly faster than the Earth with the other slightly slower, resulting

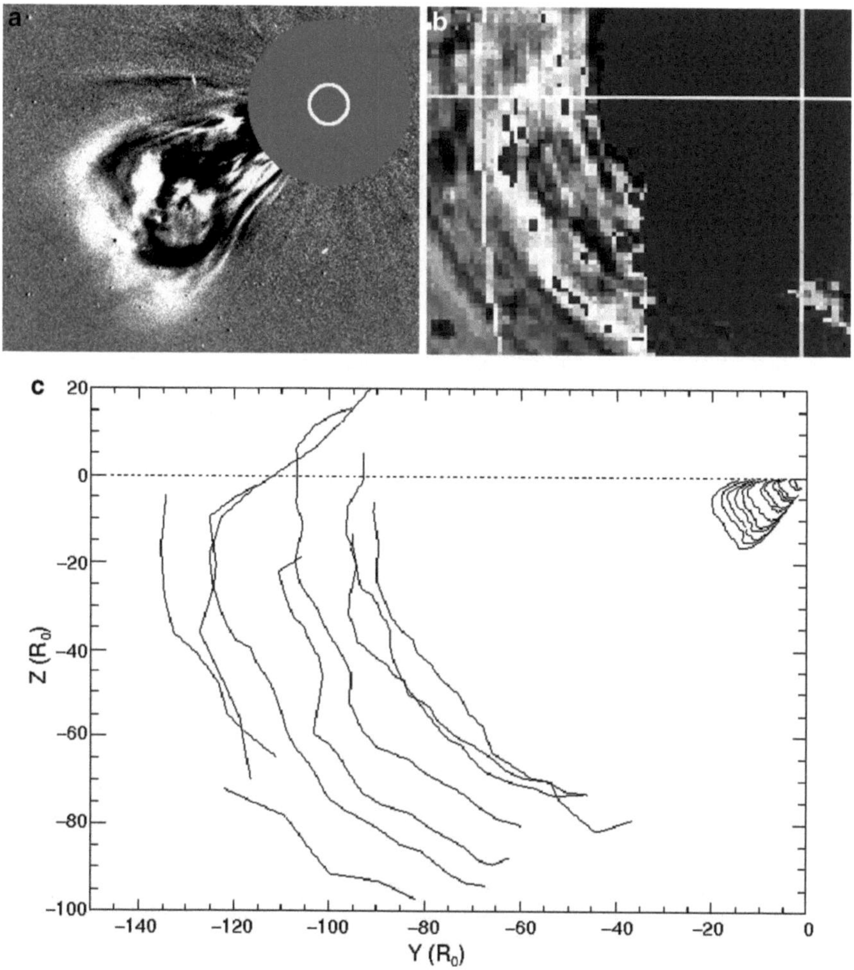

Fig. 2.13 Images of a CME observed by LASCO and SMEI in February 2004. (**a**) LASCO/C3 image obtained on 2004/02/15 at 08:18 UT. The *white circle* represents the solar surface and the *grey disk* is the occulter. (**b**) SMEI image obtained on 2004/02/16 at 07:01 UT. The horizontal and vertical lines cross at the location of the Sun. (**c**) Three-dimensional reconstruction of the leading edge of the CME combining the entire sequence of leading edge measurements from LASCO and SMEI. Because this event was very close to the sky plane, there is no Sun–Earth component for this event [123]

in leading and lagging spacecraft in the ecliptic plane. Figure 2.14 shows the location of each spacecraft at various times during the mission. The angular separation between the spacecraft and the Sun–Earth line grows by around 22.5° per year.

Along with providing continuous in-situ measurements of interplanetary transients, allowing a study of the longitudinal structure of ICMEs [123, 127], the *STEREO* imagers allow a three-dimensional image of solar structures much in the same way as depth is perceived using our two eyes. As the spacecraft become

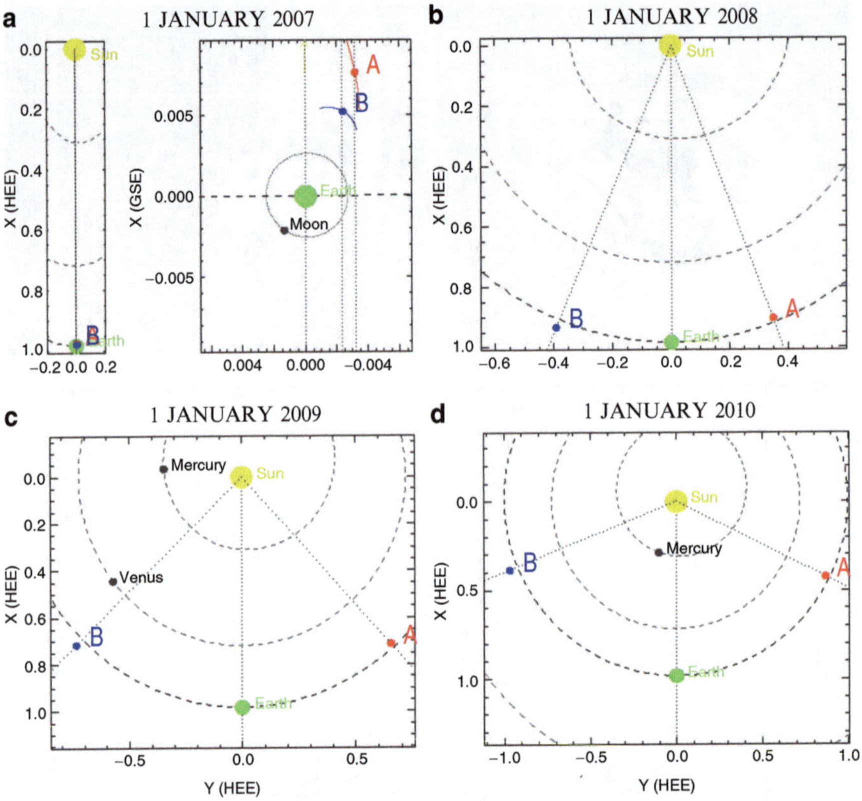

Fig. 2.14 The location of the *STEREO* spacecraft on 1 January of (**a**) 2007, (**b**) 2008, (**c**) 2009, (**d**) 2010 (from the STEREO website). The *coloured circles* indicate the following: Yellow = Sun, green = Earth, red = *STEREO-A*, blue = *STEREO-B*. The angular separation between the spacecraft and the Sun–Earth line grows by around 22.5° per day (Images provided courtesy of the "Where is STEREO" tool (NASA/GSFC))

separated further, three-dimensional reconstructions became possible with different instruments, first with the EUV low corona structures [5, 174], then the coronagraphs [125, 190], and finally the heliospheric imagers [127]. The results from one such reconstruction are shown in Fig. 2.15.

The *STEREO* spacecraft continue to move apart and are functioning to date. Eventually they will pass each other on the far side of the Sun and return from the opposite direction.

2.8 The Continuing Role of Past Missions

It is helpful to remind the reader that many of the missions launched in earlier years continue to function and play a role in our understanding of space and CMEs to this day. The *Voyagers*, for example, continue to monitor the outer regions of the

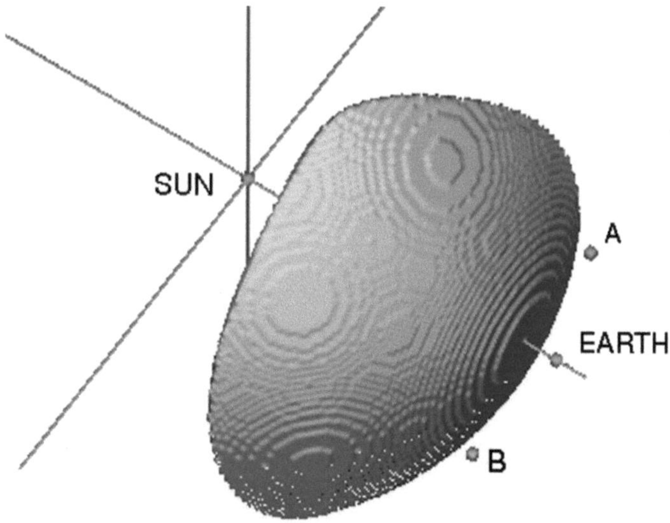

Fig. 2.15 Three-dimensional reconstruction of the leading edge of an ICME observed by SMEI and both *STEREO* in November 2007. The *grey surface* represents the leading edge of the ICME, and A and B are the locations of *STEREO-A* and *-B* at the time of the event. The location of the Earth and Sun are also indicated [127]

heliosphere, and they are reaching the edges of the heliosphere [42]. By the year 2000, they were over 58 AU from the Sun and were still capable of observing ICMEs even there. At those distances, ICMEs tend to merge with other dense regions (such as corotating interaction regions or other ICMEs), the combination of which are called merged interaction regions (MIRs [23], see Sect. 9.7). Richardson et al. [209] studied a single event from the Sun to the *WIND* spacecraft at 1 AU to *Ulysses* at 5 AU then to *Voyager 2* at 58 AU. Similar studies include the "Bastille Day" CME by Burlaga et al. [24] and a series of events the following year [257]. A review of Voyager observations of MIRs involving CIRs can be found in Lazarus et al. [167].

Along with the *Voyager* observations, new publications continue to emerge from ongoing missions dating from the 70s, such as *IMP-8* [72, 172]. Also, analysis of data from spacecraft no longer operating continue to yield new scientific results, such as those from *Helios* [72, 148, 175], *ISEE-3* [57, 213], *Solwind* [36, 157] and *SMM* [18, 181].

2.9 Summary

To summarise, CMEs have been detected using a large variety of instruments and techniques. Directly:

1. Using white light coronagraphs that detect the light that is Thomson scattered from the free electrons in the CME,

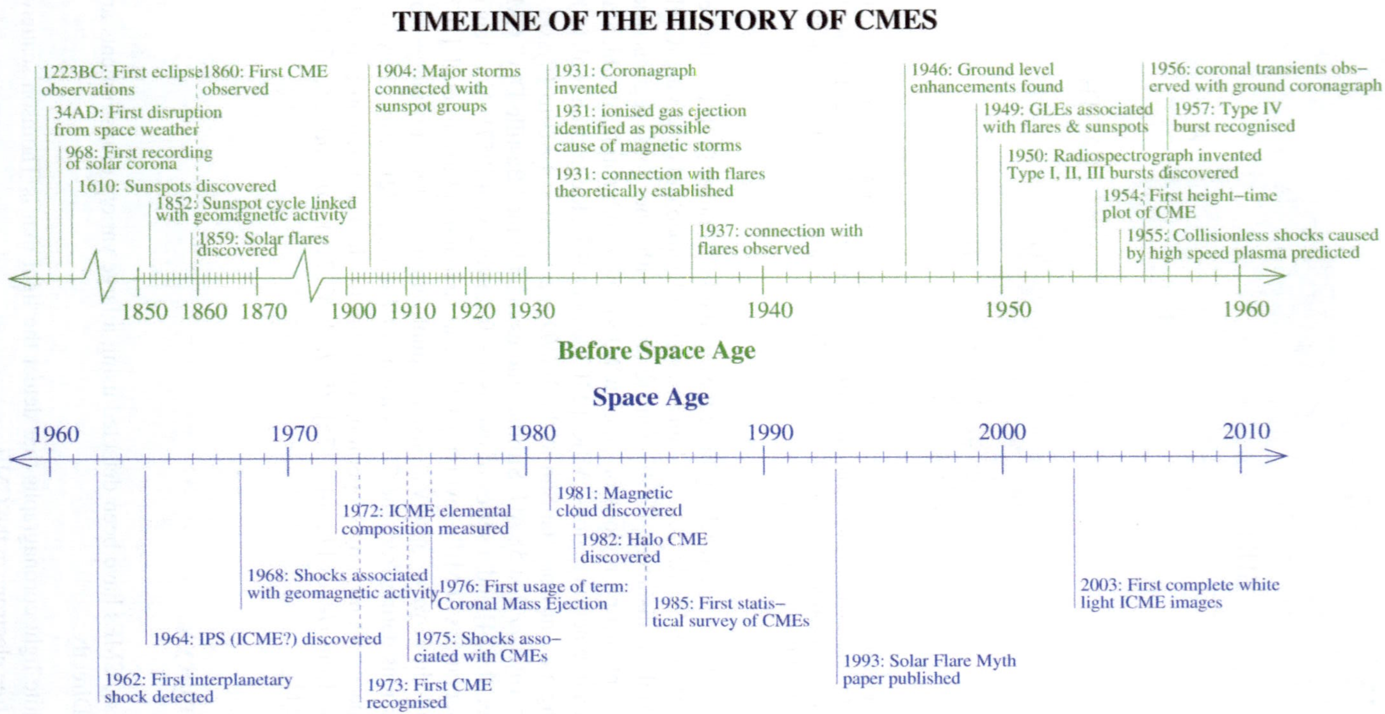

Fig. 2.16 Timeline of the significant events that have led to an enhancement of our understanding of CMEs. It has been divided into before and during the space age (green and blue respectively)

2. Directly measuring properties of the ICME as it passes by in-situ spacecraft,
3. Measuring the changes in longwave radio signals from distant sources as the ICME passes between them and the Earth (IPS),

And through investigation of the secondary effects of CME launch and propagation:

1. Solar flares, observed in visible light, EUV, x-ray,
2. Erupting prominences/disappearing filaments, observed in visible light and EUV,
3. Other solar surface eruptions, such as post-eruptive arcades and coronal dimming,
4. Solar energetic particles accelerated by the shock in the interplanetary medium from the CME,
5. Type II and Type IV radio bursts, driven by the CME shock.

Figure 2.16 shows a timeline of the significant events that have led to an enhancement of our understanding of CMEs. The passage from ground-based to space-based observations is indicated, but the importance of the work leading up to the space age cannot be overstated. It seems clear that even by the time of the emergence of the first spacecraft around 1960, our understanding of interplanetary transients and the interplanetary medium had a firm foundation.

2.9.1 The Future

As this is a history chapter, I will only briefly mention those missions planned for the future. Further details on the current plans for these missions can be found in Sect. 3.6.

On 11 February 2010, a new solar observatory which many regard as the next generation of *SOHO* was launched. The *Solar Dynamics Observatory* (*SDO*) is an geosynchronous orbit and contains a suite of instruments for solar observation, including white light, UV and EUV imagers. Other planned missions for the 2010–2020 decade include NASA's Solar Sentinels and Solar Probe, NASA/ESA's Solar Orbiter and JAXA's Solar-C all tentatively planned for launch in 2018.

References

1. Acton, L., Tsuneta, S., Ogawara, Y., Bentley, R., Bruner, M., Canfield, R., Culhane, L., Doschek, G., Hiei, E., Hirayama, T.: Science **258**, 618–625 (1992).
2. Anderson, D.P., Cobb, J., Korpela, E., Lebofsky, M., Werthimer, D.: Comm. ACM **45**, 56–61 (2002).
3. Antiochos, S.K.: Astrophys. J. **502**, L181 (1998).
4. Armstrong, J.A., Coles, W.A.: J. Geophys. Res. **77**, 4602–4610 (1972).
5. Aschwanden, M.J., Wülser, J.-P., Nitta, N.V., Lemen, J.R.: Astrophys. J. **679**, 827–842 (2008).
6. Aulanier, G., DeLuca, E.E., Antiochos, S.K., McMullen, R.A., Golub, L.: Astrophys. J. **540**, 1126–1142 (2000).

7. Balogh, A., Smith, E.J., Tsurutani, B.T., Southwood, D.J., Forsyth, R.J., Horbury, T.S., Science **268**, 1007–1010 (1995).
8. Bame, S.J., Asbridge, J.R., Feldman, W.C., Fenimore, E.E., Gosling, J.T.: Solar Phys. **62**, 179–201 (1979).
9. Bame, S.J., Asbridge, J.R., Hundhausen, A.J., Strong, I.B.: J. Geophys. Res. **73**, 5761–5767 (1968).
10. Benz, A.O., Grigis, P.C.: Solar Phys. **210**, 431–444 (2002).
11. Bertaux, J.L., Kyrölä, E., Quémerais, E., Pellinen, R., Lallement, R., Schmidt, W., Berthé, M., Dimarellis, E., Goutail, J.P., Taulemesse, C., Bernard, C., Leppelmeier, G., Summanen, T., Hannula, H., Huomo, H., Kehlä, V., Korpela, S., Leppälä, K., Strömmer, E., Torsti, J., Viherkanto, K., Hochedez, J.F., Chretiennot, G., Peyroux, R., Holzer, T.: Solar Phys. **162**, 403–439 (1995).
12. Biesecker, D., Lamy, P., St. Cyr, O.C., Llebaria, A., Howard, R.: Icarus **157**, 323–348 (2002).
13. Boggs, S.E., Coburn, W., Kalemci, E.: Astrophys. J. **638**, 1129–1139 (2002).
14. Boischot, A.: C. R. Acad. Sci. **244**, 1326 (1957).
15. Borrini, G., Gosling, J.T., Bame, S.J., Feldman, W.C.: J. Geophys. Res. **87**, 7370–7378 (1982).
16. Borrini, G., Gosling, J.T., Bame, S.J., Feldman, W.C.: Solar Phys. **83**, 367–378 (1983).
17. Brueckner, G.E., Howard, R.A., Koomen, M.J., Korendyke, C.M., Michels, D.J., Moses, J.D., Socker, D.G., Dere, K.P., Lamy, P.L., Llebaria, A., Bout, M.V., Schwenn, R., Simnett, G.M., Bedford, D.K., Eyles, C.J.: Solar Phys. **162**, 357–402 (1995).
18. Burkepile, J.T., Hundhausen, A.J., Stanger, A.L., St. Cyr, O.C., Seiden, J.A.: J. Geophys. Res. **109**, doi:10.1029/2003JA010149 (2004).
19. Burlaga, L.F.: J. Geophys. Res. **93**, 7217–7224 (1988).
20. Burlaga, L.F., Behannon, K.W.: Solar Phys. **81**, 181–192 (1982).
21. Burlaga, L.F., Klein, L., Sheeley, N.R., Jr., Michels, D.J., Howard, R.A., Koomen, M.J., Schwenn, R., Rosenbauer, H.: Geophys. Res. Lett. **9**, L1317–L1320 (1982).
22. Burlaga, L.F., Lepping, R.P., Jones, J.A.: In Russell, C.T., Priest, E.R., Lee, L.C. (eds.) Geophys. Monog. Ser. **58**, p.373, AGU, Washington DC (1990).
23. Burlaga, L.F., McDonald, F.B., Goldstein, M.L., Lazarus, A.J.: J. Geophys. Res. **90**, 12127–12039 (1985).
24. Burlaga, L.F., Ness, N.F., Richardson, J.D., Lepping, R.P.: Solar Phys. **204**, 399–411 (2001).
25. Burlaga, L., Sittler, E., Mariani, F., Schwenn, R.: J. Geophys. Res. **86**, 6673–6684 (1981).
26. Burton, M.E., Smith, E.J., Goldstein, B.E., Balogh, A., Forsyth, R.J., Bame, S.J.: Geophys. Res. Lett **19**, L1287–L1289 (1992).
27. Cane, H.V., Kahler, S.W., Sheeley, N.R., Jr.: J. Geophys. Res. **91**, 13321–13329 (1986).
28. Cane, H.V., Richardson, I.G.: J. Geophys. Res. **108**, doi:10.1029/2002JA009817 (2003).
29. Cane, H.V., Richardson, I.G., St. Cyr., O.C.: J. Geophys. Res. **102**, 7075–7086 (1997).
30. Carrington, R.C.: Mon. Not. R. Astron. Soc. **20**, 13–15 (1860).
31. Carrington, R.C.: Observations of Spots on the Sun, Williams and Norgate, London (1863).
32. Chapman, S.: J. Geophys. Res. **55**, 361 (1950).
33. Chapman, S., Ferraro, V.C.A.: Terr. Mag. Atmos. Electr. **36**, 77–97 (1931a).
34. Chapman, S., Ferraro, V.C.A.: Terr. Mag. Atmos. Electr. **36**, 171–186 (1931b).
35. Chapman, S., Ferraro, V.C.A.: Terr. Mag. Atmos. Electr. **37**, 147–156 (1932).
36. Cliver, E.W., Ling, A.G.: Astrophys. J. **556**, 432–437 (2001).
37. Coles, W.A.: Space Sci. Rev. **21**, 411–425 (1978).
38. Cook, J.W., Socker, D.G., Korendyke, C.M., Howard, R.A., Brueckner, G.E., Karovska, M., Wood, B.E.: Adv. Space Res. **25**, 1883–1886 (2000).
39. Couzens, D.A., King, J.H.: Interplanetary Medium Data Book – Supplement 3A, 1977–1985, NSSDC (1986).
40. Crooker, N.U.: J. Atmos. Solar Terr. Phys. **62**, 1071–1085 (2000).
41. Culhane, J.L., Harra, L.K., James, A.M., Al-Janabi, K., Bradley, L.J., Chaudry, R.A., Rees, K., Tandy, J.A., Thomas, P., Whillock, M.C.R., Winter, B., Doschek, G.A., Korendyke, C.M., Brown, C.M., Myers, S., Mariska, J., Seely, J., Lang, J., Kent, B.J., Shaughnessy, B.M., Young, P.R., Simnett, G.M., Castelli, C.M., Mahmoud, S., Mapson-Menard, H., Probyn, B.J., Thomas, R.J., Davila, J., Dere, K., Windt, D., Shea, J., Hagood, R., Moye, R., Hara, H., Watanabe, T., Matsuzaki, K., Kosugi, T., Hansteen, V., Wikstol, Ø.: Solar Phys. **243**, 19–61 (2007).

42. Decker, R.B., Krimigis, S.M., Roelof, E.C., Hill, M.E., Armstrong, T.P., Gloeckler, G., Hamilton, D.C., Lanzerotti, L.J.: Science **309**, 2020–2024 (2005).
43. Delaboudiniére, J.-P., Artzner, G.E., Brunaud, J., Gabriel, A.H., Hochedez, J.F., Millier, F., Song, X.Y., Au, B., Dere, K.P., Howard, R.A., Kreplin, R., Michels, D.J., Moses, J.D., Defise, J.M., Jamar, C., Rochus, P., Chauvineau, J.P., Marioge, J.P., Catura, R.C., Lemen, J.R., Shing, L., Stern, R.A., Gurman, J.B., Neupert, W.M., Maucherat, A., Clette, F., Cugnon, P., van Dessel, E.L.: Solar Phys. **162**, 291–312 (1995).
44. Delinger, J.H.: Terr. Mag. Atmos. Electr. **42**, 49–53 (1937).
45. Demastus, H.L., Wagner, W.J., Robinson, R.D.: Solar Phys. **100**, 449–459 (1973).
46. Dennison, P.A., Hewish, A.: Nature **213**, 343–346 (1967).
47. De Vaucouleurs, G.: Astronomical Photography, New York: MacMillan (1961).
48. Dryer, M.: Space Sci. Rev. **33**, 233–275 (1982).
49. Dryer, M.: Space Sci. Rev. **67**, 363–419 (1994).
50. Dryer, M., Smith, Z.K., Endrud, G.H., Wolfe, J.H.: Cosmic Electrodyn. **3**, 184–207 (1972).
51. Dungey, J.W.: Phil. Mag. **44**, 725 (1953).
52. Dungey, J.W.: In Lehnert, B (ed.), Proc IAU Symp. **6**, 135 (1958).
53. Emslie, A.G., Kucharek, H., Dennis, B.R., Gopalswamy, N., Holman, G.D., Share, G.H., Vourlidas, A., Forbes, T.G., Gallagher, P.T., Mason, G.M., Metcalfe, T.R., Mewaldt, R.A., Murphy, R.J., Schwartz, R.A., Zurbuchen, T.H.: J. Geophys. Res. **109**, doi:10.1029/2004JA010571 (2004).
54. Eyles, C.J., Simnett, G.M., Cooke, M.P., Jackson, B.V., Buffington, A., Hick, P.P., Waltham, N.R., King, J.M., Anderson, P.A., Holladay, P.E.: Solar Phys. **217**, 319–347 (2003).
55. Farrugia, C.J., Scudder, J.D., Freeman, M.P., Janoo, L., Lu, G., Quinn, J.M., Arnoldy, R.L., Torbert, R.B., Burlaga, L.F., Ogilvie, K.W., Lepping, R.P., Lazarus, A.J., Steinberg, J.T., Gratton, F.T., Rostoker, G.: J. Geophys. Res. **103**, 17261–17278 (1998).
56. Fenimore, E.E.: Astrophys. J. **235**, 245–257 (1980).
57. Feroci, M., Hurley, K., Duncan, R.C., Thompson, C.: Astrophys. J. **549**, 1021–1038 (2001).
58. Fisk, L.A.: J. Geophys. Res. **101**, 15547–15554 (1996).
59. Forbush, S.E.: Phys. Rev. **70**, 771–772 (1946).
60. Forbush, S.E., Gill, P.S., Vallarta, M.S.: Rev. Mod. Phys. **21**, 44–48 (1949).
61. Fraunhofer, J.: Denkschriften der koeniglichen Academie der Wissenschaften zu München **5**, 193–226 (1817).
62. Fröhlich, C., Romero, J., Roth, H., Wehrli, C, Andersen, B.N., Appourchaux, T., Domingo, V., Telljohann, U., Berthomieu, G., Delache, P., Provost, J., Toutain, T., Crommelynck, D.A., Chevalier, A., Fichot, A., Däppen, W., Gough, D., Hoeksema, T., Jiménez, A., Gómez, M.F., Herreros, J.M., Cortés, T.R., Jones, A.R., Pap, J.M., Willson, R.C.: Solar Phys. **162**, 101–128 (1995).
63. Gabriel, A.H., Grec, G., Charra, J., Robillot, J.-M., Roca Cortés, T., Turck-Chiéze, S., Bocchia, R., Boumier, P., Cantin, M., Cespédes, E., Cougrand, B., Crétolle, J., Damé, L., Decaudin, M., Delache, P., Denis, N., Duc, R., Dzitko, H., Fossat, E., Fourmond, J.-J., García, R.A., Gough, D., Grivel, C., Herreros, J.M., Lagardére, H., Moalic, J.-P., Pallé, P.L., Pétrou, N., Sanchez, M., Ulrich, R., van der Raay, H.B.: Solar Phys. **162**, 61–99 (1995).
64. Gergely, T.E., Kundu, M.R.: Solar Phys. **34**, 433–446 (1974).
65. Gilbert, W.: De Magnete, Peter Short, London (1600).
66. Giovanelli, R.G.: Mon. Not. R. Astron. Soc. **107**, 338 (1947).
67. Gloeckler, G., Fisk, L.A., Hefti, S., Schwadron, N.A., Zurbuchen, T.H., Ipavich, F.M., Geiss, J., Bochsler, P., Wimmer-Schweingruber, R.F.: Geophys. Res. Lett. **26**, L157–L160 (1999).
68. Gloeckler, G., Geiss, J., Roelof, E.C., Fisk, L.A., Ipavich, F.M., Ogilvie, K.W., Lanzerotti, L.J., von Steiger, R., Wilken, B.: J. Geophys. Res. **99**, 17637–17643 (1994).
69. Goff, C.P., van Driel-Gesztelyi, L., Harra, L.K., Matthews, S.A., Mandrini, C.H.: Astron. Astrophys. **434**, 761–771 (2005).
70. Gold, T.: In van de Hulst, J.C., Burgers, J.M. (eds.), Gas Dynamics of Cosmic Clouds, p.103, North-Holland, New York (1955).

71. Golub, L., Deluca, E., Austin, G., Bookbinder, J., Caldwell, D., Cheimets, P., Cirtain, J., Cosmo, M., Reid, P., Sette, A., Weber, M., Sakao, T., Kano, R., Shibasaki, K., Hara, H., Tsuneta, S., Kumagai, K., Tamura, T., Shimojo, M., McCracken, J., Carpenter, J., Haight, H., Siler, R., Wright, E., Tucker, J., Rutledge, H., Barbera, M., Peres, G., Varisco, S.: Solar Phys. **243**, 63–86 (2007).
72. González-Esparza, A.: Space Sci. Rev. **97**, 197–200 (2001).
73. Gopalswamy, N., Hanaoka, Y.: Astrophys. J. **498**, L179–L182 (1998).
74. Gosling, J.T.: In Russell, C.T., Priest, E.R., Lee, L.C. (eds.) Geophys. Monog. Ser. **58**, p.343, AGU, Washington DC (1990).
75. Gosling, J.T.: J. Geophys. Res. **98**, 18937–18949 (1993).
76. Gosling, J.T.: Proc. Solar Wind 11 – SOHO 16, Whistler, Canada (2005).
77. Gosling, J.T.: Coronal Mass Ejections in the solar wind at high solar latitudes: an overview, Presented at the Third SOHO Workshop, Estes Park, CO (1994).
78. Gosling, J.T., Asbridge, J.R., Bames, S.J., Hundhausen, A.J., Strong, I.B.: J. Geophys. Res. **73**, 43–50 (1968).
79. Gosling, J.T., Asbridge, J.R., Bame, S.J., Feldman, W.C., Zwickl, R.D.: J. Geophys. Res. **85**, 3431–3434 (1980).
80. Gosling, J.T., Bame, S.J., McComas, D.J., Phillips, J.L., Scime, E.E., Pizzo, V.J., Goldstein, B.E., Balogh, A.: Geophys. Res. Lett. **21**, L237–L240 (1994).
81. Gosling, J.T., Hildner, E., MacQueen, R.M., Munro, R.H., Poland, A.I., Ross, C.L.: J. Geophys Res. **79**, 4581–4587 (1974).
82. Gosling, J.T., Hildner, E., MacQueen, R.M., Munro, R.H., Poland, A.I., Ross, C.L.: Solar Phys. **40**, 439–448 (1975).
83. Gosling, J.T., Hildner, E., MacQueen, R.M., Munro, R.H., Poland, A.I., Ross, C.L.: Solar Phys. **48**, 389–397 (1976).
84. Gosling, J.T., Skoug, R.M., McComas, D.J., Smith, C.W.: J. Geophys. Res. **110**, A01107, doi:10.1029/20043A010809 (2005a).
85. Gosling, J.T., Skoug, R.M., McComas, D.J., Smith, C.W.: Geophys. Res. Lett. **32**, L05105, doi:10.1029/2005GL022406 (2005b).
86. Graham, G.: Phil. Trans. R. Soc. Lond. **383**, 96–107 (1724).
87. Greaves, W.M.H, Newton, H.W.: Mon. Not. R. Astron. Soc. **88**, 556–567 (1928a).
88. Greaves, W.M.H, Newton, H.W.: Mon. Not. R. Astron. Soc. **89**, 84–92 (1928b).
89. Hale, G.E.: Astrophys. J. **73**, 379–412 (1931).
90. Handy, B.N., Acton, L.W., Kankelborg, C.C., Wolfson, C.J., Akin, D.J., Bruner, M.E., Caravalho, R., Catura, R.C., Chevalier, R., Duncan, D.W., Edwards, C.G., Feinstein, C.N., Freeland, S.L., Friedlander, F.M., Hoffmann, C.H., Hurlburt, N.E., Jurcevich, B.K., Katz, N.L., Kelly, G.A., Lemen, J.R., Levay, M., Lindgren, R.W., Mathur, D.P., Meyer, S.B., Morrison, S.J., Morrison, M.D., Nightingale, R.W., Pope, T.P., Rehse, R.A., Schrijver, C.J., Shine, R.A., Shing, L., Tarbell, T.D., Title, A.M., Torgerson, D.D., Golub, L., Bookbinder, J.A., Caldwell, D., Cheimets, P.N., Davis, W.N., DeLuca, E.E., McMullen, R.A., Amato, D., Fisher, R., Maldonado, H., Parkinson, C.: Solar Phys. **187**, 229–260 (1999).
91. Harra, L.K., Hara, H., Imada, S., Young, P.R., Williams, D.R., Sterling, A.C., Korendyke, C., Attrill, G.D.R.: Publ. Astron. Soc. Japan **59**, S801–S806 (2007).
92. Harrison, R.A.: Astron. Astrophys. **162**, 283–291 (1986).
93. Harrison, R.A.: Solar Phys. **166**, 441–444 (1996).
94. Harrison, R.A., Hildner, E., Hundhausen, A.J., Sime, D.G., Simnett, G.M.: J. Geophys. Res. **95**, 917–937 (1990).
95. Harrison, R.A., Sawyer, E.C., Carter, M.K., Cruise, A.M., Cutler, R.M., Fludra, A., Hayes, R.W., Kent, B.J., Lang, J., Parker, D.J., Payne, J., Pike, C.D., Peskett, S.C., Richards, A.G., Gulhane, J.L., Norman, K., Breeveld, A.A., Breeveld, E.R., Al Janabi, K.F., McCalden, A.J., Parkinson, J.H., Self, D.G., Thomas, P.D., Poland, A.I., Thomas, R.J., Thompson, W.T., Kjeldseth-Moe, O., Brekke, P., Karud, J., Maltby, P., Aschenbach, B., Bruninger, H., Kúhne, M., Hollandt, J., Siegmund, O.H.W., Huber, M.C.E., Gabriel, A.H., Mason, H.E., Bromage, B.J.I.: Solar Phys. **162**, 233–290 (1995).

96. Harrison, R.A., Sime, D.G.: J. Geophys. Res. **94**, 2333–2344 (1989).
97. Harrison, R.A., Simnett, G.M.: Adv. Space Res. **4**, 199–202 (1984).
98. Harrison, R.A., Waggett, P.W., Bentley, R.D., Phillips, K.J.H., Bruner, M., Dryer, M., Simnett, G.M.: Solar Phys. **97**, 387–400 (1985).
99. Hartman, J.: Astrophys. J. **9**, 69–85 (1899).
100. Henke, T., Woch, J., Mall, U., Livi, S., Wilkin, R., Schwenn, R., Gloeckler, G., von Stieger, R., Forsyth, R.J., Balogh, A.: Geophys. Res. Lett. **25**, L3465–L3468 (1998).
101. Henke, T., Woch, J., Schwenn, R., Mall, U., Gloeckler, G., von Stieger, R., Forsyth, R.J., Balogh, A.: J. Geophys. Res. **106**, 10597–10613 (2001).
102. Hetherington, B.: A Chronicle of Pre-Telescopic Astronomy, John Wiley and Sons (1996).
103. Hewish, A.: New Scientist **118**, 46–50 (1988).
104. Hewish, A.: Solar Phys. **116**, 195–198 (1988).
105. Hewish, A., Bravo, S.: Solar Phys. **106**, 185–200 (1986).
106. Hewish, A., Scott, P.F., Wills, D.: Nature **203**, 1214–1217 (1964).
107. Hewish, A., Symonds, M.D.: Planet. Space. Sci. **17**, 313–320 (1969).
108. Hick, P., Jackson, B.V.: Adv. Space Res. **14**, 135–138 (1994).
109. Hick, P., Jackson, B.V., Schwenn, R.: Astron. Astrophys **244**, 242–250 (1991).
110. Hildner, E.: Proc. AIAA Meeting (1974).
111. Hirshberg, J., Bame, S.J., Robbins, D.E.: Solar Phys. **23**, 467–486 (1972).
112. Holman, G.D.: In: The RHESSI Homepage, available via NASA/GSFC. http://hesperia.gsfc.nasa.gov/hessi/ramaty.html. Cited 10 November 2008.
113. Horbury, T., Balogh, A., Forsyth, R.J., Smith, E.J.: Ann. Geophys. **13**, 105–107 (1995).
114. Houminer, Z.: Nature Phys. Sci. **231**, 165–167 (1971).
115. Houminer, Z.: Planet. Space. Sci. **21**, 1617–1624 (1973).
116. Houminer, Z., Hewish, A.: Planet. Space. Sci. **20**, 1703–1716 (1972).
117. Houminer, Z., Hewish, A.: Planet. Space. Sci. **22**, 1041–1042 (1974).
118. Houminer, Z., Hewish, A.: Planet. Space Sci. **36**, 301–306 (1988).
119. Hovestadt, D., Hilchenbach, M., Bürgi, A., Klecker, B., Laeverenz, P., Scholer, M., Grünwaldt, H., Axford, W.I., Livi, S., Marsch, E., Wilken, B., Winterhoff, H.P., Ipavich, F.M., Bedini, P., Coplan, M.A., Galvin, A.B., Gloeckler, G., Bochsler, P., Balsiger, H., Fischer, J., Geiss, J., Kallenbach, R., Wurz, P., Reiche, K.-U., Gliem, F., Judge, D.L., Ogawa, H.S., Hsieh, K.C., Möbius, E., Lee, M.A., Managadze, G.G., Verigin, M.I., Neugebauer, M.: Solar Phys. **162**, 441–481 (1995).
120. Howard, R.A., Koomen, M.J., Michels, D.J., Tousey, R., Dewiler, C.R., Roberts, D.E., Seal, R.T., Whitney, J.T., Hansen, R.T., Hansen, S.F., Garcia, C.J., Yasukawa, E.: World Data Center A, Report UAG 48A (1975).
121. Howard, R.A., Michels, D.J., Sheeley, N.R., Jr., Koomen, M.J.: Astrophys. J. **263**, L101–L104 (1982).
122. Howard, R.A., Sheeley, N.R., Jr., Michels, D.J., Koomen, M.J.: J. Geophys. Res. **90**, 8173–8191 (1985).
123. Howard, T.A., Fry, C.D., Johnston, J.C., Webb, D.F.: Astrophys. J. **667**, 610–625 (2007).
124. Howard, T.A., Tappin, S.J.: Astron. Astrophys. **440**, 373–383 (2005).
125. Howard, T.A., Tappin, S.J.: Solar Phys. **252**, 373–383 (2008).
126. Howard, T.A., Tappin, S.J.: Space Sci. Rev. **147**, 31–54 (2009a).
127. Howard, T.A., Tappin, S.J.: Space Sci. Rev. **147**, 89–110 (2009b).
128. Howard T.A., Webb, D.F., Tappin, S.J., Mizuno, D.R., Johnston, J.C.: J. Geophys. Res. **111**, doi:10.1029/2005JA011349 (2006).
129. Hudson, H.S.: Solar Phys. **113**, 1–9 (1987).
130. Hudson, H.S., Lemen, J.R., St. Cyr, O.C., Sterling, A.C., Webb, D.F.: Geophys. Res. Lett. **25**, L2481–L2484 (1998).
131. Hudson, H.S, Haisch, B., Strong, K.T.: J. Geophys. Res. **100**, 3473–3477 (1995).
132. Hundhausen, A.J.: In Sonnet, C.P., Coleman, P.J., Wilcox, J.M. (eds.), Solar Wind, NASA Spec. Publ., SP-308, p.393 (1972a).
133. Hundhausen, A.J.: Coronal Expansion and Solar Wind, Springer-Verlag, New York, 1972b.

134. Hundhausen, A.J.: In Pizzo, V.J., Holzer, T.E., Sime, D.G. (eds.), Proc. 6th Int. Solar Wind Conf., p.192 HAO CO (1987).
135. Hundhausen, A.J., Bame, S.J., Montgomery, M.D.: J. Geophys. Res. **75**, 4631–4642 (1970).
136. Hundhausen, A.J., Burkepile, J.T., St. Cyr, O.C.: J. Geophys. Res. **99**, 6543–6552 (1994).
137. Hundhausen, A.J., Sawyer, C.B., House, L., Illing, R.M.E., Wagner, W.J.: J. Geophys. Res. **89**, 2639–2646, (1984).
138. Illing, R.M.E., Hundhausen, A.J.: J. Geophys. Res. **90**, 275–282 (1985).
139. Ivanov, K.G.: Astronomicheskii Zhurnal **48**, 998 (1971).
140. Ivanov, K.G.: Soviet Astron. **17**, 94 (1973).
141. Jackson, B.V.: Solar Phys. **95**, 363–370 (1985).
142. Jackson, B.V.: Adv. Space Res. **6**, 307–310 (1986).
143. Jackson, B.V.: Adv. Space Res. **9**, 69–74 (1989).
144. Jackson, B.V.: Proc. 3rd COSPAR Coll. (Solar Wind 7), 623–634 (1992).
145. Jackson, B.V., Buffington, A., Hick, P.P., Altrock, R.C., Figueroa, S., Holladay, P.E., Johnston, J.C., Kahler, S.W., Mozer, J.B., Price, S., Radick, R.R., Sagalyn, R., Sinclair, D., Simnett, G.M., Eyles, C.J., Cooke, M.P., Tappin, S.J., Kuchar, T., Mizuno, D., Webb, D.F., Anderson, P.A., Keil, S.L., Gold, R.E., Waltham, N.R.: Solar Phys. **225**, 177–207 (2004).
146. Jackson, B.V., Buffington, A., Hick, P.P., Wang, X., Webb, D.F.: J. Geophys. Res. **111**, doi:10.1029/2004JA010942 (2006).
147. Jackson, B.V., Froehling, H.R.: Astron. Astrophys. **299**, 885–892 (1995).
148. Jackson, B.V., Hick, P.P.: Solar Phys. **211**, 345–356 (2002).
149. Jackson, B.V., Howard, R.A., Sheeley, N.R., Jr., Michels, D.J., Koomen, M.J., Illing, R.M.E.: J. Geophys. Res. **90**, 5075–5081 (1985).
150. Jackson, B.V., Leinert, C.: J. Geophys. Res. **90**, 10759–10764 (1985).
151. Jackson, B.V., Rompolt, B., Švestka, Z.: Solar Phys. **115**, 327–343 (1988).
152. Jacobs, I.: In: About the World Wide Web Consortium, available via W3C. http://www.w3.org/Consortium/. Cited 29 April 2008.
153. Janardhan, P., Balasubramanian, V., Ananthakrishnan, S.: In Wilson, A. (ed.), Proc. 31st ESLAB Symp., Noordwijk, The Netherlands, ESA SP-415, p.177 (1997).
154. Jones, R., Canals, A., Breen, A., Fallows, R., Lawrence, G.: Proc. 1st EGU Sci. Assem., Nice, France, EGU-A-00452 (2004).
155. Joselyn, J.A., McIntosh, P.S.: J. Geophys. Res. **86**, 4555–4564 (1981).
156. Kahler, S.W.: Ann. Rev. Astron. Astrophys. **30**, 113–141 (1992).
157. Kahler, S.W., Reames, D.V. Sheeley, N.R., Jr.: Astrophys. J. **562**, 558–565 (2001).
158. Kaiser, M.L., Kucera, T.A., Davila, J.M., St. Cyr, O.C., Guhathakurta, M., Christian, E.: Space Sci. Rev. **136**, 5–16 (2008).
159. Kanekal, S.G., Baker, D.N., Blake, J.B., Klecker, B., Mewaldt, R.A., Mason, G.M.: J. Geophys. Res. **104**, 24885–24894 (1999).
160. King, J.H.: Interplanetary magnetic field data 1963–1974, NSSDC (1975).
161. Klein, L.W., Burlaga, L.F.: J. Geophys. Res. **87**, 613–624 (1982).
162. Kohl, J.L., Esser, R., Gardner, L.D., Habbal, S., Daigneau, P.S., Dennis, E.F., Nystrom, G.U., Panasyuk, A., Raymond, J.C., Smith, P.L., Strachan, L., van Ballegooijen, A.A., Noci, G., Fineschi, S., Romoli, M., Ciaravella, A., Modigliani, A., Huber, M.C.E., Antonucci, E., Benna, C., Giordano, S., Tondello, G., Nicolosi, P., Naletto, G., Pernechele, C., Spadaro, D., Poletto, G., Livi, S., von der Lühe, O., Geiss, J., Timothy, J.G., Gloeckler, G., Allegra, A., Basile, G., Brusa, R., Wood, B., Siegmund, O.H.W., Fowler, W., Fisher, R., Jhabvala, M.: Solar Phys. **162**, 313–356 (1995).
163. Koomen, M., Howard, R.A., Hansen, R., Hansen, S.: Solar Phys. **34**, 447–452 (1974).
164. Korreck, K.E., Zurbuchen, T.H., Lepri, S.T., Raines, J.M.: Astrophys. J. **659**, 773–779 (2007).
165. Lazarus, A.J., Binsack, J.H.: Ann. I. Q. S. Y. **3**, 378–385 (1969).
166. Lazarus, A.J., Ogilvie, K.W., Burlaga, L.F.: Solar Phys. **13**, 232–239 (1970).
167. Lazarus, A.J., Richardson, J.D., Decker, R.B., McDonald, F.B.: Space Sci. Rev. **89**, 53–59 (1999).
168. Leamon, R.J., Canfield, R.C., Jones, S.L., Lambkin, K., Lundberg, B.J., Pevtsov, A.A.: J. Geophys. Res. **109**, doi:10.1029/2003JA010324 (2004).

169. Leinert, C., Link, H., Pitz, E., Salm, N., Kluppelberg, D.: Raumfahrtforschung **19**, 264–267 (1975).
170. Lepping, R.P., Berdichevsky, D.: Recent Res. Devel. Geophys. **3**, 77–96 (2000).
171. Lepping, R.P., Jones, J.A., Burlaga, L.F.: J. Geophys. Res. **95**, 11957–11965 (1990).
172. Lepping, R.P., Wu, C.-C., McClernan, K.: J. Geophys. Res. **108**, doi:10.1029/2002JA009640 (2003).
173. Lepri, S.T., Zurbuchen, T.H., Fisk, L.A., Richardson, I.G., Cane, H.V., Gloeckler, G.: J. Geophys. Res. **106**, 29231–29238 (2001).
174. Liewer, P.C., De Jong, E.M., Hall, J.R., Howard, R.A., Thompson, W.T., Culhane, J.L., Bone, Laura, van Driel-Gesztelyi, L.: Solar Phys. **256**, 57–72 (2009).
175. Liu, Y., Richardson, J.D., Belcher, J.W.: Planet. Space Sci. **53**, 3–17 (2005).
176. Lynch, B.J., Gruesbeck, J.R., Zurbuchen, T.H.: J. Geophys. Res. **108**, doi:10.1029/2005JA011137 (2005).
177. Lynch, B.J., Zurbuchen, T.H., Fisk, L.A., Antiochos, S.K.: J. Geophys. Res. **108**, doi:10.1029/2002JA009591 (2003).
178. Lyot, M.B.: Mon. Not. R. Astron. Soc. **99**, 578–590 (1939).
179. MacIntosh, S.W.: Astrophys. J. **693**, 1306–1309 (2009).
180. MacQueen, R.M.: Phil. Trans. R. Soc. Lond. A. **297**, 605–620 (1980).
181. MacQueen, R.M., Burkepile, J.T., Holzer, T.E., Stanger, A.L., Spence, K.E.: Astrophys. J. **549**, 1175–1182 (2001).
182. MacQueen, R.M., Csoeke-Poeckh, A., Hildner, E., House, L., Reynolds, R., Stanger, A., Tepoel, H., Wagner, W.: Solar Phys. **65**, 91–107 (1980).
183. Marubashi, K.: In Crooker, N., Joselyn, J.A., Feynman, J. (eds.), Geophys. Monog. Ser. **99**, p.147, AGU, Washington DC (1997).
184. Maunder, E.W.: Mon. Not. R. Astron. Soc. **65**, 2–34 (1904).
185. McComas, D.J., Riley, P., Gosling, J.T., Balogh, A., Forsyth, R.: J. Geophys. Res. **103**, 1955–1967 (1998).
186. McLean, D.J., Labrum, N.R. (eds): Solar radiophysics: Studies of emission from the sun at metre wavelengths, Cambridge Univ. Press, Cambridge and New York (1985).
187. Meyer, P., Parker, E.N., Simpson, J.A.: Phys. Rev. **104**, 768–783 (1956).
188. Michels, D.J., Howard, R.A., Koomen, M.J., Sheeley, N.R., Jr.: In Kundu, M.R., Gergely, T.E. (eds.), Radio Physics of the Sun, p.439, D. Reidel, Hingham MA (1980).
189. Michels, D.J., Sheeley, N.R., Jr., Howard, R.A., Koomen, M.J., Schwenn, R., Mulhauser, K.H., Rosenbauer, H.: Adv. Space Res. **4**, 311–321 (1984).
190. Mierla, M., Davila, J., Thompson, W., Inhester, B., Srivastava, N., Kramar, M., St. Cyr, O.C., Stenborg, G., Howard, R.A.: Solar Phys. **252**, 385–396 (2008).
191. Mierla, M., Schwenn, R., Teriaca, L., Stenborg G., Podlipnik, B.: Adv. Space Res. **35**, 2199–2203 (2005).
192. Mitchell, A.C.: Terr. Magn. Atmosph. Electr. **37**, 105–146 (1932).
193. Morrison, P.: Phys. Rev. **95**, 646 (1954).
194. Moullard, O., Burgess, D., Salem, C., Mangeney, A., Larson, D.E., Bale, S.D.: J. Geophys. Res. **106**, 8301–8314 (2001).
195. Müller-Mellin, R., Kunow, H., Fleißner, V., Pehlke, E., Rode, E., Röschmann, N., Scharmberg, C., Sierks, H., Rusznyak, P., McKenna-Lawlor, S., Elendt, I., Sequeiros, J., Meziat, D., Sanchez, S., Medina, J., Del Peral, L., Witte, M., Marsden, R., Henrion, J.: Solar Phys. **162**, 483–504 (1995).
196. Munro, R.H., Sime, D.G.: Solar Phys. **97**, 191–201 (1985).
197. Newton, H.W.: Observatory **62**, 317–326 (1939).
198. Newton, H.W.: Mon. Not. R. Astron. Soc. **103**, 244–257 (1943).
199. Ogilvie, K.W., Burlaga, L.F.: Solar Phys. **8**, 422–434 (1969).
200. Parker, E.N.: J. Geophys. Res. **62**, 509–520 (1957).
201. Petschek, H.E.: In Hess, W.N. (ed.), Physics of Solar Flares, NASA SP-50 (1964).
202. Plunkett, S.P., Brueckner, G.E., Dere, K.P., Howard, R.A., Koomen, M.J., Korendyke, C.M., Michels, D.J., Moses, J.D., Moulton, N.E., Paswaters, S.E., St. Cyr, O.C., Socker, D.G., Wang, D., Simnett, G.M., Bedford, D.K., Biesecker, D.A., Eyles, C.J., Tappin, S.J., Schwenn, R., Lamy, P.L., Llebaria, A.: Solar Phys. **175**, 699–718 (1997).

203. Pudovkin, M.I.: J. Geophys. Res. **100**, 7917–7919 (1995).
204. Qiu, J., Hu, Q., Howard, T.A., Yurchyshyn, V.B.: Astrophys. J. **659**, 758–772 (2007).
205. Ranyard, C.A.: Mem. R. Astron. Soc. **41**, 520 (1879).
206. Reid, G.C., Leinbach, H.: J. Geophys. Res. **64**, 1801–1805 (1959).
207. Reinard, A.A.: Astrophys. J., **682**, 1289–1305 (2008).
208. Richardson, I.G., Cane, H.V.: J. Geophys. Res. **109**, doi: 10.1029/2004JA010598 (2004).
209. Richardson, J.D., Paularena, K.I., Wang, C., Burlaga, L.F.: J. Geophys. Res., **107**, doi:10.1029/2001JA000175 (2002).
210. Richter, I., Leinert, C., Planc, B.: Astron. Astrophys **110**, 115–120 (1982).
211. Rickett, B.J.: Solar Phys. **43**, 237–247 (1975).
212. Riley, P., Linker, J.A., Lionello, R., Mikiċ, Odstrcil, D., Hidalgo, M.A., Cid, C., Hu, Q., Lepping, R.P., Lynch, B.J., Rees, A.: J. Atmos. Solar. Terr. Phys. **66**, 1321–1331 (2004).
213. Rodríguez-pacheco, J., Cid, C., Blanco, J.J., Sequeiros, J.: Solar Phys. **213**, 121–145 (2003).
214. Roelof, E.C., Simnett, G.M., Tappin, S.J.: Astron. Astrophys. **316**, 481–486 (1996).
215. Rust, D.M., Kumar, A.: Astrophys. J. **464**, L199–L202 (1996).
216. Sabine, E.: Phil. Trans. R. Soc. Lond. **142**, 103–124 (1852).
217. Sanderson, T.R., Marsden, R.G., Heras, A.M., Wenzel, K.-P., Anglin, J.D., Balogh, A., Forsyth, R.: Geophys. Res. Lett. **19**, L1263–L1266 (1992).
218. Scherrer, P.H., Bogart, R.S., Bush, R.I., Hoeksema, J.T., Kosovichev, A.G., Schou, J., Rosenberg, W., Springer, L., Tarbell, T.D., Title, A., Wolfson, C.J., Zayer, I., MDI Engineering Team: Solar Phys. **162**, 129–188 (1995).
219. Schröder, W., Wiederkehr, K-H.: J. Atmos. Solar Terr. Phys. **63**, 1649–1660 (2001).
220. Schwabe, S.H.: Astronomische Nachrichten **20**, 234–235 (1843).
221. Schwenn, R., Rosenbauer, H., Muehlhaeuser, K.-H.: Geophys. Res. Lett. **7**, 201–204 (1980).
222. Schwenn, R., Watanabe, T., Kakinuma, T., Kojima, M.: Proc. Nagoya Univ. Res. Inst. Atmospherics **36**, 11–28 (1989).
223. Sime, D.G., MacQueen, R.M., Hundhausen, A.J.: J. Geophys. Res. **89**, 2113–2121 (1984).
224. Sime, D.G., Hundhausen, A.J.: J. Geophys. Res. **92**, 1049–1055 (1987).
225. Simnett, G.M.: Astron. Astrophys. **416**, 759–764 (2004).
226. Simnett, G.M., Harrison, R.A.: Adv. Space Res. **4**, 279–282 (1984).
227. Simnett, G.M., Harrison, R.A.: Solar Phys. **99**, 291–311 (1985).
228. Sisko, G., Crooker, N.U., Clauer, C.R.: Adv. Space Res. **38**, 173–179 (2006).
229. Sonnet, C.P., Colburn, D.S., Davis, L., Smith, E.J., Coleman, P.J.: Phys. Rev. Lett. **13**, 153–156 (1964).
230. Stanger, A.L.: SMM C/P CME Event: 14 April 1980 — Day of Year (DOY): 105 Available via HAO. http://smm.hao.ucar.edu/smm/smmcp_events/1980apr14.html. Cited 29. August 2000
231. Stern, D.P.: Rev. Geophys. **40**, 1–30 (2002).
232. Sterling, A.C., Hudson, H.S.: Astrophys. J. **491**, L55–L58 (1997).
233. Sterling, A.C., Hudson, H.S., Thompson, B.J., Zarro, D.M.: Astrophys. J. **532**, 628–647 (2000).
234. Švestka, Z.F: Solar Phys. **160**, 53–56 (1995).
235. Švestka, Z.F., Jackson, B.V., Howard, R.A., Sheeley, N.R., Jr.: Solar Phys. **122**, 131–143 (1989).
236. Sweet, P.A.: In Lehnert, B. (ed.), Proc. IAU Symp. **6**, 123 (1958).
237. Tappin, S.J.: PhD thesis, Univ. Cambridge (1984).
238. Tappin, S.J.: Solar Phys. **233**, 233–248 (2006).
239. Tappin, S.J., Buffington, A., Cooke, M.P., Eyles, C.J., Hick, P.P., Holladay, P.E., Jackson, B.V., Johnston, J.C., Kuchar, T., Mizuno, D., Mozer, J.B., Price, S., Radick, R.R., Simnett, G.M., Sinclair, D., Waltham, N.R., Webb, D.F.: Geophys. Res. Lett **31**, doi:10.1029/2003GL018766 (2004).
240. Tappin, S.J., Hewish, A., Gapper, G.R.: Planet. Space Sci. **31**, 1171–1176 (1983).
241. Tappin, S.J., Howard, T.A.: Space Sci. Rev. **147**, 55–87 (2009).
242. Taylor, H.E.: Solar Phys. **6**, 320–334 (1969).

243. Torsti, J., Valtonen, E., Lumme, M., Peltonen, P., Eronen, T., Louhola, M., Riihonen, E., Schultz, G., Teittinen, M., Ahola, K., Holmlund, C., Kelhä, V., Leppälä, K., Ruuska, P., Strömmer, E.: Solar Phys. **162**, 505–531 (1995).
244. Tousey, R.: In Rycroft, M.J., Runcorn, S.K. (eds.), Space Research XIII, Akademie-Verlag, Berlin p.713–730 (1973).
245. Trella, M., Greenfield, M., Herring, E.L., Credland, J., Freeman, H.R., Laine, R., Kilpatrick, W., Machi, D., Reth, A., Smith, A., In: SOHO Mission Interruption Join NASA/ESA Investigation Board: Final Report, available via ESA. http://sohowww.estec.esa.nl/whatsnew/SOHO_final_report.html. Cited 31 August 1998.
246. Tsuneta, S., Acton, L., Bruner, M., Lemen, J., Brown, W., Caravalho, R., Catura, R., Freeland, S., Jurcevich, B., Morrison, M., Obawara, Y., Hirayama, T., and Owens, J.: Solar Phys. **136**, 37–61 (1991).
247. Tsuneta, S., Ichimoto, K., Katsukawa, Y., Nagata, S., Otsubo, M., Shimizu, T., Suematsu, Y., Nakagiri, M., Noguchi, M., Tarbell, T., Title, A., Shine, R., Rosenberg, W., Hoffmann, C., Jurcevich, B., Kushner, G., Levay, M., Lites, B., Elmore, D., Matsushita, T., Kawaguchi, N., Saito, H., Mikami, I., Hill, L.D., Owens, J.K.: Solar Phys. **249**, 167–196 (2008).
248. Tsurutani, B.T., Gonzalez, W.D., Lakhina, G.S., Alex, S.: J. Geophys. Res. **108**, 1268, doi:10.1029/2002JA009504 (2002).
249. Utton, T.: A Super Highway for Aliens?, The Daily Mail, London, January 15 (2003).
250. Vaquero, J.M.: Adv. Space Res. **40**, 929–941 (2007).
251. Vitkevich, V.V., Vlasov, V.I.: Soviet Astron. **13**, 669–676 (1970).
252. Vlasov, V.I.: Geomag. Aeron. **21**, 324–326 (1981).
253. Vlasov, V.I.: Geomag. Aeron. **22**, 446–450 (1982).
254. Vlasov, V.I., Shishov, V.I., Shisova, T.D.: Geomag. Aeron. **25**, 211–214 (1985).
255. von Steiger, R., Schwadron, N.A., Fisk, L.A., Geiss, J., Gloeckler, G., Hefti, S., Wilken, B., Wimmer-Schweingruber, R.-F., Zurbuchen, T.-H.: J. Geophys. Res. **105**, 27217–27236 (2000).
256. Vršnak,B., Warmuth, A., Temmer, M., Veronig, A., Magdalenic, J., Hillaris, A., Karlicky, M.: Astron. Astrophys. **448**, 739–752 (2006).
257. Wang, C., Richardson, J.D.: Geophys. Res. Lett **29**, doi:10.1029/2001GL014472 (2002).
258. Watanabe, T.: Proc. COSPAR, Plenary Meet. (1977).
259. Watanabe, T., Kakinuma, T.: Publ. Astron. Soc. Japan **24**, 459–467 (1972).
260. Watanabe, T., Kakinuma, T., Kojima, M., Shibasaki, K.: J. Geophys. Res. **78**, 8364–8366 (1973).
261. Webb, D.F.: In IAU Colloq. 133: Eruptive Solar Flares, pp.234–247 (1992).
262. Webb, D.F.: Proc. 37th COSPAR Sci. Assem., Montréal, Canada., p.3444 (2008).
263. Webb, D.F., Cheng, C.-C., Dulk, G.A., Edberg, S.J., Martin, S.F., McKenna-Lawlor, S., McLean, D.J.: In Sturrock, P.A (ed.), Solar Flares: A Monograph From Skylab Workshop II, Colo. Assoc. Univ. Press, Boulder CO, p.471 (1980).
264. Webb, D.F., Mizuno, D.R., Buffington, A., Cooke, M.P., Eyles, C.J., Fry, C.D., Gentile, L.C., Hick, P.P., Holladay, P.E., Howard, T.A., Hewitt, J.G., Jackson, B.V., Johnston, J.C., Kuchar, T.A., Mozer, J.B., Price, S., Radick, R.R., Simnett, G.M., Tappin, S.J.: J. Geophys. Res. **111**, doi:10.1029/2006JA011655 (2006).
265. Webb, D.F., Jackson, B.V.: J. Geophys. Res. **95**, 20641–20661 (1990).
266. Webb, D.F., Jackson, B.V., Hick, P., Schwenn, R., Bothmer, V., Reames, D.: Adv. Space Res. **13**, 71–74 (1993).
267. Wild, J.P., McCready, L.L.: Aust. J. Sci. Res. **3**, 387–398 (1950).
268. Wild, J.P., Murray, J.D., Rowe, W.C.: Aust. J. Phys. **7**, 439–459 (1954a).
269. Wild, J.P., Roberts, J.A., Murray, J.D.: Nature **173**, 532–534 (1954b).
270. Wild, J.P., Sheridan, K.V., Trent, G.H.: In Bracewell, R.N. (ed.), Proc. Paris Symp. Radio Astron., IAU/URSI, Stanford Univ. Press, Stanford, CA, p.176 (1959).
271. Wild, J.P., Smerd, S.F., Weiss, A.A.: Ann. Rev. Astron. Astrophys. **1**, 291–366 (1963).
272. Wilhelm, K., Curdt, W., Marsch, E., Schühle, U., Lemaire, P., Gabriel, A., Vial, J.-C., Grewing, M., Huber, M.C.E., Jordan, S.D., Poland, A.I., Thomas, R.J., Khne, M., Timothy, J.G., Hassler, D.M., Siegmund, O.H.W.: Solar Phys. **162**, 189–231 (1995).

273. Wilson, R.M., Hildner, E.: Solar Phys. **91**, 169–180 (1984).
274. Wilson, R.M., Hildner, E.: J. Geophys. Res. **91**, 5867–5872 (1986).
275. Woo, R.: J. Geophys. Res. **98**, 18999–19004 (1993).
276. Woo, R., Armstrong, J.W.: Nature **292**, 608–610 (1981).
277. Woo, R., Armstrong, J.W.: Nature **304**, 756 (1983).
278. Woo, R., Armstrong, J.W., Sheeley, N.R., Jr., Howard, R.A., Michels, D.J., Koomen, M.J., Schwenn, R.: J. Geophys. Res. **90**, 154–162 (1985).
279. Yashiro, S., Gopalswamy, N., Michalek, G., St. Cyr, O.C., Plunkett, S.P., Rich, N.B., Howard, R.A.: J. Geophys. Res. **109**, doi:10.1029/2003JA010282 (2004).
280. Zhang, J., Wang, J.: Astrophys. J. **554**, 474–487 (2001).
281. Zhang, G., Burlaga, L.F.: J. Geophys. Res. **93**, 2511–2518 (1988).
282. Zhao, X.-P.: J. Geophys. Res. **97**, 15051–15055 (1992).
283. Zirker, J.B.: Total Eclipses of the Sun, Princeton Univ. Press (1995).

Chapter 3
Summary of Spacecraft

In Chap. 2, instruments contributing to our understanding of CMEs were discussed. These were mostly in the form of spacecraft, many of which were not purely dedicated solar missions. With the variety of methods employed in the study along with the multitude of missions and instruments, one cannot be blamed for becoming confused when linking research efforts with instruments. For an understanding of the history of solar physics over the last few decades, it is important to follow the hardware (perhaps more important than other disciplines[1]), so in this chapter a review of the important spacecraft is presented.

The list in the following pages is in chronological order (of launch date) and presents a very brief summary of basic information relevant to each spacecraft, past, present and future. Only those spacecraft that provided important advancements in our understanding of CMEs are presented, and these milestones are listed with each summary. We do not include the *GOES* spacecraft, but a complete listing of these spacecraft can be found in the Wikipedia entry for *GOES* (http://en.wikipedia.org/wiki/List_of_GOES_satellites). NOAA provides a website of the status of *GOES* (http://www.oso.noaa.gov/goestatus/) and a summary of the Space Environment Monitor (SEM) can be found at http://ngdc.noaa.gov/stp/satellite/goes/index/html.

[1] It has been my observation that observational solar physicists tend to be associated with instruments rather than ideas. If, for example, you were to show me a publication I could often identify the instruments used in the study simply from the names of the authors on the publication. As with most areas of science, developments tend to be associated with individuals. Hence, to a certain extent, to understand the development of ideas in solar physics it is important to follow the hardware.

T. Howard, *Coronal Mass Ejections: An Introduction*, Astrophysics and Space Science Library 376, DOI 10.1007/978-1-4419-8789-1_3,

3.1 The Early Space Age: 1960–1969

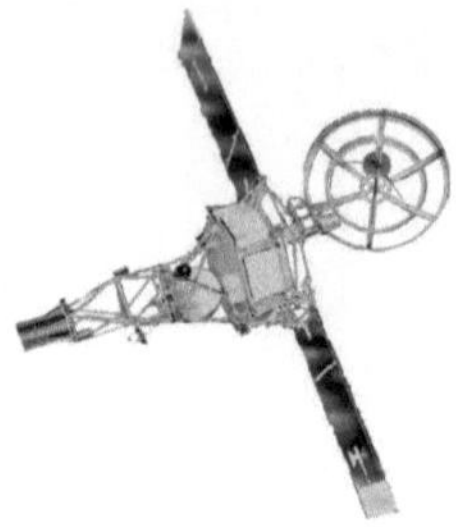

Name: *Mariner 2*.
Institution: NASA (USA).
Purpose: Venus flyby.
Launch Date: 27 August 1962.
End Date: Last transmission received 3 January 1963, remains in heliocentric orbit.
Instruments: Microwave and infrared radiometer, fluxgate magnetometer, ionisation chamber, particle and cosmic dust detector, solar plasma spectrometer.
Orbit: Heliocentric.
CME Milestones: First confirmed observation of the solar wind. First recorded interplanetary shock.
Reference: Sonnet [38].

Name: *Vela 3* (pair).
Institution: US Air Force (USA).
Purpose: To monitor worldwide compliance with the 1963 nuclear test ban treaty.
Launch Date: 17 July 1964.
Instruments: X-ray, neutron and gamma-ray detectors, electrostatic analysers.
Orbit: High Earth orbit ($\sim$18.5 Earth radii).
CME Milestone: Further observations of near-Earth interplanetary shocks.
Reference: Gosling et al. [9].

Name: *Pioneer 6*.
Institution: NASA/JPL (USA).
Purpose: Scientific research on the interplanetary medium.
Launch Date: 16 December 1965.
End Date: Primary transmitter failed on 15 December 1996. Mission ended December 2004. Maintains telemetry contact to date.
Instruments: Fluxgate magnetometer, plasma and cosmic-ray instruments, radio wave experiment.
Orbit: Heliocentric.
CME Milestone: First association between interplanetary shock and IPS ICME.
Reference: Siddiqi [36].

Name: *Pioneer 9*.
Institution: NASA/JPL (USA).
Purpose: Scientific research on the interplanetary medium.
Launch Date: 8 November 1968.
End Date: Lost contact on 19 May 1983. Spacecraft officially declared inactive following another failed communication attempt on 3 March 1987.
Instruments: Fluxgate magnetometer, plasma analyzer, cosmic-ray detectors, radio wave experiment, electric field detector and cosmic dust detector.
Orbit: Heliocentric.
CME Milestone: First association between CME and ICME. Shocks compared with IPS observations.
Reference: Siddiqi [36].

3.2 The Discovery of CMEs: 1970–1979

Name: *Orbiting Solar Observatory 7* (*OSO-7*).
Institution: NASA/GSFC (USA).
Purpose: Solar scientific research. Primary mission was to study the solar cycle in UV and x-ray.
Launch Date: 29 September 1971.
End Date: Re-entered Earth 9 July 1974.
Instruments: UV, x-ray and gamma-ray imagers, white light coronagraph.
Orbit: Low Earth Orbit.
CME Milestone: First recorded CME.
Reference: Koomen et al. [18].

Name: *Skylab*.
Institution: NASA/MSFC (USA).
Purpose: Manned space station, ATM: scientific solar physics research.
Launch Date: 14 May 1973.
End Date: 11 July 1979. Early re-entry due to space weather.
Instruments: The Apollo Telescope Mount (ATM): EUV spectrograph, x-ray telescope, coronagraph.
Orbit: Low Earth Orbit.
CME Milestone: First statistical survey of CMEs – first characteristics catalogued.
Reference: MacQueen et al. [23].

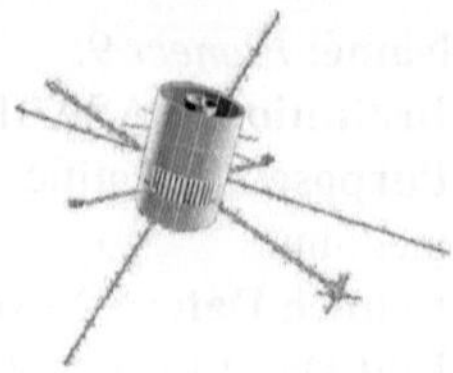

Name: *Interplanetary Monitoring Platform 8* (*IMP-8*).
Institution: NASA (USA).
Purpose: Scientific research of magnetic fields, plasmas, and energetic charged particles of the Earth's magnetotail and magnetosheath, and the near-Earth solar wind.
Launch Date: 26 October 1973.
End Date: Retired October 2001, but continues to function to date.
Instruments: Fluxgate magnetometer, ion and electron detector.
Orbit: Elliptical High Earth orbit (~25–45 Earth radii).
CME Milestone: Has provided significant long-term information on interplanetary shocks, ICMEs and on particles accelerated by shocks.
Reference: Papitashvili [32].

Name: *Helios* (pair).
Institution: Max Planck Institute (Germany).
Purpose: Scientific research on the interplanetary medium and zodiacal light.
Launch Date: *Helios 1*: 10 December 1974. *Helios 2*: 15 January 1976.
End Date: Mission concluded in 1982 but spacecraft remain in their orbits.
Instruments: Plasma experiment (E1), fluxgate (E2, E3) and search coil (E4) magnetometer, plasma wave experiment (E5A), x-ray and cosmic ray experiment (E6, E7), electron and ion spectrometer (E8), zodiacal light experiment (E9), micro-meteoroid analyser (E10).
Orbit: Eccentric heliocentric.
CME Milestone: First white light observation of ICMEs, in-situ information of the solar wind close to the Sun.
Reference: Leinert et al. [19].

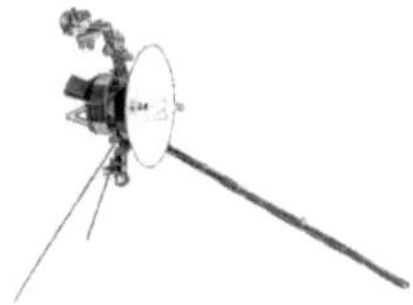

Name: *Voyager* (pair).
Institution: NASA/JPL (USA).
Purpose: Exploration of outer planets.
Launch Date: *Voyager 2*: 20 August 1977. *Voyager 1*: 5 September 1977.
End Date: Both continue to function to date and have enough power until at least 2025.
Instruments: Magnetometer (MAG), solar wind and energetic particle instruments (PLS, LECP), plasma wave and cosmic ray instruments (PWS, CRS), imagers (ISS), UV and IR spectrometers (UVS, IRIS), photopolarimeter.
Orbit: No orbit, spacecraft are on a trajectory away from the Sun.
CME Milestone: Early observations of magnetic clouds and ICMEs, and provided information on ICMEs at very large distances from the Sun.
Reference: Angrum et al. [3].

Name: *International Sun–Earth Explorer 3* (*ISEE-3*).
Institution: NASA/GSFC (USA).
Purpose: Scientific research of interplanetary medium.
Launch Date: 12 August 1978.
End Date: Became the *International Cometary Explorer* (*ICE*) in 1982, and flew through Comet P/Giacobini-Zinner on 11 September 1985. Is now in a heliocentric orbit due to return to the vicinity of Earth/Moon in August 2014.
Instruments: X-ray and gamma ray spectrometer, solar wind particle and spectrum analyser, energetic particle detector, vector helium magnetometer.
Orbit: The L1 Lagrange point.
CME Milestone: First spacecraft to observe CMEs from the L1 point (otherwise known as halo orbit).
References: Ogilvie et al. [30, 31].

Name: *P78-1* (*Solwind*).
Institution: Naval Research Laboratory (NRL)/US Air Force (USA).
Purpose: Scientific solar corona and x-ray research.
Launch Date: 24 February 1979.
End Date: 13 September 1985 (shot down during a US Air Force ASM-135 ASAT test).
Instruments: x-ray monitor, white light coronagraph.
Orbit: Sun-synchronous, Low Earth Orbit 600 km altitude.
CME Milestone: Early comprehensive statistical survey of CME properties. First halo CME recorded.
Reference: Michels et al. [25].

3.3 Our Understanding of the Sun and Interplanetary Medium Develops: 1980–1989

Name: *Solar Maximum Mission* (*SMM*).
Institution: NASA/GSFC (USA).
Purpose: Scientific research on the Sun.
Launch Date: 14 February 1980.
End Date: Malfunctioned on January 1981 (HXIS failed at that time), repaired by crew aboard space shuttle *Challenger* (STS-41C) in April 1984, re-entered Earth's orbit 2 December 1989.
Instruments: Radiometer (ACRIM), gamma-ray and x-ray spectrometers (HXRBS, GRS), hard x-ray spectroscopic imager (HXIS), soft x-ray and EUV spectrographic imagers (XRP, BCS/FCS, UVSP), white light coronagraph/polarimeter (C/P).
Orbit: Low Earth orbit.
CME Milestone: Intensive study of solar flares. Comprehensive statistical study of CME properties (over 1,000 CMEs observed). Classic three part CME identified.
Reference: MacQueen et al. [22].

3.4 The *SOHO* Era: 1990–1999

Name: *Ulysses*.
Institution: NASA/ESA (USA, Europe).
Purpose: To investigate the interplanetary medium at high heliographic latitudes and at large distances from the Sun.
Launch Date: 6 November 1990.
End Date: Lost normal telemetry in February 2007. X-band failure in February 2008 resulted in the deactivation of the spacecraft on 30 June 2009. It continues in its orbit.
Instruments: Magnetometer (VHM/FGM), solar wind plasma and ion detectors (SWOOPS), suprathermal ion detectors (SWICS) radio and plasma wave instruments (URAP), energetic ion and cosmic ray instrument (HISCALE, EPAC, COSPIN), x-ray and gamma-ray detectors (GRB), dust (DUST), coronal-sounding and gravitational wave experiments.
Orbit: Near polar heliocentric orbit ($\sim$1–5 AU).
CME Milestone: First in-situ measurements of polar regions of the heliosphere, ICMEs observed at large distances from the Sun and at high latitudes.
Reference: Longdon [21], Angrum [4].

Name: *Yohkoh* (Solar-A).
Institution: JAXA (Japan).
Purpose: Scientific investigation of the Sun, particularly solar flares.
Launch Date: 31 August 1991.
End Date: Ceased operations 14 December 2001. Re-entered 12 September 2005.
Instruments: soft and hard x-ray imagers (SXT, HXT), x-ray and gamma-ray spectrometers (WBS, BCS).
Orbit: Low Earth orbit.
CME Milestone: Comparison of high-resolution x-ray structure with CME signatures. CME-related "sigmoid" identified.
Reference: Ogawara [27], Ogawara et al. [28], McKenzie [24].

Name: SPARTAN 201.
Institution: NASA (USA).
Purpose: Scientific research on the solar corona.
Launch Date: 8 April 1993, 9 September 1994, 7 September 1995, 19 November 1997, 1 November 1998.
End Date: Flew five times on the space shuttle (STS-56, STS-64, STS-69, STS-87, STS-95) and brought back to Earth. Failed to deploy on STS-87. End of service life on 7 November 1998.
Instruments: EUV spectrometer, white light coronagraph.
Orbit: Low Earth orbit.
CME Milestone: Provided coronal information during the data gap between *SMM* and *SOHO*.
Reference: Guhathakurta et al. [10], Gurman and Fisher [11].

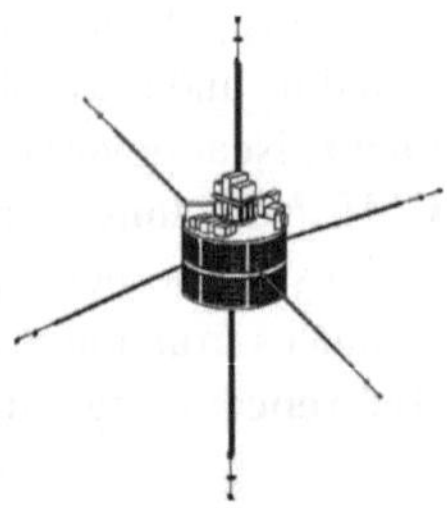

Name: *WIND*.
Institution: NASA/GSFC (USA).
Purpose: To study the near-Earth solar wind and magnetosphere.
Launch Date: 1 November 1994.
End Date: Continues to function to date.
Instruments: Radio and plasma wave experiment (WAVES), energetic and solar wind particle detectors (EPACT, SWE, SWICS), fluxgate magnetometer (MFI), gamma-ray detectors (TGRS).
Orbit: High Earth Orbit: Original mission: Eccentric Earth orbit spending most of its time in the upstream solar wind region (out to around 100–200 Earth radii), New (from November 1998): near polar eccentric Earth orbit ($\sim$10–80 Earth radii).
CME Milestone: Provides ongoing vital information on ICME occurrence, structure and composition, allowing comparison with current CME observing instruments on board *SOHO* and *STEREO*.
Reference: Ogilvie [29].

Name: *Solar and Heliospheric Observatory* (*SOHO*).
Institution: NASA/ESA (USA/Europe).
Purpose: Scientific study of the Sun.
Launch Date: 2 December 1995.
End Date: Malfunctioned on 25 June 1998, restored mid-September 1998. (LASCO) C1 failed at that time. Continues to function to date.
Instruments: EUV imagers and spectrometers (EIT, CDS, SUMER, UVCS), three white light coronagraphs (LASCO), Doppler imager (MDI), Helioseismology instruments (MDI, GOLF, VIRGO), in-situ particle detectors (CELIAS, COSTEP, ERNE), neutral particle instrument (SWAN).
Orbit: The L1 Lagrange point.
CME Milestone: Most comprehensive study of CMEs to date (over 10^4 observed), statistics over solar cycle, clear observations of the largest number of Earth-directed (halo) CMEs and clear connection with space weather. Concluded the Solar Flare Myth by conclusively bringing the role of the CME into the foreground.
Reference: Domingo et al. [6], SOHO [37].

Name: *Advanced Composition Explorer* (*ACE*).
Institution: Caltech (USA).
Purpose: Scientific research of interplanetary medium.
Launch Date: 25 August 1997.
End Date: Continues to function to date.
Instruments: Magnetometer (MAG), solar wind instruments (SWEPAM, SWICS/SWIMS), energetic particle instruments (ULEIS, EPAM, SEPICA, SIS, CRIS).
Orbit: The L1 Lagrange point.
CME Milestone: Ongoing information on ICME occurrence, structure and composition, allowing comparison with current CME observing instruments aboard *SOHO* and *STEREO*. Magnetic reconnection with ICMEs.
Reference: Stone et al. [39], Christian and Davis [5].

Name: *Transition Region And Coronal Explorer* (*TRACE*).
Institution: NASA/SMEX (USA).
Purpose: Scientific research on the transition region and corona of the Sun.
Launch Date: 1 April 1998.
End Date: Final observing sequence in 21 June 2010.
Instruments: UV and EUV imager.
Orbit: Sun-synchronous polar Low Earth Orbit.
CME Milestone: CME helicity and associated solar "surface" activity.
Reference: Handy et al. [12], Hurlburt [14].

3.5 The Next Generation: 2000–2009

Name: *(Ramaty) High Energy Solar Spectroscopic Imager* (*(R)HESSI*).
Institution: NASA/GSFC (USA).
Purpose: Solar flare observer.
Launch Date: 5 February 2002.
End Date: Continues to function to date.
Instruments: Hard x-ray imager and spectrometer.
Orbit: Low Earth Orbit (600 km).
CME Milestone: High energy solar "surface" activity.
Reference: Lin et al. [20], Holman [13].

Name: *P98-2* (*Coriolis*).
Institution: Naval Research Laboratory (NRL)/US Air Force (USA).
Purpose: Proof of concept heliospheric imager, ocean weather studies.
Launch Date: 6 January 2003.
End Date: Continues to function to date.
Instruments: White light whole sky imager (SMEI), microwave polarimetery instrument (WINDSAT).
Orbit: Sun-synchronous polar Low Earth Orbit (800 km).
CME Milestone: First widefield (all-sky) white light images of ICMEs. First statistical survey of ICME images.
Reference: Eyles et al. [8], Mozer [26].

Name: *Hinode* (Solar-B).
Institution: JAXA (Japan).
Purpose: Solar observations.
Launch Date: 22 September 2006.
End Date: Continues to function to date.
Instruments: Visible light and soft x-ray imagers (SOT, XRT), EUV spectrometer (EIS).
Orbit: Sun-synchronous Low Earth Orbit (600 km).
CME Milestone: Solar activity. CME onset data provision.
Reference: Ichimoto et al. [15], Adams and Davis [1].

Name: *Solar-TErrestrial RElations Observatory* (*STEREO*) (pair).
Institution: NASA/GSFC (USA).
Purpose: Solar and heliospheric investigation.
Launch Date: 25 October 2006.
End Date: Continues to function to date.
Instruments: Three instrument suites, SECCHI: EUV imager (EUVI) and white light coronagraphs (COR) and heliospheric imagers (HI), SWAVES: Radio burst tracker, IMPACT/PLASTIC: Solar wind particle detector (SWEA, SEPT), suprathermal ion and electron analysers (STE, SIT), energetic particle instruments (LET, HET), magnetometer (MAG).
Orbit: Heliocentric, sharing Earth orbit but with one leading and the other lagging.
CME Milestone: First images of CMEs from different viewpoints, first images of the region between 30 and 80 $R_\odot$, first complete coverage of CME images from the Sun to 1 AU.
Reference: Kaiser et al. [16], Addison and Kaiser [2].

Name: *Solar Dynamics Observatory* (*SDO*).
Institution: NASA/GSFC/LWS (USA).
Purpose: Solar observatory.
Launch Date: 11 February 2010.
End Date: Continues to date.
Instruments: Helioseismic and magnetic imager (HMI), White light, UV, EUV imagers (AIA), EUV irradiance instrument (EVE).
Orbit: Geosynchronous: Longitude 100°, inclination 30°.
CME Milestone: Early evolution of CMEs – launch mechanism.
Reference: Pesnell and Addison [33].

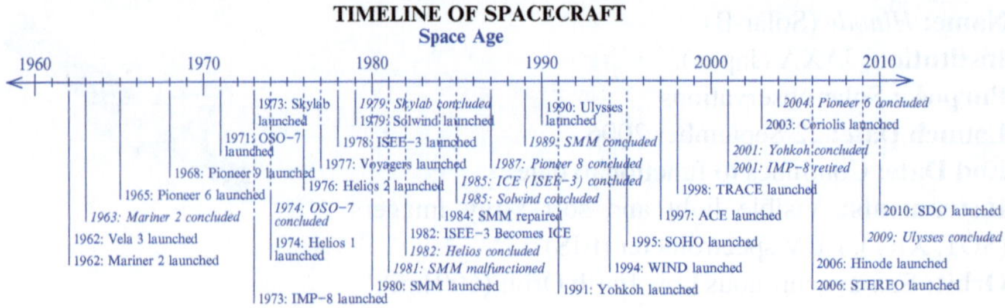

Fig. 3.1 Timeline of spacecraft launch and conclusions (*italic font*). Times of major anomalies in the spacecraft are also shown

Figure 3.1 shows the timeline for the spacecraft above, excluding the future spacecraft in the next section.

3.6 The Future

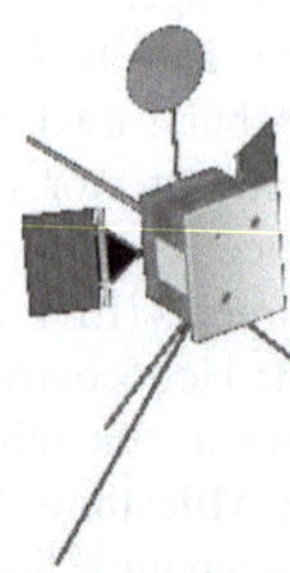

Name: *Solar Orbiter*.
Institution: ESA (Europe).
Purpose: High resolution solar imager and in-situ measurements.
Launch Date: 2018
End Date: Six year (nominal) mission planned.
Instruments: Energetic Particle Detector (EPD), Extreme Ultraviolet Imager (EUI), Magnetometer (MAG), Coronagraph (METIS/COR), Visible Imager and Magnetograph (PHI), Radio and Plasma Waves (RPW), Heliospheric Imager (SoloHI), EUV Spectrometer (SPICE), X-ray Imager (STIX), Solar Wind Plasma Analyser (SWA).
Orbit: Heliocentric, passing within 0.3 AU.
CME Milestone: High-resolution images of CMEs and in-situ measurements of ICMEs.
Reference: ESA [7].

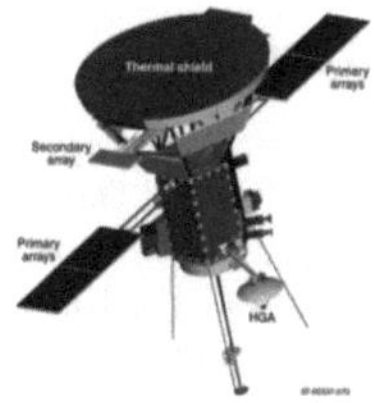

Name: *Solar Probe (Plus).*
Institution: NASA/APL (USA).
Purpose: Direct measurement of the solar corona.
Launch Date: 2018
End Date: Seven year (nominal) mission planned.
Instruments: Solar Wind Electrons Alphas and Protons Investigation (SWEAP), Wide-field Imager for Solar Probe Plus (WISPR), a magnetic and electric fields investigation (FIELDS), Integrated Science Assembly (ISIS).
Orbit: Heliocentric spiral in the ecliptic, 24 solar encounters, passing within 0.16 AU (35 $R_\odot$).
CME Milestone: In-situ measurements of CMEs as they appear in coronagraphs.
Reference: Rumberg [34].

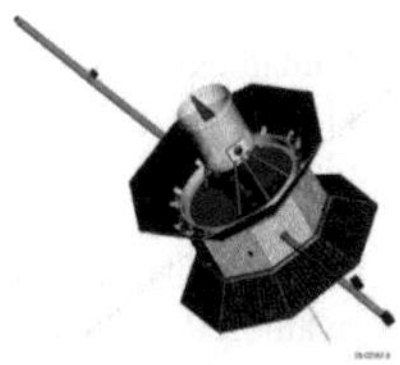

Name: *Solar Sentinels* (four spacecraft).
Institution: NASA/APL (USA).
Purpose: Simultaneous multiple measurements of interplanetary medium.
Launch Date: 2018?
End Date: Five year (nominal) mission planned.
Instruments: Energetic particle (LICA, STE, EPI, HICA, SEPQ) and solar wind (SWI, SWE, SWComp) detectors, DC and search coil magnetometers (MAG, SCM), radio experiment (Waves), neutron, x-ray, gamma-ray spectrometer (NS, XRI/GRS).
Orbit: Heliocentric, passing within 0.3 AU.
CME Milestone: Multiple in-situ measurements of ICMEs.
Reference: Kirby [17].

Name: *Solar-C.*
Institution: JAXA (Japan).
Purpose: Solar observation.
Launch Date: 2018?
End Date: Six year (nominal) mission planned.
Instruments: Plan A: Photospheric and chromospheric Dopplergram, polarimeter, EUV/x-ray imager, possible in-situ instruments and coronagraph. Plan B: Near IR-visible-UV imagers, EUV-x-ray imager.
Orbit: Plan A: Out of the ecliptic, 30–45° inclination. Plan B: Sun synchronous.
CME Milestone: Plan A: First out of the ecliptic images of CMEs.
Reference: Tsuneta [35].

References

1. Adams, M., Davis, J.M.: In: The Hinode (Solar-B) Webpage, available via NASA/MSFC. http://solarb.msfc.nasa.gov/.CitedApril102009.
2. Addison, K., Kaiser, M.L.: In: The STEREO Webpage, available via NASA/GSFC. http://stereo.gsfc.nasa.gov/.Cited25March2009.
3. Angrum, A., Medina, E., Sedlacko, D.: In: The Voyager Webpage, available via NASA/JPL. http://voyager.jpl.nasa.gov/.Cited21January2009.
4. Angrum, A., Sedlacko, D.: In: The Ulysses webpage, available via NASA/JPL. http://ulysses.jpl.nasa.gov/index.html.Cited30September2008.
5. Christian, E.R., Davis, A.J.: In: The Advanced Composition Explorer (ACE) Webpage, available via SRL/Caltech. http://www.srl.caltech.edu/ACE/.Cited2October2006.
6. Domingo, V., Fleck, B., Poland, A.I.: Solar Phys. **162**, 1–37 (1995).
7. ESA: The Solar Orbiter Webpage, available via ESA. http://sci.esa.int/science-e/www/area/index.cfm?fareaid=45.CitedMarch2010.
8. Eyles, C.J., Simnett, G.M., Cooke, M.P., Jackson, B.V., Buffington, A., Hick, P.P., Waltham, N.R., King, J.M., Anderson, P.A., Holladay, P.E.: Solar Phys. **217**, 319–347 (2003).
9. Gosling, J.T., Asbridge, J.R., Bame, S.J., Hundhausen, A.J., Strong, I.B.: J. Geophys. Res. **73**, 43–50 (1968).
10. Guhathakurta, M., Fisher, R.R., Holzer, T.E., Sime, D.G.: Bull. Amer. Astron. Soc. **25**, 1213 (1993).
11. Gurman, J.B., Fisher, R.R.: In: SPARTAN 201: NASA's mission to explore the Sun's corona, available via NASA/GSFC. http://umbra.nascom.nasa.gov/spartan/.Cited28July2002.
12. Handy, B.N., Acton, L.W., Kankelborg, C.C., Wolfson, C.J., Akin, D.J., Bruner, M.E., Caravalho, R., Catura, R.C., Chevalier, R., Duncan, D.W., Edwards, C.G., Feinstein, C.N., Freeland, S.L., Friedlander, F.M., Hoffmann, C.H., Hurlburt, N.E., Jurcevich, B.K., Katz, N.L., Kelly, G.A., Lemen, J.R., Levay, M., Lindgren, R.W., Mathur, D.P., Meyer, S.B., Morrison, S.J., Morrison, M.D., Nightingale, R.W., Pope, T.P., Rehse, R.A., Schrijver, C.J., Shine, R.A., Shing, L., Tarbell, T.D., Title, A.M., Torgerson, D.D., Golub, L., Bookbinder, J.A., Caldwell, D., Cheimets, P.N., Davis, W.N., DeLuca, E.E., McMullen, R.A., Amato, D., Fisher, R., Maldonado, H., Parkinson, C.: Solar Phys. **187**, 229–260 (1999).
13. Holman, G.D.: In: The RHESSI Homepage, available via NASA/GSFC. http://hesperia.gsfc.nasa.gov/hessi/.Cited10November2008.
14. Hurlburt, N.: In: The TRACE Webpage, available via Lockheed Martin Missile and Space. http://trace.lmsal.com/.CitedJune2000
15. Ichimoto, K., the Solar-B Team: J. Korean Astron. Soc. **38**, 307–310 (2005).
16. Kaiser, M.L., Kucera, T.A., Davila, J.M., St. Cyr, O.C., Guhathakurta, M., Christian, E.: Space Sci. Rev. **136**, 5–16 (2008).
17. Kirby, K.W.: Solar Sentinels: Mission Study Report, available via APL. http://sentinels.gsfc.nasa.gov/Sentinels_MSR.pdf.CitedFebruary2008.
18. Koomen, M.J., Detwiler, C.R., Brueckner, G.E., Cooper, H.W., and Tousey, R.: Appl. Opt. **14**, 743–751 (1975).
19. Leinert, C., Link, H., Pitz, E., Salm, N., Kluppelberg, D.: Raumfahrtforschung **19**, 264–267 (1975).
20. Lin, R.P., Dennis, B.R., Hurford, G.J., Smith, D.M., Zehnder, A., Harvey, P.R., Curtis, D.W., Pankow, D., Turin, P., Bester, M., Csillaghy, A., Lewis, M., Madden, N., van Beek, H.F., Appleby, M., Raudorf, T., McTiernan, J., Ramaty, R., Schmahl, E., Schwartz, R., Krucker, S., Abiad, R., Quinn, T., Berg, P., Hashii, M., Sterling, R., Jackson, R., Pratt, R., Campbell, R.D., Malone, D., Landis, D., Barrington-Leigh, C.P., Slassi-Sennou, S., Cork, C., Clark, D., Amato, D., Orwig, L., Boyle, R., Banks, I.S., Shirey, K., Tolbert, A.K., Zarro, D., Snow, F., Thomsen, K., Henneck, R., McHedlishvili, A., Ming, P., Fivian, M., Jordan, John, Wanner, Richard, Crubb, Jerry, Preble, J., Matranga, M., Benz, A., Hudson, H., Canfield, R.C., Holman, G.D., Crannell, C., Kosugi, T., Emslie, A.G., Vilmer, N., Brown, J.C., Johns-Krull, C., Aschwanden, M., Metcalf, T., Conway, A.: Solar Phys. **210**, 3–32 (2002).

21. Longdon, N.: The Ulysses data book. A summary of the technical elements of the Ulysses spacecraft and its scientific payload., ESA, Paris (1990).
22. MacQueen, R.M., Csoeke-Poeckh, A., Hildner, E., House, L., Reynolds, R., Stanger, A., Tepoel, H., Wagner, W.: Solar Phys. **65**, 91–107 (1980).
23. MacQueen, R.M., Eddy, J.A., Gosling, J.T., Hildner, E., Munro, R.H., Newkirk, G.A., Jr., Poland, A.I., Ross, C.L.: Astrophys. J. **87**, L85 (1974).
24. McKenzie, D.: In: Yohkoh Solar Observatory Webpage, available via Montana State University. http://solar.physics.montana.edu/sxt/.Cited29November2006.
25. Michels, D.J., Howard, R.A., Koomen, M.J., Sheeley, N.R.Jr.: In Kundu, M.R., Gergely, T.E. (eds.), Radio Physics of the Sun, p.439, D. Reidel, Hingham MA (1980).
26. Mozer, J.: In: The Solar Mass Ejection Imager (SMEI) Webpage, available via NSO. http://smei.nso.edu/.Cited29March2004.
27. Ogawara,Y.: Solar Phys. **113**, 361–370 (1987).
28. Ogawara,Y., Takano,T., Kato,T., Kosugi,T., Tsuneta,S., Wtanabe,T., Kondo,I., Uchida,Y.: Solar Phys. **136**, 1–16 (1991).
29. Ogilvie, K.W.: In: The WIND Webpage, available via NASA/GSFC. http://www-istp.gsfc.nasa.gov/istp/wind/.Cited2005.
30. Ogilvie, K.W., Durney, A.C., von Rosenvinge, T.T.: IEEE Trans. Geosci. Electron., GE-16, 151–153, (1978).
31. Ogilvie, K.W., von Rosenvinge, T.T, Durney, A.C.: Science **198**, 131–138 (1977).
32. Papitashvili, N.E.: In: IMP-8 Project Information, available via NASA/GSFC. http://spdf.gsfc.nasa.gov/imp8/project.html.Cited2000.
33. Pesnell, D, Addison, K.: In: The Solar Dynamics Observatory Webpage, available via NASA/GSFC. http://sdo.gsfc.nasa.gov/.Cited2April2009.
34. Rumberg, J.: In: The Solar Probe Webpage, available via NASA/GSFC. http://solarprobe.gsfc.nasa.gov/Cited24July2008.
35. Tsuneta, S.: JAXA Solar-C mission, In: Proc. Solar-C Sci. Def. Meet., JAXA/ISAS (2008).
36. Siddiqi, A.A.: Deep Space Chronicle: A Chronology of Deep Space and Planetary Probes 1958–2000, NASA Monog. Aerosp. His. **24** (2002).
37. Solar and Heliospheric Observatory, In: The SOHO Webpage, available via NASA. http://sohowww.nascom.nasa.gov/.Cited2010.
38. Sonnet, C.P.: Space Sci. Rev. **2**, 751–777 (1963).
39. Stone, E.C., Frandsen, A.M., Mewaldt, R.A., Christian, E.R., Margolies, D., Ormes, J.F., Snow, F.: Space Sci. Rev. **86**, 1–22 (1998).

Chapter 4
How We Observe CMEs

Coronal mass ejections are observed in visible white light by coronagraphs which block out the light from the photosphere. This is made possible because CMEs are comprised of plasma, so they contain large numbers of free electrons. The white light we observe is originally from the photosphere, and is scattered off these electrons via a process called Thomson scattering.

In order to completely understand the evolution of CMEs it is necessary to understand the meaning of changes in their appearance in coronagraph and heliospheric images. Are these due to physical changes in the CME itself, or do they reflect changes in the observed intensity of the CME due to the Thomson scattering physics? Do we continue to observe the same component of a CME as it moves through the sky, or do different parts of the structure become more detectable as a results of the geometry and scattering? How do projection effects come into play? Can we use the CME image and our knowledge of how the light reaches us to identify physical characteristics of the CME itself (such as mass, trajectory, three-dimensional structure)? With careful analysis of the white light images and a detailed understanding of the Thomson scattering physics and projection geometry, it is possible to extract a great deal of information about the CME from their white light images.

This chapter addresses the analysis surrounding how CMEs and ICMEs are observed in white light. We begin with the basics of Thomson scattering theory and build up a picture of a CME from a line-of-sight integration to a volume of electrons. The objective is to identify how the appearance of a CME image changes due to the physics surrounding its appearance, particularly at large distances from the Sun. The effects of geometry on the appearance of CMEs and ICMEs are discussed in the next chapter.

A large component of this chapter is a modified version of the review paper by Howard and Tappin [2], but the theory of Thomson scattering is detailed in much earlier work, such as Minnaert [5] and van de Hulst [8]. The most popular work for describing the physics of Thomson scattering in the corona today is *A Guide to the Solar Corona* by Billings [1].[1] Because of some notational problems in this

[1] Note that all of these early works pre-date the discovery of the CME.

T. Howard, *Coronal Mass Ejections: An Introduction*, Astrophysics and Space Science Library 376, DOI 10.1007/978-1-4419-8789-1_4,

popular text, where the confusion arises we return to the work of Minnaert as we did in Howard and Tappin [2]. This chapter may be considered a consolidation of the work of Jackson [4], Billings [1], Tappin [7], Vourlidas and Howard [9], and Howard and Tappin [2]. The reader is encouraged to read these texts for a more complete understanding of CME detection and measurement.

4.1 Thomson Scattering Fundamentals

The idea that scattered photospheric light could be a major contributor to the observed solar corona predates the discovery of the electron. Schuster [6] integrated the incident radiation over the visible region of the Sun and derived the variation of polarisation with scattering angle [6]. Fifty years later, Minnaert [5] extended the analysis to include a quantitative scattering cross-section [5].

Thomson scattering is a special case of the general theory of the scattering of electromagnetic radiation by charged particles. Put simply, if electromagnetic radiation is incident on a free particle of charge e and mass m_e, the particle will be accelerated. As it accelerates, it too will emit radiation, and in Thomson scattering the momentum transfer from the photon to the electron is ignored, so the frequency of the scattered light is the same as the incident. This is the (Thomson) scattered radiation which will be emitted in all directions as demonstrated in Fig. 4.1a.

So, the scattered light is in a pattern symmetrical about the direction of the incident wave. Since the electric field of an electromagnetic wave is always perpendicular to the direction of propagation, the acceleration of the electron will be confined to the plane perpendicular to that direction. If an observer lies at a scattering angle (χ) of 0° or 180°, then the scattered light from the electron will be unpolarised because the electron will appear to be displaced equally in all directions (Fig. 4.1b). If, however, the observer lies at $\chi = 90°$ (Fig. 4.1d), then the electron will appear to move only in a linear motion perpendicular to the incident light. In that case the observer will see linearly polarised scattered light, but the maximum magnitude of the electric field will be the same. For an observer at an intermediate angle (Fig. 4.1c), the electric field in the plane of both the observer and the incident direction is shortened, resulting in a scattered light that is partially polarised. Hence, the intensity of the component seen as transverse to the incident beam is isotropic, while the intensity of the component seen as parallel to the projected direction of the incident beam varies as $\cos^2 \chi$.

If the particle with charge e and mass m_e moves with speeds that are small compared with the speed of light c, then the acceleration or radiation field $\mathbf{E}_a$ is given [4] by

$$\mathbf{E_a} = \frac{e}{4\pi\varepsilon_0 c}\left[\frac{\hat{\mathbf{n}}\times(\hat{\mathbf{n}}\times(\mathbf{a}/c))}{R}\right], \tag{4.1}$$

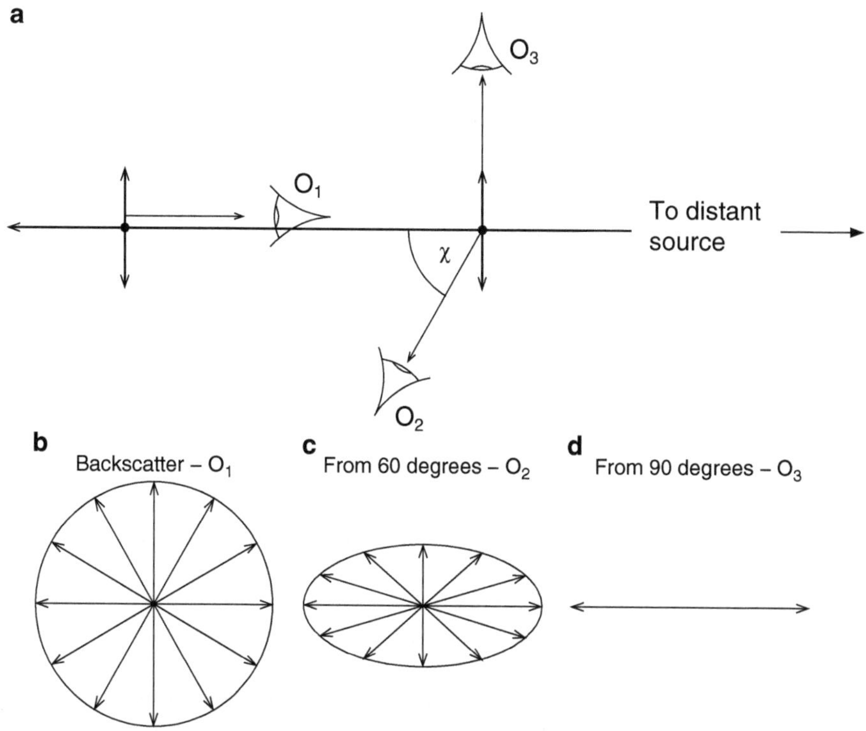

Fig. 4.1 Schematic demonstrating how the angular variation in Thomson scattering arises (**a**) the conceptual set up, the scattering angle χ is shown for the oblique observer (O_2), (**b–d**) the scattered electric vectors as seen by observers at O_1 at $\chi = 180°$, O_2 at $\chi = 60°$ and O_3 at $\chi = 90°$ respectively [2]. (Reproduced with kind permission of Springer Science and Business Media)

where ε_0 is the permittivity of free space (we are assuming the charge is moving in a vacuum or near vacuum), R is the distance traveled by the particle in a given time, $\hat{\mathbf{n}}$ is a unit vector in the direction of R and $\mathbf{a}$ is the acceleration vector at the same time. The energy flux at this time is the Poynting vector $\mathscr{P}$, defined by

$$\mathscr{P} = \mathbf{E} \times \mathbf{H} = \frac{1}{\mu_0}\mathbf{E} \times \mathbf{B}, \tag{4.2}$$

where $\mathbf{E}$ is the electric field, $\mathbf{B}$ is the magnetic flux density and μ_0 is the permeability of free space. Now, since the speed of light is defined by $c = 1/\sqrt{\mu_0\varepsilon_0}$ and the relationship between the vectors $\mathbf{E}$ and $\mathbf{B}$ is such that $\mathbf{E} \times \mathbf{B} = (1/c)|\mathbf{E}^2|\hat{\mathbf{n}}$, we can rewrite (4.2) as

$$\mathscr{P} = \varepsilon_0 c^2 \left(\frac{1}{c}|\mathbf{E}|^2\hat{\mathbf{n}}\right) = \varepsilon_0 c|\mathbf{E}|^2\hat{\mathbf{n}}. \tag{4.3}$$

Then by the definition of $\mathbf{E}_a$ and R, the power P radiated per unit solid angle ω is

$$\begin{aligned}\frac{\mathrm{d}P}{\mathrm{d}\omega} &= \varepsilon_0 c|R\mathbf{E}_a|^2 \\ &= \varepsilon_0 c\left(\frac{e}{4\pi\varepsilon_0 c}\right)^2 |\hat{\mathbf{n}}\times(\hat{\mathbf{n}}\times\mathbf{a}/c)|^2 \\ &= \frac{\varepsilon_0}{c}\left(\frac{e}{4\pi\varepsilon_0 c}\right)^2 |\hat{\mathbf{n}}\times(\hat{\mathbf{n}}\times\mathbf{a})|^2. \end{aligned} \quad (4.4)$$

This can be re-written in terms of the polarisation of the outgoing state (scattered beam) $\mathscr{E}$:

$$\frac{\mathrm{d}P}{\mathrm{d}\omega} = \frac{\varepsilon_0}{c}\left(\frac{e}{4\pi\varepsilon_0 c}\right)^2 |\mathscr{E}^*\cdot\mathbf{a}|^2, \quad (4.5)$$

where * denotes the complex conjugate.

Now, the acceleration **a** is provided by the incident wave, which has an electric field in the form of a plane wave,

$$E(\mathbf{x},t) = \mathscr{E}_0 E_0 \mathrm{e}^{(\mathbf{k}_0\cdot\mathbf{x}-\omega_0 t)i}, \quad (4.6)$$

where $\mathscr{E}_0$ is the polarisation vector of the incoming state (incident waves), $\mathbf{k}_0$ is the propagation vector, E_0 is the incident energy, ω_0 is the angular speed and $\mathbf{x}$ is the displacement vector. The electron is accelerated by an electric field, so we can now combine Newton's second law ($\mathbf{F} = m_e\mathbf{a}$) with Coulomb's law ($\mathbf{F} = e\mathbf{E}$) and find

$$E(\mathbf{x},t) = \mathscr{E}_0 \frac{e}{m_e} E_0 \mathrm{e}^{(\mathbf{k}_0\cdot\mathbf{x}-\omega_0 t)i}. \quad (4.7)$$

By assuming that the particle moves a small distance during a wave cycle (much smaller than a wavelength) we have the time average of $|\mathbf{a}|^2$ as $\frac{1}{2}\mathrm{Re}\{\mathbf{a}\cdot\mathbf{a}^*\}$. So the average power per solid angle becomes

$$\overline{\frac{\mathrm{d}P}{\mathrm{d}\omega}} = \frac{1}{2}\frac{e^2}{m_e^2}E_0^2\frac{\varepsilon_0}{c}\left(\frac{e}{4\pi\varepsilon_0 c}\right)^2 |\mathscr{E}^*\cdot\mathscr{E}_0|^2. \quad (4.8)$$

Now we introduce a new term for convenient use for scattering. This is the scattering cross-section $\mathrm{d}\sigma/\mathrm{d}\omega$, which is defined by Jackson [4] as

$$\frac{\mathrm{d}\sigma}{\mathrm{d}\omega} = \frac{\text{Energy radiated/unit time/unit solid angle}}{\text{Incident energy flux/unit area/unit time}}. \quad (4.9)$$

The incident energy flux (denominator) is the Poynting vector $\mathscr{P}$ (defined in (4.2)) for the incident wave when it has been time-averaged, that is

$$\text{Incident energy flux} = \frac{1}{2}\varepsilon_0 c E_0^2, \tag{4.10}$$

so (4.8) becomes:

$$\begin{aligned}\frac{d\sigma}{d\omega} &= \frac{e^2}{m_e^2 c^2}\left(\frac{e}{4\pi\varepsilon_0 c}\right)^2 |\mathscr{E}^* \cdot \mathscr{E}|^2 \\ &= \left(\frac{e^2}{4\pi\varepsilon_0 m_e c^2}\right)^2 |\mathscr{E}^* \cdot \mathscr{E}|^2.\end{aligned} \tag{4.11}$$

So now we need to express $|\mathscr{E}^* \cdot \mathscr{E}|^2$ in terms of the scattering geometry. Following Jackson once again, we resolve the scattered polarisation vector $\mathscr{E}$ into two components: $\mathscr{E}_1$ in the plane containing $\hat{\mathbf{n}}$ and $\mathbf{k}_0$; and $\mathscr{E}_2$ in the plane orthogonal to $\mathscr{E}_1$. This is shown in Fig. 4.2. The angle between the incident wave vector $\mathbf{k}_0$ and the scattered wave vector $\hat{\mathbf{n}}$ is the *scattering angle*, denoted χ.

Using the geometry in Fig. 4.2 we see that

$$\mathscr{E}_1 = (\cos\chi\cos\psi, \cos\chi\sin\psi, -\sin\chi), \tag{4.12}$$

$$\mathscr{E}_2 = (-\sin\psi, \cos\psi, 0). \tag{4.13}$$

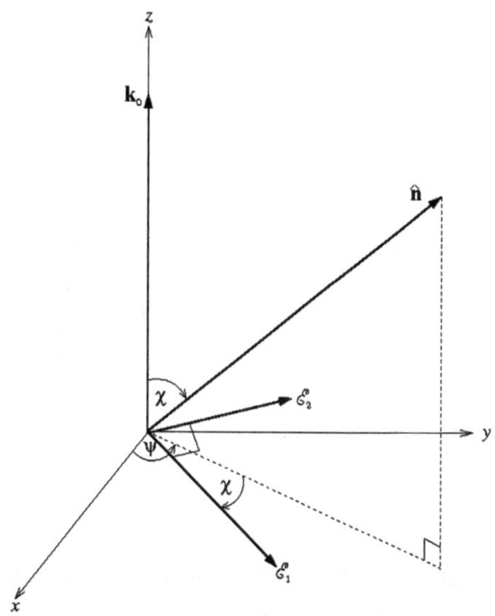

Fig. 4.2 Diagram showing the geometry where the scattered polarisation beam $\mathscr{E}$ is resolved into two components: $\mathscr{E}_1$ in the plane containing $\hat{\mathbf{n}}$ and $\mathbf{k}_0$; and $\mathscr{E}_2$ in the plane orthogonal to $\mathscr{E}_1$. The separation between the direction of the incident wave $\mathbf{k}_0$ and $\hat{\mathbf{n}}$ is the scattering angle χ and the angle between the x-axis and the $\hat{\mathbf{n}}$ vector is ψ. (Modified from Jackson [4])

According to this coordinate system, the incident wave $\mathscr{E}_0$ is represented by a unit vector in the x and y directions. We can thus resolve the scattered polarised beam by adding the two components from both axes. When performing the vector products the reader is reminded that we are dealing with incoherent power quantities, and so we add the squared components rather than adding the components before performing the square.

So, for the case of the x component of the incident wave:

$$\begin{aligned}|\mathscr{E}^* \cdot \mathscr{E}_0|_x^2 &= \left|\left(\mathscr{E}_1^2 + \mathscr{E}_2^2\right) \cdot \left(\mathscr{E}_{0x}^2\right)\right| \\ &= |((\cos\chi\cos\psi)^2 + (-\sin\psi)^2, (\cos\chi\sin\psi)^2 + (\cos\psi)^2, (-\sin\chi)^2) \\ &\quad \cdot(1,0,0)| \\ &= \cos^2\chi\cos^2\psi + \sin^2\psi, \end{aligned} \tag{4.14}$$

and for the y component:

$$\begin{aligned}|\mathscr{E}^* \cdot \mathscr{E}_0|_y^2 &= \left|\left(\mathscr{E}_1^2 + \mathscr{E}_2^2\right) \cdot (0,1,0)\right| \\ &= \cos^2\chi\sin^2\psi + \cos^2\psi.\end{aligned} \tag{4.15}$$

The total contribution is then the addition of these two components, taking care to use units of amplitude rather than peak-to-peak power:

$$\begin{aligned}|\mathscr{E}^* \cdot \mathscr{E}_0|^2 &= \frac{1}{2}\left(|\mathscr{E}^* \cdot \mathscr{E}_0|_x^2 + |\mathscr{E}^* \cdot \mathscr{E}_0|_y^2\right) \\ &= \frac{1}{2}(\cos^2\chi\cos^2\psi + \sin^2\psi + \cos^2\chi\sin^2\psi + \cos^2\psi) \\ &= \frac{1}{2}((sin^2\psi + \cos^2\psi) + \cos^2\chi(\cos^2\psi + \sin^2\psi)) \\ &= \frac{1}{2}(1 + \cos^2\chi),\end{aligned} \tag{4.16}$$

so the scattering cross-section (4.11) becomes

$$\frac{d\sigma}{d\omega} = \frac{1}{2}\left(\frac{e^2}{4\pi\varepsilon_0 m_e c^2}\right)^2 (1 + \cos^2\chi). \tag{4.17}$$

Now, by integrating over all solid angles we can then derive the total cross-section σ_T as

$$\sigma_T = \frac{8\pi}{3}\left(\frac{e^2}{4\pi\varepsilon_0 m_e c^2}\right)^2 = \frac{8\pi}{3}r_e^2 = 6.65 \times 10^{-29}\mathrm{m}^2, \tag{4.18}$$

where r_e is the classical electron radius. A useful alternative value is the differential cross-section for perpendicular scattering σ_e:

$$\sigma_e = \frac{e^4}{(4\pi\varepsilon_0)^2 m_e^2 c^4} = r_e^2 = 7.95 \times 10^{-30}\mathrm{m}^2\mathrm{sr}^{-1}. \tag{4.19}$$

This is the Thomson cross-section used by Billings [1], but he quotes it as having units of sr^{-1} rather than the correct cm^2sr^{-1} (or m^2sr^{-1}). This is probably the major cause of the confusion surrounding his derivations.

The Thomson theory can be used if the following conditions are met:

1. The coherence length of the radiation is small compared with the separation of the particles. In the case of incoherent sources means that the wavelength is small compared with the particle separation. If this is not met (e.g. radio waves) then the theory of coherent scattering applies;
2. The energy of the photons is negligible compared with the rest mass energy of the scattering particles. If this is not met (e.g. for x-rays) then the theory of Compton scattering applies.

For optical wavelengths and the densities encountered in the corona and solar wind, both of these conditions are comfortably met so we can safely use the Thomson theory throughout.

4.2 Application to the Solar Corona

The preceding analysis applies only to the scattering of light from a point source by a single electron or small packet of electrons. Unfortunately, from the perspective of the solar corona the photosphere is not a point source, so it is necessary to integrate the scattering over light from the visible disk of the Sun. This was first attempted by Schuster [6], and Minnaert [5] included the fact that the scatterers are electrons (unknown in Schuster's time) and also added solar limb-darkening. They showed how the tangential and radial electric vector components of Thomson-scattered radiation from a single electron in the solar corona may be calculated in terms of a small number of measurable parameters.

4.2.1 Resolving into Useful Components

The most conceptually difficult aspect of the integration over the visible photosphere is expressing the polarisation components from an element of the photosphere in terms of a common coordinate system. Here, we follow the example of Billings [1], who resolved the radius vector joining the centre of the Sun into parallel and perpendicular components, and dealt with each component separately. Consider the situation illustrated in Fig. 4.3a (adapted from the left-hand panel of Fig. 6.1 in Billings [1]). The y-axis is defined as the radius vector from the centre of the Sun (S) through the scattering point Q, the x-axis is defined such that the observer is in the xy-plane, and the z-axis completes the orthogonal set. The observer is located such that the line of sight from the observer through Q makes an angle χ with the radius vector SQ.[2]

[2] This is the same χ defined earlier as the scattering angle.

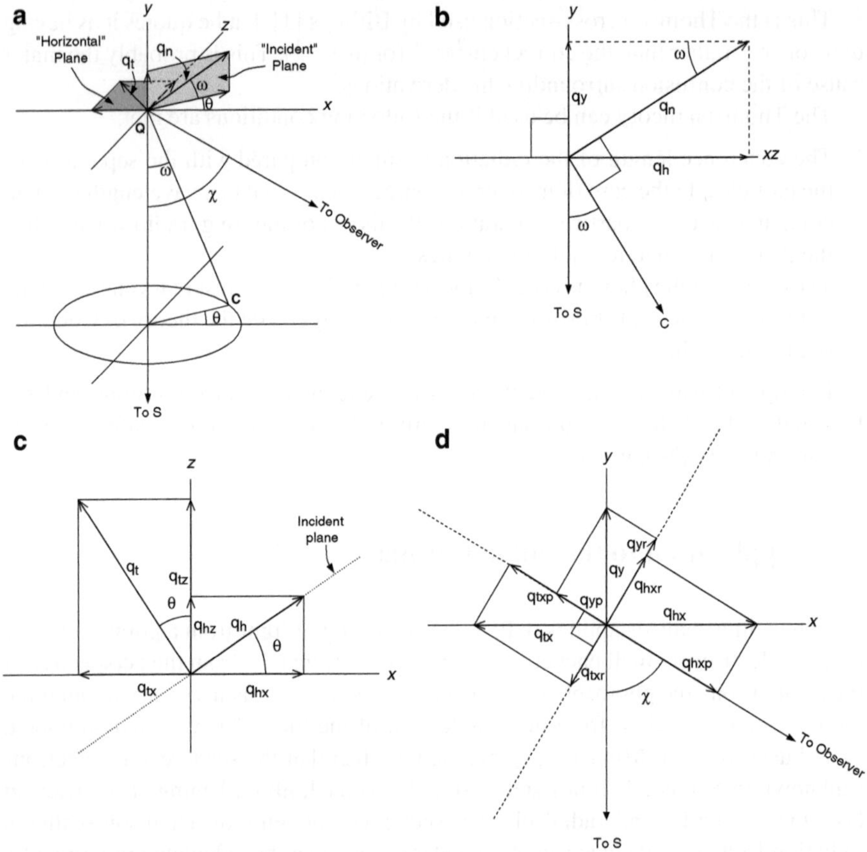

Fig. 4.3 The geometry involved in converting the polarisation components from an element of the photosphere to a common coordinate system. (**a**) A general overview, modified from Fig. 6.1 of Billings [1], (**b**) the CQS, or "incident" plane, resolved into *xyz* components, (**c**) the *xy* or "horizontal" plane, and (**d**) the *xy* or "emergent" plane [2]. (Reproduced with kind permission of Springer Science and Business Media)

We now introduce a point C on the photosphere and angles ω and θ, such that the angle CQS is ω, and the plane CQS makes an angle θ with the *xy*-plane. The light emitted from the element of the photosphere at C is unpolarised, so we may divide it into two arbitrary equal perpendicular polarisation components. Let's follow Billings [1] and divide it into one component in the *xz*-plane and one in the CQS plane, and let's call them the"horizontal" and "incident" planes respectively. We'll denote them as q_t, and q_n, as the *xz*-plane is tangential to the solar surface, and the incident plane is normal to it. Note that the two planes are not orthogonal to each other – q_n is elevated at an angle ω to q_t.

The first step is to transform q_t and q_n into *xyz* coordinates. In the "incident" plane (Fig. 4.3b), q_n is at an angle ω to the *xz*-plane, so it resolves into one component along the *y*-axis and one in the *xz*-plane, or

$$q_y = q_n \sin^2 \omega$$
$$q_h = q_n \cos^2 \omega, \quad (4.20)$$

since the decomposition is properly done with the electric vectors, rather than the intensities we are currently working with.

Turning to the "horizontal" plane (Fig. 4.3c), we can transform q_t and q_h into their x and z components. The "incident" plane crosses the "horizontal" plane at an angle θ to the x-axis, so q_h makes an angle θ with the x-axis and q_t is at θ to the z-axis. Hence q_h and q_t are resolved into

$$q_{hx} = q_h \cos^2 \theta = q_n \cos^2 \omega \cos^2 \theta$$
$$q_{hz} = q_h \sin^2 \theta = q_n \cos^2 \omega \sin^2 \theta, \quad (4.21)$$

and

$$q_{tx} = q_t \sin^2 \theta$$
$$q_{tz} = q_t \cos^2 \theta. \quad (4.22)$$

The next step is to transform the x and y components into the observer's coordinates in the xy-plane (Billings calls this the "emergent" plane). In this plane each of the x and y components are resolved into two further components: perpendicular and parallel to the observer's line of sight. These are labeled r for "radial" (since they are parallel to the solar radius vector) and p for "parallel" (Fig. 4.3d).

Using the geometry in Fig. 4.3d and Eqs. 4.20–4.22, we may resolve the y and x components q_y and q_x into their r and p components:

$$q_{yr} = q_y \sin^2 \chi = q_n \sin^2 \omega \sin^2 \chi$$
$$q_{yp} = q_y \cos^2 \chi = q_n \sin^2 \omega \cos^2 \chi, \quad (4.23)$$

$$q_{hxr} = q_{hx} \cos^2 \chi = q_n \cos^2 \omega \cos^2 \theta \cos^2 \chi$$
$$q_{hxp} = q_{hx} \sin^2 \chi = q_n \cos^2 \omega \cos^2 \theta \sin^2 \chi, \quad (4.24)$$

and

$$q_{txr} = q_{tx} \cos^2 \chi = q_t \sin^2 \theta \cos^2 \chi$$
$$q_{txp} = q_{tx} \sin^2 \chi = q_t \sin^2 \theta \sin^2 \chi. \quad (4.25)$$

Finally we gather together the terms transverse to the radius vector and those parallel to it. Remember that q_t and q_n are equal in magnitude, so let's call these simply q hereafter. The transverse terms are those with electric vectors on the z axis (q_{hz} and q_{tz}) so we add these components to determine the total transverse component q_T:

$$q_T = q_{hz} + q_{tz} = q(\cos^2 \theta + \cos^2 \omega \sin^2 \theta). \quad (4.26)$$

The radial terms are those in the "emergent" plane and perpendicular to the observer's line of sight, namely q_{yr}, q_{hxr} and q_{txr}. We add these to determine the total radial component q_R:

$$q_R = q_{yr} + q_{hxr} + q_{txr} = q(\sin^2\omega\sin^2\chi + \cos^2\omega\sin^2\theta\cos^2\chi + \cos^2\theta\cos^2\chi). \tag{4.27}$$

4.2.2 Incident and Scattered Intensity

Now that the incident and emergent terms have been resolved into components containing exclusively q, θ, ω and χ, it is time to consider the received intensity and the scattered radiation.

The intensity q[3] *received* at Q from the element of photosphere around the point C is given by

$$q = I\sin\theta d\theta d\omega = -I d\theta d(\cos\omega), \tag{4.28}$$

where I is the emitted intensity from the photosphere in units of power per unit area per unit solid angle. Note that to be strictly correct we ought to write I as a function of ω and Ω, in order to accommodate for limb-darkening.

The total *scattered* radiation is determined by integrating Eqs. (4.26) and (4.27) over the visible portion of the photosphere, using the geometry shown in Fig. 4.4. That is, we perform two double integrations, the first with respect to θ across the range $\theta = 0 - 2\pi$ and the second with respect to $\cos\omega$ from $\cos\Omega$ to 1. That is:

$$I_T = \frac{\sigma_e}{2z^2}\int_{\cos\Omega}^{1}\int_0^{2\pi} I(\cos^2\theta + \cos^2\omega\sin^2\theta)d\theta d(\cos\omega), \tag{4.29}$$

and

$$I_R = \frac{\sigma_e}{2z^2}\int_{\cos\Omega}^{1}\int_0^{2\pi} I(\sin^2\omega\sin^2\chi + \cos^2\omega\sin^2\theta\cos^2\chi + \cos^2\theta\cos^2\chi)d\theta d(\cos\omega), \tag{4.30}$$

where z is the distance from Q to the observer.[4] It is here we depart from the formalism of Billings and return to Minnaert. Note that just one electron is being considered (hence there is no N_e term) and that since we wish to avoid the use of power per unit solid angle, the distance to the observer (z) has been included.

The integration over θ can be easily accomplished by applying the trigonometric identity $\sin^2\alpha + \cos^2\alpha \equiv 1$ to the $\cos^2\chi$ and $\sin^2\omega$ terms. Equations (4.29) and (4.30) then become:

$$I_T = \frac{\pi\sigma_e}{2z^2}\int_{\cos\Omega}^{1} I(1 + cos^2\omega)d(\cos\omega), \tag{4.31}$$

[3] This can be either q_t or q_n, it doesn't matter as the equation is the same.

[4] This is **not** the same z as in Fig. 4.3 and the one we used in the coordinate deconstruction.

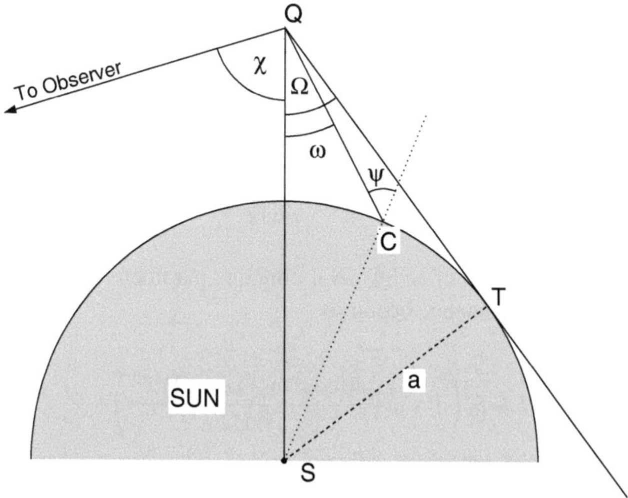

Fig. 4.4 The Sun and a nearby scattering point Q (modified from Fig. 6.2 of Billings [1]). The shaded *grey circle* represents the Sun with its centre at S and radius a, T is the point where the scattered point vector crosses the Sun at a tangent and Ω is the angle between the tangent from Q and the SQ vector. The line of sight (To Observer) has been added and the angle between it and the SQ vector is shown as χ. The geometry of a ray from a point C is also shown, including the angles ψ and ω [2]. (Reproduced with kind permission of Springer Science and Business Media)

and

$$I_R = \frac{\pi\sigma_e}{2z^2}\int_{\cos\Omega}^{1} I[(1+\sin^2\chi)+\cos^2\omega(1-3\sin^2\chi)]\mathrm{d}(\cos\omega). \tag{4.32}$$

Before we perform the second integration, there are two modifications we need to apply. First, when integrating over $\cos\omega$, it is more convenient to work with a term that allows the χ dependency to be factored out. Since

$$I_P = I_T - I_R = -\frac{\pi\sigma_e}{2z^2}\int_{\cos\Omega}^{1} I\sin^2\chi(1-3\cos^2\omega)\mathrm{d}(\cos\omega), \tag{4.33}$$

it is better to work in terms of $I_P = I_T - I_R$ rather than I_R alone. Second, we must factor in the contributions of limb-darkening. The following expression has been used by both Minnaert and Billings for limb darkening:

$$I = I_0(1-u+u\cos\psi), \tag{4.34}$$

and that is good enough for us. In this expression, the limb-darkening coefficient is denoted u and is a function of wavelength, and ψ is the angle between the emitted radiation and the radius vector. From Fig. 4.4 we can show that

$$\frac{\sin\psi}{\sin\omega} = \frac{|\mathbf{SQ}|}{a}, \tag{4.35}$$

and

$$a = |\mathbf{SQ}| \sin \Omega, \tag{4.36}$$

which allows us to produce the following expression relating ψ to ω and Ω:

$$\sin \psi = \frac{\sin \omega}{\sin \Omega}. \tag{4.37}$$

This allows the conversion of (4.34) so it contains parameters already in our integrations. Equation (4.34) hence becomes

$$I = I_0 \left(1 - u + \frac{u(\cos^2 \omega - \cos^2 \Omega)^{1/2}}{\sin \Omega} \right). \tag{4.38}$$

So now the limb-darkening parameters can be factored into the two integrals (4.31) and (4.33), which become

$$\begin{aligned} I_T = I_0 \frac{\pi \sigma_e}{2z^2} \Bigg[(1-u) \int_{\cos \Omega}^{1} (1 + cos^2 \omega) \mathrm{d}(\cos \omega) \\ + u \int_{\cos \Omega}^{1} \frac{(\cos^2 \omega - \cos^2 \Omega)^{1/2}}{\sin \Omega} (1 + cos^2 \omega) \mathrm{d}(\cos \omega) \Bigg] \\ \equiv I_0 \frac{\pi \sigma_e}{2z^2} \left[(1-u)C + uD \right], \end{aligned} \tag{4.39}$$

and

$$\begin{aligned} I_P = -I_0 \frac{\pi \sigma_e}{2z^2} \sin^2 \chi \Bigg[(1-u) \int_{\cos \Omega}^{1} (1 - 3\cos^2 \omega) \mathrm{d}(\cos \omega) \\ + u \int_{\cos \Omega}^{1} \frac{(\cos^2 \omega - \cos^2 \Omega)^{1/2}}{\sin \Omega} (1 - 3\cos^2 \omega) \mathrm{d}(\cos \omega) \Bigg] \\ \equiv I_0 \frac{\pi \sigma_e}{2z^2} \sin^2 \chi \left[(1-u)A + uB \right]. \end{aligned} \tag{4.40}$$

The integrals within each have been given the coefficients A, B, C and D, which can finally be integrated to give

$$\begin{aligned} A = \int_{\cos \Omega}^{1} (1 + cos^2 \omega) \mathrm{d}(\cos \omega) \\ = \cos \Omega \sin^2 \Omega, \end{aligned} \tag{4.41}$$

$$B = \int_{\cos\Omega}^{1} \frac{(\cos^2\omega - \cos^2\Omega)^{1/2}}{\sin\Omega}(1+cos^2\omega)\mathrm{d}(\cos\omega)$$
$$= -\frac{1}{8}\left[1 - 3\sin^2\Omega - \frac{\cos^2\Omega}{\sin\Omega}(1+3\sin^2\Omega)\ln\left(\frac{1+\sin\Omega}{\cos\Omega}\right)\right], \quad (4.42)$$

$$C = \int_{\cos\Omega}^{1}(1-3\cos^2\omega)\mathrm{d}(\cos\omega)$$
$$= \frac{4}{3} - \cos\Omega - \frac{\cos^3\Omega}{3}, \quad (4.43)$$

$$D = \int_{\cos\Omega}^{1} \frac{(\cos^2\omega - \cos^2\Omega)^{1/2}}{\sin\Omega}(1-3\cos^2\omega)\mathrm{d}(\cos\omega)$$
$$= \frac{1}{8}\left[5 + \sin^2\Omega - \frac{\cos^2\Omega}{\sin\Omega}(5-\sin^2\Omega)\ln\left(\frac{1+\sin\Omega}{\cos\Omega}\right)\right]. \quad (4.44)$$

The total intensity of the light scattered from the electrons is

$$I_{tot} = (I_T + I_R) = 2I_T - I_P, \quad (4.45)$$

where I_0 is the intensity of the source (in this case, the Sun) as a power per unit area (of the photosphere) per unit solid angle. The coefficients A, B, C and D are generally known as the van de Hulst coefficients after the author who modified the coefficients in order to reduce the number of tabulated functions needed (van de Hulst [8]). The terms were actually originally introduced by Minnaert [5], and in the modern computer era the modifications imposed by van de Hulst are no longer required.

For modern calculations involving CMEs and Thomson scattering, workers refer to the equations as defined by Billings [1]. All the necessary equations are conveniently provided on a single page (p. 150) so there is little need to trawl through the theory. Unfortunately, Billings defines his scattered intensities I_T, I_R and I_P in units of power per unit solid angle, while I_0 is actually power *per unit area* per unit solid angle. Thus his scattered intensities differ from those by Minnaert (and from those presented here and by Howard and Tappin [2]) by a factor of z^2. I regard this as an oversight at best (and error at worst) by Billings, as it is not meaningful to translate the relations into units of solar brightness. The good news is that the z^2 term cancels out when the integrations are performed over large areas, so the many workers who have used Billings to perform their calculations can be reassured that the results obtained have not been incorrect. I do not see the need to continue to propagate this error into the literature, and so here we have returned to the theory as originally intended and recommend the reader do the same.

4.3 Moving Away from the Sun

In the previous section we have derived the equations necessary to determine the total intensity scattered from a single electron in the solar corona by Thomson scattering. Those applying this theory to CMEs today use the SolarSoft routine `eltheory.pro`, which applies the Billings version of (4.39–4.40), (4.41–4.44), and (4.45) which in the Billings text include a density term N_e. Hence the van de Hulst coefficients are central to the application of Thomson scattering to determine CME properties. In this section, we investigate how the theory can be applied to distances further from the Sun into the heliosphere.

It is important to note that the van de Hulst coefficients only contain a single independent variable: Ω, the angular separation of the solar radius relative to the electron at point Q (Fig. 4.4). So what happens when Q is at large distances from the Sun, when Ω becomes small and the Sun tends toward a point source? This has two effects on Q: the incident radiation becomes more collimated; and it also becomes fainter. Both of these effects are contained in the van de Hulst coefficients. The fall-off in incident intensity is represented by coefficients tending towards $1/R^2$, while the collimation is seen in the decreasing difference between A and C and between B and D. Figure 4.5 shows the variation of the four coefficients with distance from the Sun, out to a distance of 0.1 AU ($\sim$21.5 $R_\odot$). It is important to note that the coefficients tend toward $1/R^2$ much sooner than would be expected: well inside 0.1 AU where the Sun would still have an apparent diameter greater than 5°. This means that *even in the corona, from around 2 $R_\odot$ outwards*, the Thomson scattering physics may be approximated with the Sun as a point source.

For any line of sight (LOS), the maximum scattered intensity is the point along that LOS that is closest to the Sun. This is *not* due to the scattering physics itself, but rather because the closest point has the largest density and maximum incident intensity[5] along the LOS. Geometrically, this is the point where the LOS is normal to the solar radial vector (i.e. when $\chi = 90°$). It was shown by Vourlidas and Howard [9] that the locus of all points where $\chi = 90°$ form a sphere with the Sun-observer line as the diameter (their Fig. 1). They called it the "Thomson surface" and showed how the scattered brightness of a single electron varied when projected radially from the Sun. They stated that the only important parameter in determining the relative brightness was the distance of the electron from the Thomson surface, and made some predictions on the trajectory of ICMEs most likely to be detected by heliospheric imagers. While I disagree with some of their conclusions, I do believe that the foundations they laid in extending the Thomson scattering physics into the heliosphere are sound – they just didn't take it far enough or with sufficient depth.

[5] Both density and incident intensity fall off with an approximate $1/R^2$ factor as we move away from the Sun.

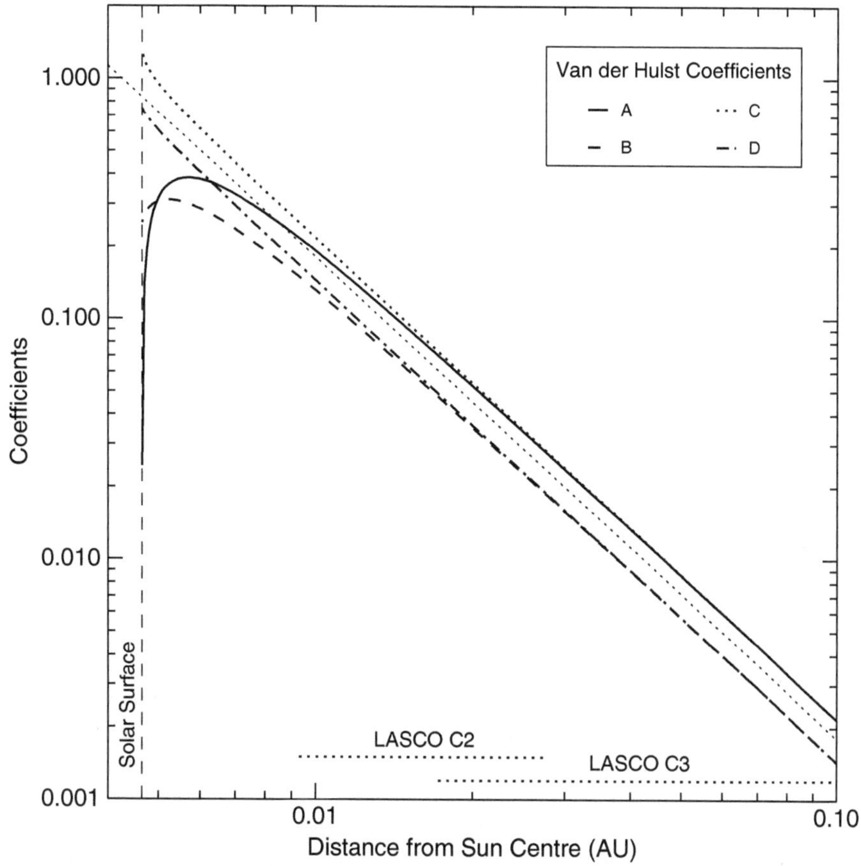

Fig. 4.5 The four van de Hulst coefficients, plotted as functions of R, the distance from the centre of the Sun [2]. Overlayed on these plots is a light dotted trace with a slope of −2. As this is a log–log plots, this line indicates a curve with a trend of $1/R^2$. Hence when the coefficients have a trend which is parallel with this line, they also have a $1/R^2$ trend. This begins to happen at distances beyond a few solar radii. For comparison, the sky plane projection of the LASCO/C2 and C3 fields of view are indicated. (Reproduced with kind permission of Springer Science and Business Media)

As demonstrated by Howard and Tappin [2], The total scattered intensity received at a location on a LOS is governed by three separate terms:

1. The scattering efficiency, which is a *minimized* on the Thomson surface.
2. The incident intensity, which is *maximized* on the Thomson surface. This is because that is where the LOS is closest to the Sun.
3. The electron density in the scattering region, which is *maximized* on the Thomson surface. This is also because the LOS is closest to the Sun, and because the solar wind density falls off with distance from the Sun.

The separate contributions of efficiency and intensity (Points 1 and 2) are rarely appreciated because they are bundled together in the expressions for the scattered intensity (4.39), (4.40) and (4.45).

When these three terms are combined the point at which the scattered light is maximized remains that at which the LOS is orthogonal to the Sun-Q vector, i.e. when on the Thomson surface. This is because the incident intensity and density effects are far more significant than scattering efficiency. However, minimisation of the scattering efficiency does result in a partial cancelling, which causes a spread of observed intensity when moving to large distances from the Sun. This spreading is also greater at larger distances from the Sun. As stated by Howard and Tappin:

> Hence... the distance of the observed volume from the Thomson surface is less significant when making measurements based on the assumption of Thomson scattering, but at the cost of increased uncertainty in identification of the location of the observed unit volume. (p. 43 [2]).

4.4 Integrated Lines of Sight

When one observes an object in space, one does not observe the object exclusively, but rather everything between the object and the observer is observed together collectively. The result is a two-dimensional projection representing an integration of everything along the direction observed. A single direction along which one observes is called a line of sight, and the resulting two-dimensional projection of everything is called an integrated line of sight. Our brains later attempt to introduce depth by identifying a number of so-called "cues" that it is programmed to recognise as an indicator of depth. It is important then to remember that the original image contains no depth, but rather is a collection of integrated lines of sight.

When observing CMEs we are presented with the same problem of integrated lines of sight, except here it is far more difficult to identify depth in the resulting image. The theory of Thomson scattering regarding CMEs is presented as a line-of-sight integral, but this is not strictly true. It is in fact an integral though the cone of the observer's (instrument's) point spread function. In other words, it is not restricted to a single line integral, but rather is a function of the collecting area and beam size of the instrument.

To illustrate this process, consider the diagram in Fig. 4.6. Here we have an idealised detector with a surface area δA and beam size $\delta\omega$. The coordinate system is defined such that the z-axis is along the LOS and the xy plane is the plane of the sky relative to the observer. Let's introduce a unit volume of plasma located at distance z from the detector, of dimensions (dx,dy,dz). At the detector, this volume element contributes a power of

$$\mathrm{d}P = I_0 G_x(z) N_e(x,y,z)\,\mathrm{d}x\,\mathrm{d}y\,\mathrm{d}z\,\delta A, \tag{4.46}$$

where I_0 is the source intensity, N_e is the electron number density in the volume element and G_x is the scattering expression associated with I_T, I_R, I_P or I_{tot} (where x can be any parameter) as described in (4.39), (4.40) and (4.45). So, substituting each expression we have

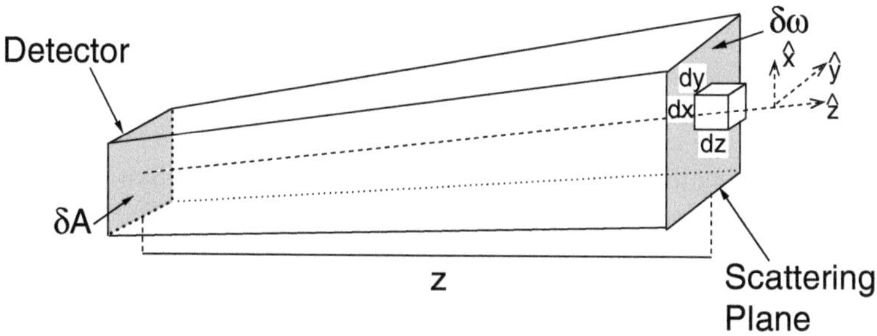

Fig. 4.6 Diagram demonstrating the configuration for the LOS integral calculations [2]. A unit volume of dimensions (dx,dy,dz) is at a distance Z from the observer. The coordinate system is defined such that the z-axis is along the line of sight and the x-y plane is in the plane of the sky. The detector at the observer location has an area of δA and its field of view at Z has area $\delta\omega$. (Reproduced with kind permission of Springer Science and Business Media)

$$G_T = \frac{\pi\sigma_e}{2z^2}[(1-u)C + uD],$$

$$G_P = \frac{\pi\sigma_e}{2z^2}[(1-u)A + uB],$$

$$G_R = G_T - G_P,$$

and

$$G_{tot} = G_T + G_R = 2G_T - G_P,$$

where each of the G parameters are functions of the distance z from the detector (i.e. $G_x = G_x(z)$).

Now, if the beam from the detector observes a solid angle of sky $\delta\omega$ and we allow an incrementation of the distance z by a small amount dz, then the total power received by the detector from the part of the LOS between z and $z + \mathrm{d}z$ is

$$P_{\mathrm{rec}}(z)\,\mathrm{d}z = \iint_{\delta\omega} I_0 N_e G \mathrm{d}x\,\mathrm{d}y\,\mathrm{d}z\delta A. \tag{4.47}$$

Also, the element of solid angle dω at a distance z observed by the detector is

$$\mathrm{d}\omega = \frac{\mathrm{d}x\,\mathrm{d}y}{z^2},$$

which implies

$$\iint_{\delta\omega} \mathrm{d}x\mathrm{d}y = z^2\delta\omega,$$

so (4.47) becomes

$$P_{\mathrm{rec}}(z)\,\mathrm{d}z = \iint_{\delta\omega} I_0 N_e z^2 G\delta A \mathrm{d}z\mathrm{d}\omega, \tag{4.48}$$

or

$$P_{\text{rec}}(z)\,\mathrm{d}z = I_0 N_e z^2 G \delta A \mathrm{d}z \delta\omega. \tag{4.49}$$

At this point we can convert the power into intensity using units of solar surface brightness by setting the power that would be received from the Sun by the observing instrument $I_0 \delta A \delta\omega$ equal to 1. Equation (4.49) then becomes

$$I_{\text{rec}}(z)\,\mathrm{d}z = N_e z^2 G \mathrm{d}z. \tag{4.50}$$

(4.50) represents the intensity contribution to a detector of a single element on the LOS. We must now integrate along the entire LOS to determine the whole contribution. The total intensity integrated along the LOS received by the detector is thus

$$I_{\text{rec}} = \int_0^\infty N_e z^2 G \mathrm{d}z. \tag{4.51}$$

Here is the simple line-of-sight integral shown in Billings [1], which assumes that the electron density does not vary significantly across the beam of the instrument.

When observing CMEs, it is generally assumed that the entire intensity is located on the Thomson surface, so the integral in (4.51) is reduced to

$$I_{\text{rec}} = z_T^2 N_e G, \tag{4.52}$$

where z_T is the distance from the observer to the Thomson surface and N_e and G are both functions of z_T (i.e. $N_e = N_e(z_{\mathrm{T}})$ and $G = G(z_{\mathrm{T}})$). The spreading of the signal created by the partial cancelling by the minimising of the scattering efficiency results in this assumption being applicable even across fairly large distances from the Thomson surface. As mentioned before, this allows a gain in accuracy in the determination of parameters such as density, but a loss of information about the location of the scattering volume itself. Figure 4.7 demonstrates this, where the geometry for a single LOS is shown in Fig. 4.7a and the scattered intensity distribution as a function of distance for a number of lines of sight are shown in Fig. 4.7b. In Fig. 4.7b, the intensity distributions have been normalised by

$$\frac{z^2 N_e G}{\int_0^\infty z^2 N_e G \mathrm{d}z}. \tag{4.53}$$

As shown in Fig. 4.7b, the peaks in the contributions to the intensity occur on the Thomson surface as expected. However of note is the width of the peak, which becomes broader with increasing elongation. This indicates that as a volume of electrons moves away from the Thomson surface, its scattered light intensity will drop off at a less rapid rate when it is at larger distances from the Sun.

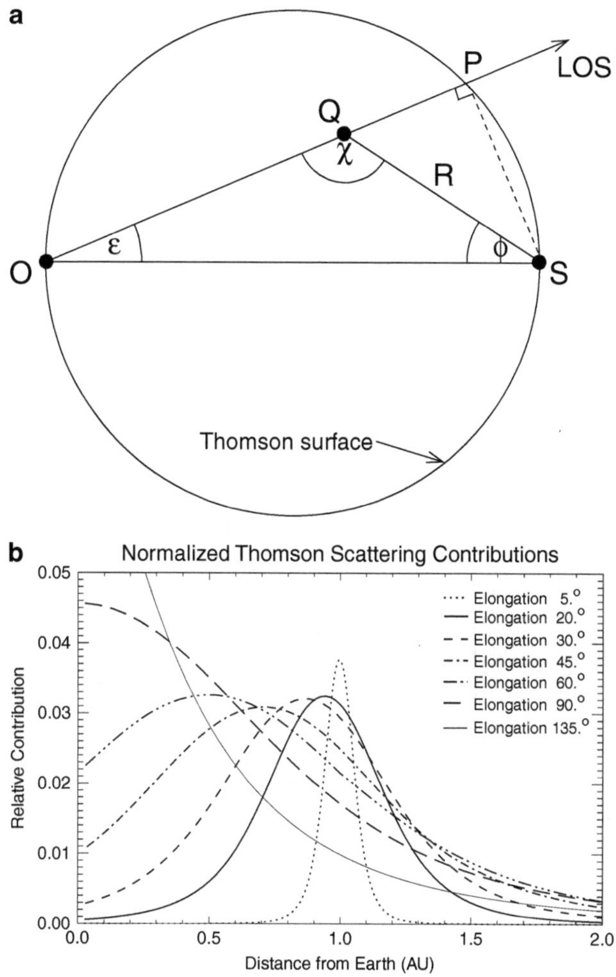

Fig. 4.7 (**a**) Diagram of the relevant vectors and angles to the plots below. Shown are the LOS, Sun-Q vector SQ and Sun-observer line SO. The elongation ε and Sun-observer packet angle ϕ are indicated along with the location of the Thomson surface (the LOS always meets the Thomson surface at point P). (**b**) Normalised scattered intensity vs distance from the observer in AU [2]. (Reproduced with kind permission of Springer Science and Business Media)

4.5 Reconstructing the CME/ICME

To this point we have considered the scattering from the perspective of a single element only, first the scattering of a single electron or volume of electrons, and then as an integrated LOS along a single direction. The CME itself, however, is a much larger volume, and as such represents the combination of a large number of electrons and lines of sight. Thus, it is necessary to treat the CME as a complete structure, and consider this structure as it becomes an ICME.

From a mathematical perspective, the density of a CME (the part observed by a coronagraph and heliospheric imager) can be represented as a function multiplying the background solar wind. It could also be an additive function, but in this case let's choose the multiplicative density formulation, as did Howard and Tappin [2] and earlier workers [3]. The total intensity along an arbitrary LOS (denoted with the subscript *i*) can then be expressed in the following modified version of (4.51):

$$I_{\mathrm{rec}_i} = \int_0^\infty N_e z^2 F_i G_i \,\mathrm{d}z, \tag{4.54}$$

where F_i is the CME function along the *i*th LOS. An example from Howard and Tappin [2] which demonstrates the inclusion of this multiplicative function is discussed below.

Consider the structure shown in Fig. 4.8a, which is the simplest form of CME structure: a spherical shell. In this case the simulated ICME has an angular width of ~90°, a uniform thickness of 0.05 AU, and is at a time when the leading edge of the ICME is ~0.5 AU from the Sun. The ICME density is a fractional change F of the baseline electron density, which in this case is assumed to be uniform and twice the base level. Finally the ICME central axis is offset from the Sun-observer line.

Three lines of sight are shown at elongations 0° (the Sun-observer line), 10° and 20° and their functions are shown in Fig. 4.8b. The ICME has a value of 2 across its thickness and the LOS function has a value of 1 everywhere else. As shown in Fig. 4.8a, the thickness of the ICME through each LOS is determined by the angle at which the LOS passes through the ICME. Note that close to the Sun, the angular difference is negligible, and so little change occurs across the LOS collective when considering CMEs. For this ICME the function has a width of 0.053 AU for the 10° elongation LOS and a width of 0.069 AU for the 20° LOS. The thickness along the 0° LOS is exactly 0.05 AU because this is a spherical shell the Sun-observer line passes normally through the ICME.

Generally, the ICME thickness along any LOS may be calculated from

$$\mathrm{d}z_r = [(r+\mathrm{d}r)^2 - p^2]^{1/2} - z_s, \tag{4.55}$$

where r is the location of the outer edge of the ICME, $\mathrm{d}r$ is the ICME thickness, p is the distance from the Sun to the Thomson surface and z_s is the distance along the LOS from the leading edge to the Thomson surface [7].

So, mathematically the observed CME is the collection of all of the LOS functions that pass through the ICME structure. These may be compiled in a number of ways, but the most useful and convenient is to reproduce the two-dimensional image representing the projection of each LOS. This is best achieved by means of

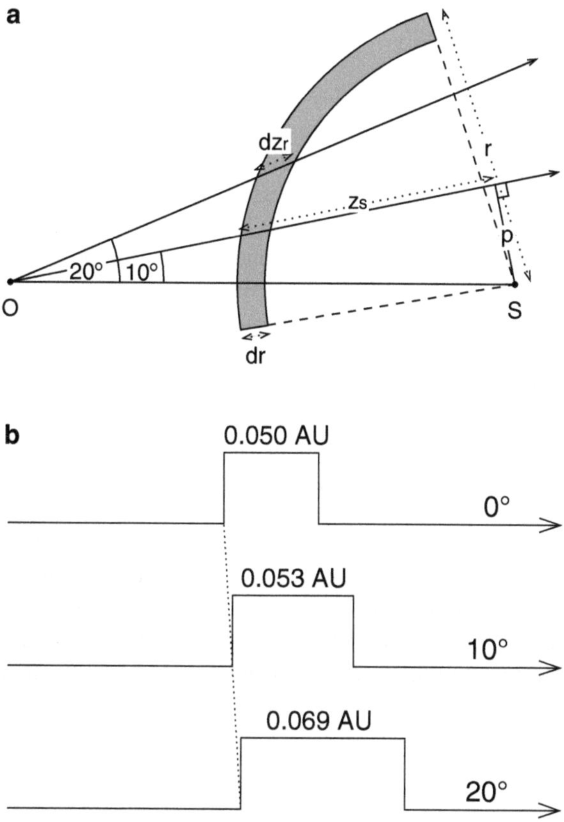

Fig. 4.8 Illustration of the CME function relative to different lines of sight. (**a**) A simple ICME spherical shell structure, with an angular width of ~90°. Here, the ICME is ~0.5 AU from the Sun. The ICME also has a uniform density of two times the base level electron density. For generality the ICME central axis is offset from the Sun-observer line. Three lines of sight are shown by way of example, at elongations 0° (the Sun-observer line), 10° and 20°. (**b**) The ICME function for each LOS. The ICME has a value of 2 across its thickness and the LOS function has a value of 1 everywhere else. The thickness of the ICME through each LOS is determined by the angle at which the LOS passes through the ICME. In this case the ICME function is wider for the 20° elongation LOS than the 10° LOS. The Sun-observer line passes normally through the ICME, and so the function on this LOS has a thickness of exactly 0.05 AU [2]. (Reproduced with kind permission of Springer Science and Business Media)

a map with a suitable projection, upon which each LOS integral is a single point. The complete map can be mathematically represented by

$$I_{\text{map}} = \int_0^N I_{\text{rec}_i} \mathrm{d}i, \tag{4.56}$$

where N is the total number of lines of sight and I_{rec_i} is the LOS intensity with the ICME superimposed, defined in (4.54). For those values of i for which the LOS does not pass through the ICME, the CME function F_i has a value of 1.

4.6 Concluding Remarks

The theoretical basis for how CMEs and ICMEs are detected in white light has a heritage dating back to before the discovery of the electron. While the theory established in the 1960s by Billings is still in use today for the analysis of CMEs (particularly for determination of CME mass), we cite a word of caution with using his notation because of a few discrepancies. These have probably not been regarded as significant by the community, as they cancel out when the CME is treated as a collective. Nonetheless, it is not helpful to continue to propagate an error when we have become aware of it.

The other problem is that the theory outlined in this chapter is spread out among a variety of literature, some of which are out of print, difficult to get hold of, and when they are obtained, difficult to understand. The paper in *Space Science Reviews* by Howard and Tappin consolidated this theory into a more accessible (and, we hope, easier to understand) layout, and it has been my intention to do the same with this chapter.

There remains one final problem associated with CME and ICME detection which must be accommodated. The geometry and location of the CME/ICME also plays a significant role in the apparent geometry of the white light images. The next chapter addresses the theory behind the geometry, and also how a careful application of this theory can result in three-dimensional structural and kinematic information about CMEs and ICMEs.

References

1. Billings, D.E.: A Guide to the Solar Corona, New York, Academic Press (1966).
2. Howard, T.A., Tappin, S.J.: Space Sci. Rev. **147**, 31–54 (2009).
3. Hundhausen, A.J.: J. Geophys. Res. **78**, 1528–1542 (1973).
4. Jackson, J.D.: Classical Electrodynamics (2nd edn.), New York, Wiley (1975).
5. Minnaert, M.: Z. Astrophys. **1**, 209–235 (1930).
6. Schuster, A.: Mon. Not. R. Astron. Soc. **40**, 35–57 (1879).
7. Tappin, S.J.: Planet. Space Sci. **35**, 271–283 (1987).
8. van de Hulst, H.C., Bull. Astron. Inst. Netherlands **11**, 135–150 (1950).
9. Vourlidas, A., Howard, R.A.: Astrophys. J. **642**, 1216–1221 (2006).

Chapter 5
Geometry of CMEs and ICMEs

In the previous chapter the theoretical framework for how CMEs are observed in white light was established. This framework was built up from a single electron scattering white light by the process of Thomson scattering, through to an integrated line of sight, and finally to a complete map of the CME through an integration of the collective of lines of sight. Upon review of that chapter it should be clear that there is an additional complication when analysing white light images of ICMEs that does not occur when analysing CMEs. Many of the assumptions we can apply when considering images close to the Sun break down when we move further away. The theory in Chap. 4 has shown, for example:

1. The behaviour of the van de Hulst coefficients (the coefficients governing the intensity of Thomson scattered light) tend toward $1/R^2$ as the Sun starts to look like a point source (Sect. 4.3);
2. The assumption that the Thomson scattered light is maximised in the plane of sky no longer applies, instead it is maximised on the Thomson surface, a sphere with diameter of the Sun-observer line (Sect. 4.3);
3. Information on the location of the scattering volume of electrons is reduced, due to a spreading of the intensity function with distance along a given LOS (Sect. 4.4);
4. The ICME function multiplying the background along a LOS (the function describing the ICME within the solar wind) changes in apparent structure and length for different lines of sight (Sect. 4.5).

We now must consider one final complication to the analysis of white light CME/ICME images: The geometry of the CME/ICME itself. Section 4.5 has dealt with this to a certain extent when regarding the complete picture of the ICME relative to the entire sky, but those of us who measure ICME parameters (such as distance from the Sun) tend to select a feature that is easy to identify. The most commonly-used feature is the leading edge, identified in images as a sharp change in intensity from the CME to the background (see Fig. 1.1). When measuring this feature it is generally assumed that the location of the leading edge on the ICME does not change, i.e. that the same part of the ICME is measured each time. This is a valid assumption for CMEs close to the Sun, but once again matters become more

T. Howard, *Coronal Mass Ejections: An Introduction*, Astrophysics and Space Science Library 376, DOI 10.1007/978-1-4419-8789-1_5,

complicated at larger distances. While more difficult to understand and analyse, additional information can be extracted about the three-dimensional structure of ICMEs (information impossible to extract from CME images), provided one is careful and thorough in their geometrical analysis. In this chapter we reveal how this can be achieved using ICME images far from the Sun, but first we consider how multiple images from different viewpoints can be used to reconstruct the three-dimensional structure of near-Sun CMEs.

5.1 CMEs Close to the Sun

Let us first consider what can be done with CME images close to the Sun. It has been demonstrated in Chap. 4 that very little three-dimensional information can be ascertained from an image of a CME from a single viewpoint. However, we *can* extract such information if we are fortunate enough to observe the same CME from different viewpoints at the same time. The SECCHI imagers on board *STEREO* allow us this ability for the first time. Recall from Chaps. 2 (Sect. 2.7.3) and 3 (Sect. 3.5) that the *STEREO* spacecraft orbit in such a way as to increase their angular distance from each other and the Sun-Earth line by around 22° a year. Hence, after around a year into the mission, they were far enough apart for projection differences in CME images to be detectable (and measurable). By measuring these differences and applying triangulation geometry, the three-dimensional location of the CME can be identified. Such a technique was employed using the COR2s by Howard & Tappin [10] on two CMEs observed in November 2007, when the *STEREO* spacecraft were 20° away from the Sun–Earth and 40° away from each other. Figure 1.2 shows the coronagraph (LASCO and COR2) images of one of these CMEs. The projection effects are clear in these images, and they are discussed briefly in Sect. 1.2.6.

The triangulation was applied using the geometry shown in Fig. 5.1, which is based on the assumptions that each observer lies in the same plane (in this case, the ecliptic) and that three measurements of the same point on the CME can be obtained. The elevation (or latitude) from the Sun of the points is the angle β and the angular separation between the projection into the equatorial plane of each observer is α. So there is a single β and three values of α are to be determined. It is also known that each α is separated by a known quantity $\Delta\alpha$, which in this case is equal to the separation angle between the spacecraft. The Position Angle (PA) of the measured point relative to each observer is Π and the projected distance of the point in the equatorial plane is r_e. Finally, the perpendicular distances from the projected point and the vectors from the Sun to observer 1 and 2 are l_1 and l_2 respectively [10].

Using this geometry, we arrive at

$$\tan\beta = \tan\Pi_1 \sin\alpha_1 = \tan\Pi_2 \sin(\alpha_1 + \Delta\alpha), \tag{5.1}$$

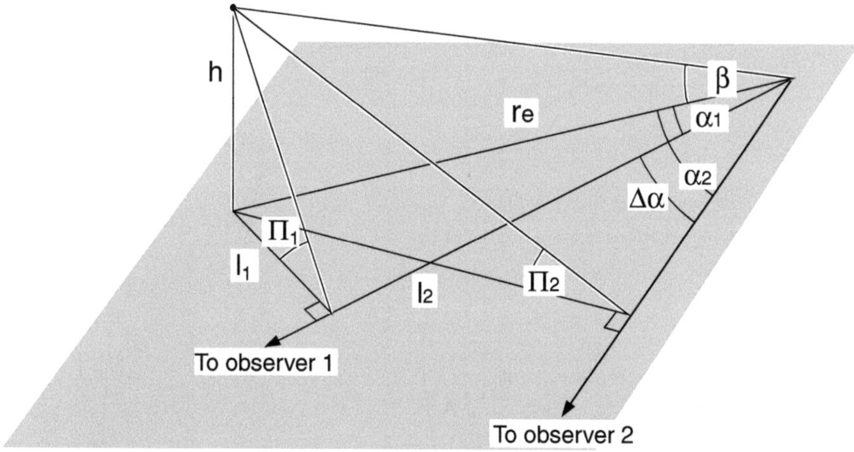

Fig. 5.1 The geometry for determination of the 3-D triangulation of the CME [10] (Reproduced with kind permission of Springer Space and Business Media)

or

$$\tan\alpha_1 = \frac{\sin\Delta\alpha}{\frac{\tan\Pi_1}{\tan\Pi_2} - \cos\Delta\alpha}. \tag{5.2}$$

Howard & Tappin [10] performed additional coordinate rotations, first into solar ecliptic, then to heliocentric using

$$\beta' = \sin^{-1}(\cos\beta\sin\delta\cos(\alpha-\eta) + \sin\beta\cos\delta), \tag{5.3}$$

and

$$\alpha' = \tan^{-1}\left(\frac{\sin\beta - \sin\beta'\cos\delta}{\cos\beta\sin(\alpha-\eta)\sin\delta}\right) - (90^\circ - \eta), \tag{5.4}$$

where δ is the tilt of the solar axis and η is the longitude of the plane containing the solar axis and the ecliptic pole, relative to the spacecraft.

So, the heliocentric coordinates of the latitude and longitude of each measured point on the CME were identified, allowing an estimation of the three-dimensional structure of each CME to be made. These reconstructions, as produced by Howard & Tappin for two events observed in November 2007 [10], are shown in Fig. 5.2. The Howard & Tappin [10] paper was the first to perform geometrical triangulation using the *STEREO*/COR2 coronagraphs. Since then a number of techniques has emerged, including forward modelling [28, 30], tie-pointing [22], inverse reconstruction [1] and polarisation ratio [23]. Other triangulation methods have also been developed [3, 17, 27]. A review of many of these techniques can be found in Mierla et al. [21]. Theirs and others' developments continue.

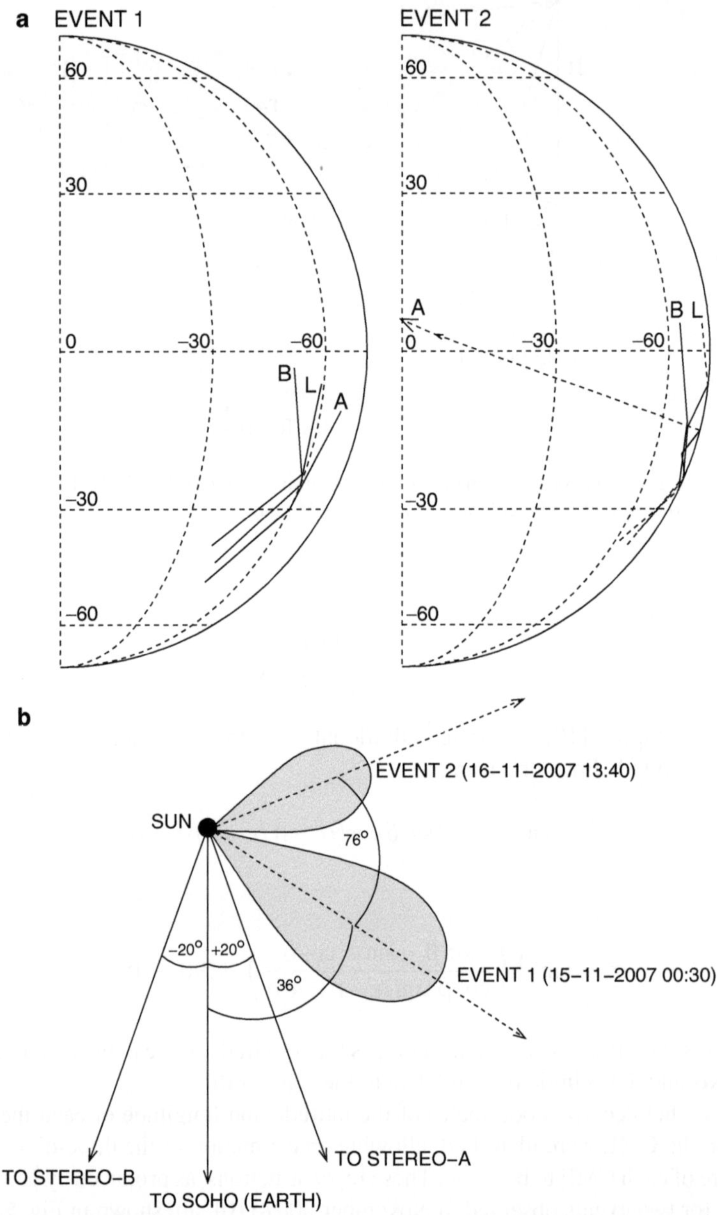

Fig. 5.2 Three-dimensional reconstruction of each CME. (**a**) The location of the CME projected onto the solar surface for (*left*) Event 1 and (*right*) Event 2, with solar latitude and longitude contours added. The three traces on each disk represents the projected location from all three spacecraft, rotated into the *SOHO* FOV. Dashed lines for Event 2 represent measurements behind the solar limb. (**b**) Each CME as viewed down onto the equatorial plane from the north. The directions to each of the three spacecraft is indicated, along with the relative distance of each CME based on the times chosen [10] (Reproduced with kind permission of Springer Science and Business Media)

5.2 Single Point

Moving further from the Sun the geometry becomes more complicated, but with careful analysis one can reveal additional information about the ICME without the need for multiple viewpoints. Before we consider the ICME as a collective structure, however, let us consider the effects of geometry on a single point as it moves away from the Sun to distances beyond 1 AU.

As mentioned previously, any measured point in the sky is in fact a projection of the true location of the point. So it is not meaningful to assign a distance from the Sun to any measured point. Instead, we use the elongation angle ε, which is the angle between the Sun-observer line and the LOS (Fig. 4.7a). The objective is to find a way to convert the elongation into units of distance, using meaningful assumptions.

5.2.1 Near the Sun

Close to the Sun, the popular approach has been to use the Point P approximation [12]. This assumes that the section of the CME on which the measured point lies can be approximated by a spherical arc with its centre at the Sun. Therefore the LOS makes a tangent with the solar radial vector at the measured point, resulting in a right-angled triangle between the Sun, the observer and the measured point. Figure 5.3 illustrates this situation.

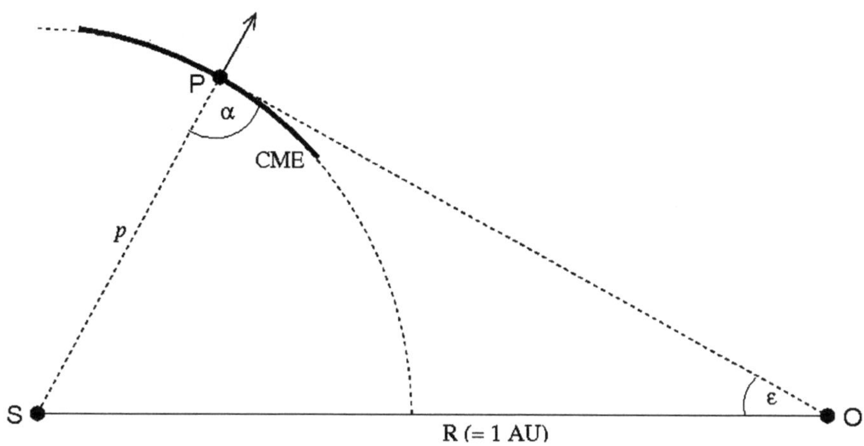

Fig. 5.3 Diagram of the geometry for the Point P approximation. S and O are the locations of the Sun and observer and the CME is represented by the arc with its centre at S. P is the point on the CME at which the measurement is made. The elongation ε is shown, along with angle SPO $= \alpha$. If we assume the transient is spherical, then PO becomes a tangent and $\alpha = 90°$ (Modified from Howard et al. [12])

The resulting triangle from the Point P approximation greatly simplifies the conversion from elongation to distance. The distance p from the Sun to the measured point P is simply

$$p = R\sin\varepsilon, \tag{5.5}$$

where R is the distance from the Sun to the observer. If the observer is near 1 AU then the distance p in units of AU is determined from simply $\sin\varepsilon$. Also, since we are close to the Sun the elongation is small, so we can also apply the trigonometric approximation $\sin\varepsilon \sim \varepsilon$. Hence, p in AU is simply approximated by ε in radians. This is the assumption applied when determining distance for CMEs observed in coronagraphs such as LASCO.

5.2.2 *Far From the Sun*

As we move further from the Sun certain assumptions begin to break down. Beyond elongations around 15° the small angle assumption becomes invalid, so we can only apply Point P via $p \sim \sin\varepsilon$. Beyond around 30° we can no longer reliably use the Point P approximation. We must then apply the complete three-dimensional geometry on the measured point.

Figure 5.4 demonstrates the geometry for a single point in 3-D space far from the Sun [7]. The coordinate system is defined such that the Earth is the x-axis, z is ecliptic northward and the yz plane lies in the plane of the sky. The Sun, observer, elongation, and other parameters are the same as in Fig. 5.3, and we have three additional parameters: the co-latitude θ and longitude ϕ, and the projection of the point P into the xy (ecliptic) plane, Q.

Using this geometry and assuming the observer is 1 AU away from the Sun (i.e. R = 1), the following relationship can be obtained:

$$\frac{1}{p} = \sin\alpha\cot\varepsilon + \cos\alpha, \tag{5.6}$$

where

$$\cos\alpha = \sin\theta\cos\phi. \tag{5.7}$$

These can be expressed in Cartesian coordinates (x,y,z) as:

$$x = \frac{\sin\theta\cos\phi}{\sin\alpha\cot\varepsilon + \cos\alpha} \tag{5.8}$$

$$y = \frac{\sin\theta\sin\phi}{\sin\alpha\cot\varepsilon + \cos\alpha} \tag{5.9}$$

$$z = \frac{\cos\theta}{\sin\alpha\cot\varepsilon + \cos\alpha}, \tag{5.10}$$

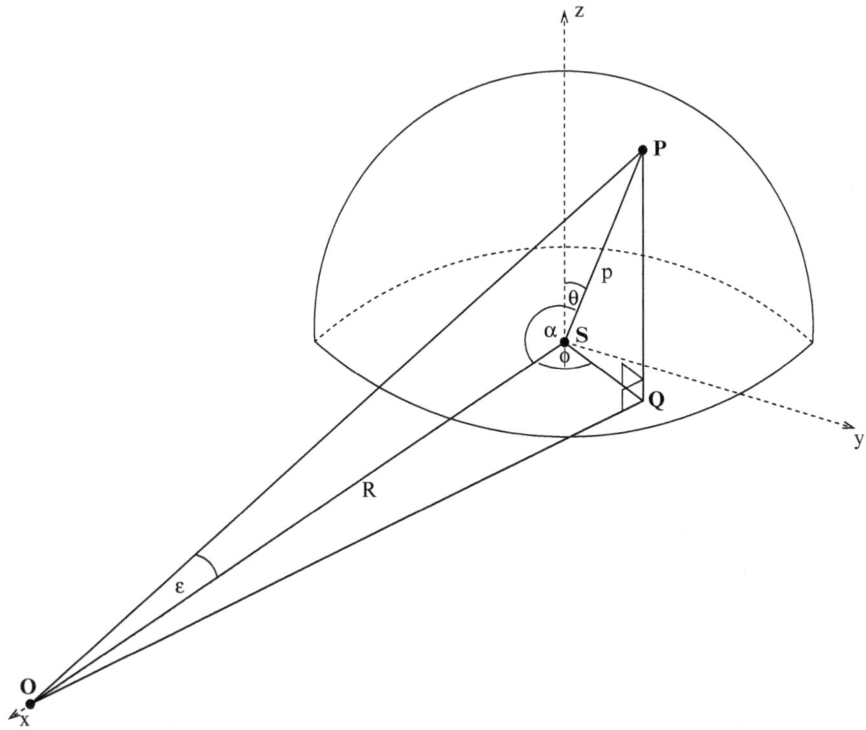

Fig. 5.4 Three-dimensional geometry of a single point far from the Sun. The coordinate system is defined such that the Earth is the x-axis, z is ecliptic northward and y completes the orthogonal set (the yz plane is the plane of the sky). The location of the measured point is P and its projection into the xy plane is Q. The distance from the Sun to the point P is p, R is the distance from the Sun S and observer O, and ε is the elongation. The co-latitude θ and longitude ϕ are shown, as is the Sun-P-Observer angle α (Modified from Howard et al. [7] and reproduced by permission of the AAS)

where (x,y,z) are in units of AU [7]. Thus we have an expression for the 3-D distance of point P from the Sun, in terms of the elongation ε and co-latitude and longitude (θ, ϕ). Likewise, we can obtain expressions for the 3-D speed and acceleration by differentiating (5.6) and assuming the observer is at 1 AU and is fixed relative to the movement of the point [8]:

$$V_{3D} = \frac{dR}{dt} = R^2 \sin\alpha \operatorname{cosec}^2\varepsilon \frac{d\varepsilon}{dt}, \tag{5.11}$$

and

$$A_{3D} = \frac{d^2R}{dt^2} = \frac{V_{3D}}{d\varepsilon/dt}\left(\frac{d^2\varepsilon}{dt^2}\right) + 2V_{3D}\left[\frac{V_{3D}}{R} - \cot\varepsilon\left(\frac{d\varepsilon}{dt}\right)\right], \tag{5.12}$$

where

$$\frac{d\varepsilon}{dt} = V_0 \sec \varepsilon, \tag{5.13}$$

and

$$\frac{d^2\varepsilon}{dt^2} = A_0 \sec \varepsilon + V_0^2 \tan \varepsilon \sec^2 \varepsilon. \tag{5.14}$$

Now let's consider the effects of projection on this travelling point. Figure 5.5 shows the effects on the elongation-time profile of a single point moving with a speed of 1,000 km/s in different directions. Each panel represents a different latitude and each curve a different longitude. The plots themselves show how the elongation of the point (which represents the projected distance) relative to an observer 1 AU away varies with time as the point moves through the heliosphere [9].

The first thing to note is that there are no linear relationships (i.e. no straight lines), even though a distance-time plot of each of these curves would be linear. The second thing is that the closer the point is to passing by the observer, the more distorted the curve becomes. In the most extreme case, the "Long = 1" curve in Fig. 5.5a, the point remains at very low elongations until around 40 h after launch

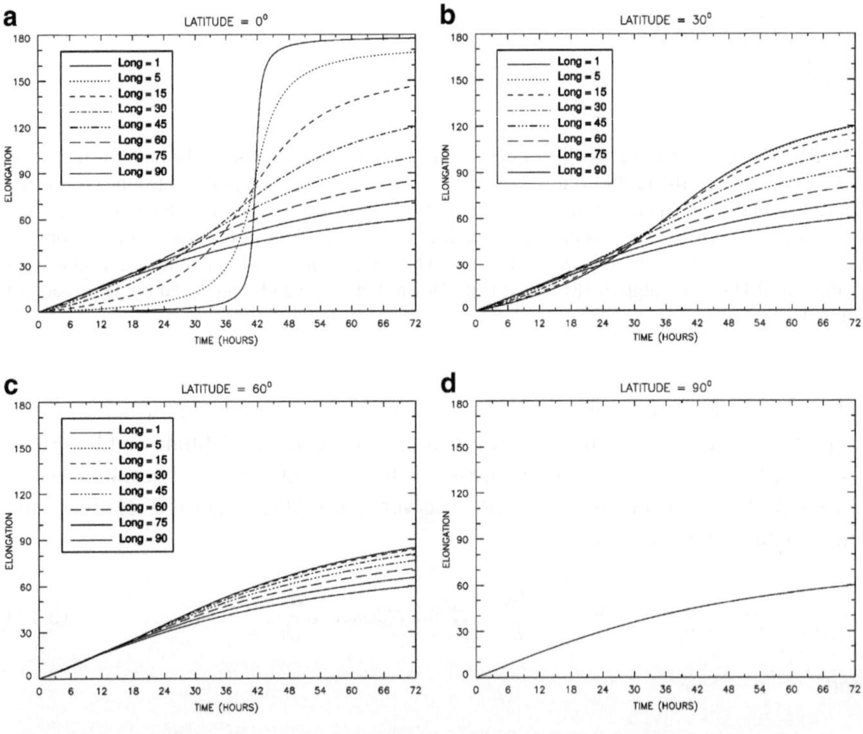

Fig. 5.5 Plots of elongation vs time (relative to an observer 1 AU away) for a single point moving with a speed of 1,000 km/s in different directions. For each plot curves represent longitudinal directions of [1, 5, 15, 30, 45, 60, 75 and 90]° from the Sun-Earth line, while each panel represents different latitudinal directions, i.e. (**a**) 0° (the equator), (**b**) 30°, (**c**) 60° and (**d**) 90° (the pole) [9]

when there is a sharp increase in elongation for a few hours, before it flattens out again near 180°.

This elongation-time trend for the point heading directly toward the observer can be explained if we note that a point moving at with a speed of 1,000 km/s will reach 1 AU in just under 42 h (41 h, 40 min). This is the time at which the curve reaches an elongation of 90°.[1] Physically this means that the point will not be observed to change in elongation until it has almost reached the observer, and once it passes it is directly behind ($\varepsilon = 180°$). For the other points it takes a longer time to reach elongations of 90° and for some it could take several weeks to get there. This is because for any trajectory other than that along the Sun-Earth line, $\varepsilon = 90°$ does not represent a distance of 1 AU from the Sun, but instead a point in the observer plane that is some considerable distance from the observer itself.

5.3 ICMEs Far From the Sun

The previous analysis applies if we were to consider the ICME as a single point. This is then appropriate for very narrow events or if we regard that a measured point on any given ICME is always at the same location on its surface. Unfortunately there are two problems with these conditions: No narrow ICME has ever been observed in a heliospheric imager; and the measured point on the surface of the ICME is not the same each time. The latter is because the relative leading edge changes as the ICME expands. Hence far from the Sun is the most complicated realm in which to analyse white light ICME images, but it is also potentially the realm with the most information about the three-dimensional structure and trajectory of the ICME.

Let us first address the complications by once again considering the ICME as a simple geometric structure. Figure 5.6 shows two basic ICME structures: an expanding spherical bubble of material moving away from the Sun (Figs. 5.6a–b); and a simple expanding spherical shell (Figs. 5.6c–d). The Sun S and observer O are fixed points, and the central axis or diameter of the ICME is arbitrarily oriented so it has a component that is observer-directed. Figs. 5.6a and 5.6c show the ICME at three different locations, at small, medium and large distances from the Sun.

Now let us consider what happens to the observed leading edge of the ICME.[2] The observed leading edge of the ICME is the point at which the LOS is a tangent to the ICME. This point, indicated by the $\times$ symbols in Figs. 5.6a and 5.6c, moves closer to the observer as the ICME expands. Thus, if one assumes the same point on the ICME is measured each time, indicated by the $\circ$ symbols, then the distance measurements of the ICME would appear to be larger than they actually are (the $+$

[1] Note that an elongation of 90° is the equivalent to the plane of the observer.

[2] Recall that the leading edge is the most common feature measured on a CME when one wishes to measure kinematic properties such as distance and speed. Also recall that the common assumption applied when making leading edge measurements is that the measured point is the same location on the CME for each successive image.

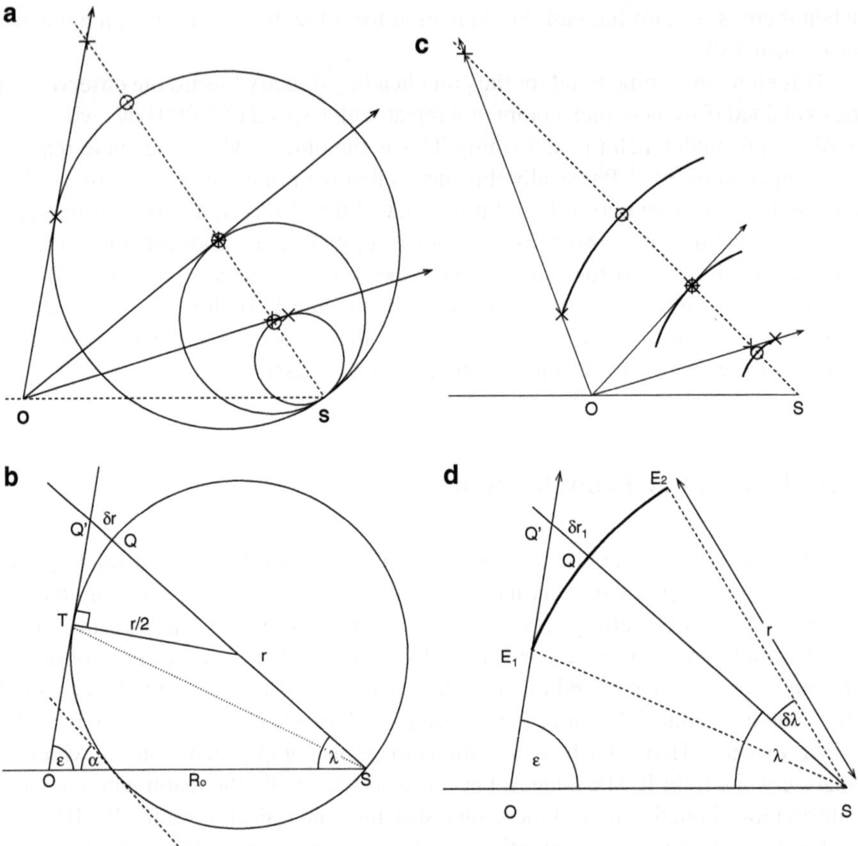

Fig. 5.6 Diagrams of basic ICME structures, (**a–b**) the expanding bubble and (**c–d**) the spherical shell with a semi-vertical angle of 30°. (**a**) and (**c**) The ICME at three different locations during its expansion, at small, medium and large distances. The tangent drawn from the observer O across the CME surface shows the location of the relative leading edge. The × symbols represent the location of the leading edge seen by the observer the ∘ symbols show the true location of the leading edge at the central location and the + symbols the inferred location of the leading edge based on the central location. (**b** & **d**) The geometry allowing the derivation of the relationship between the difference in measured distance at a given point as a function of elongation [11]. (Reproduced with kind permission of Springer Science and Business Media)

symbols). Consequently, speed measurements of ICMEs at large elongations would also appear to be larger than their true speeds [11].

Consider the geometry shown in Fig. 5.6b. The assumed measured point on the ICME is Q and the actual measured point is Q′. T′ represents the point on the sphere where the LOS is tangent and crosses the ICME, i.e. the observed leading edge. The distance of the assumed measured point from the Sun is r and the actual measured distance of this point is $r+\delta r$. This means that the difference between the measured distance and true distance is δr. Using this geometry, it can be shown that

$$r+\delta r=\frac{r}{2}(1+\mathrm{cosec}(\lambda+\varepsilon)), \tag{5.15}$$

or

$$r=\frac{2R_0\sin\varepsilon\,\mathrm{cosec}(\lambda+\varepsilon)}{1+\mathrm{cosec}(\lambda+\varepsilon)}, \tag{5.16}$$

where R_0 is the distance from the observer to the Sun. Figure 5.6b also includes the angle α which is the angle the tangent to the CME makes to the Sun-observer line. When $\alpha+\varepsilon=180°$ the leading edge of the ICME has reached the observer, so beyond this point calculations of δr are meaningless.

For the case of the simple shell (Figs. 5.6c and 5.6d) there are three situations to consider:

$$\begin{aligned}
r &= R_0\sin\varepsilon && \text{when the LOS is a tangent to the front,}\\
& && \text{i.e. } 90° - (\lambda+\delta\lambda) < \varepsilon < 90° - (\lambda-\delta\lambda);\\
r &= R_0\sin\varepsilon\,\mathrm{cosec}(\varepsilon+(\lambda-\delta\lambda)) && \text{when the LOS contacts at } \mathrm{E}_1,\\
& && \text{i.e. } \varepsilon > 90° - (\lambda-\delta\lambda);\\
r &= R_0\sin\varepsilon\,\mathrm{cosec}(\varepsilon+(\lambda+\delta\lambda)) && \text{when the LOS contacts at } \mathrm{E}_2,\\
& && \text{i.e. } \varepsilon < 90° - (\lambda+\delta\lambda).
\end{aligned} \tag{5.17}$$

The $\delta\lambda$ parameter is the semi-vertical angle of the cone swept out by the shell and $r+\delta r$ is the same as in (5.15).

Using these equations for the bubble and shell, Howard & Tappin [11] showed that for elongations where $\varepsilon+\lambda$ is near 90°, the error in distance is relatively small, but as $\varepsilon+\lambda$ approaches 0° or 180° the apparent distance quickly diverges.

It should be noted that the examples discussed above are for very simple cases of ICME structure. In reality, the structure is much more complex, but the purpose is to demonstrate the effects of geometry for large ICME-like structures. Even in their simplest forms it is clear that geometry plays a major role in their appearance, i.e. the relative location of the central axis of the ICME to that of the Sun-observer line and the angular extent.

5.4 Physically Reasonable Assumptions

Another method of estimating the structure of the CME is to develop models applying physically reasonable assumptions about its internal structure and comparing them with data. Models involving CME initiation are discussed in Chap. 8 and models involving their evolution are covered in Chap. 9, so here we discuss empirical reconstruction models based on assumptions of the magnetic structure of the CME. These assume the CME comprises of a flux rope, in the form of a magnetic cloud. Geometric triangulation techniques using this assumption and both *STEREO* exist [29, 30], but here we review an older technique using in-situ measurements of magnetic clouds.

Recall (Sect. 2.3.1.1) that magnetic clouds are highly structured magnetic field configurations that exist in many ICMEs observed by in-situ spacecraft. Armed with the knowledge that the structure of the field is a spiral, one may attempt to reproduce this structure. The popular methods that have been developed can be broken into two categories: force free and non force free. A review of the performance of many types of models from these principles can be found in Riley et al. [24].

The force free field reconstruction of magnetic clouds has been developed and utilised in a number of publications [2, 6, 15, 16, 20]. The calculations are based on the constant-α solution of Lundquist [19]. The solution is cylindrically symmetric and can be written in cylindrical coordinates as

$$B_r = 0 \tag{5.18}$$
$$B_t = B_0 J_0(\alpha R) \tag{5.19}$$
$$B_p = H B_0 J_1(\alpha R), \tag{5.20}$$

where B_r, B_t and B_t are the radial, toroidal (axial) and poloidal (azimuthal) components, B_0 is the magnetic field strength along the axis of the cylinder, $H = \pm 1$ is the handedness of the flux rope (+ indicates right-handed, – indicates left-handed) and J_0 and J_0 are the zeroth and first-order Bessel functions. These can be solved empirically by comparing with the in-situ data. The magnetic flux magnitudes Φ for each component can also be determined from

$$\Phi_t = \frac{2\pi J_1(x_{01})}{x_{01}} B_0 R_0^2 \tag{5.21}$$

$$\Phi_p = \frac{L}{x_{01}} B_0 R_0, \tag{5.22}$$

where x_{01} is the first zero of the Bessel function J_0 ($\sim$2.4048) and L is the length of the entire magnetic cloud (of the order of 1 AU).

The non force-free field, otherwise known as the Grad-Shafranov technique [4, 25] assumes a magnetic field varying with distance according to a field A, whose equal value contours represent transverse magnetic field lines. In Cartesian coordinates (x,y,z), the 2-D ($\partial/\partial z = 0$) magnetic and plasma structures are given by

$$\frac{\partial^2 A}{\partial x^2} + \frac{\partial^2 A}{\partial y^2} = -\mu_0 \frac{dP_t}{dA}, \tag{5.23}$$

where the magnetic field vector $\mathbf{B}$ is

$$\mathbf{B} = \left(\frac{\partial A}{\partial y}, -\frac{\partial A}{\partial x}, B_z(A) \right), \tag{5.24}$$

which satisfies $\nabla_t A \cdot \mathbf{B}_t = 0$ with $\mathbf{B}_t$ as the transverse field (B_x, B_y). $P_t(A)$ is the sum of the plasma pressure and magnetic pressure along the axis. Assuming the moving ICME reference frame (so the spacecraft moves through the ICME) values

Table 5.1 The advantages and disadvantages from the effects of geometry on white light images of CMEs (*left*) and ICMEs (*right*)

	CME	ICME
Advantages	Assumptions can be applied to simplify calculations of distance and mass.	Can extract 3-D information through careful analysis of the geometry.
Disadvantages	No information can be extracted on the 3-D structure or location.	Simplifying assumptions break down at large distances from the Sun.

of $A(x,0)$ and $P_t(x,0)$ are found using data from the spacecraft and (5.23) solved to obtain $A(x,y)$. From this the magnetic field structure can be formed.

Reports developing and discussing the Grad-Shafranov technique include Hu & Sonnerup [13], Hu et al. [14], Hasegawa et al. [5] and Liu et al. [18].

5.5 Conclusions

The purpose of this chapter is to alert the reader to the additional problems when dealing with CME and ICME data. Not only is an appreciation of the physics responsible for the intensity of the CME required (Chap. 4), but it is also necessary to regard the CME as a large three-dimensional structure with additional geometrical consequences. It is also important to note that the effects of geometry play a different role depending on whether one is looking at a CME (close to the Sun) or an ICME (far from the Sun). The advantages and disadvantages for each is shown in Table 5.1.

To elaborate on the advantages of working with ICME white light images, the effects of the geometry become effective for the ICME as described for the two basic structures in (5.15–5.17). These variations are not significant for CMEs. Therefore if we can estimate the basic structure of the ICME, we can effectively remove the projection effects from the image. One such technique for achieving this has demonstrated by Tappin & Howard [26], who based their model on a collective of simulated ICME leading edges using a simple assumption of ICME structure. Figure 2.15 shows one such reconstruction using this technique. Other techniques involve physically reasonable deductions about the internal structure of the CME such as its intrinsic magnetic cloud.

References

1. Antunes, A., Thernisien, A., Yahil, A.: Solar Phys. **259**, 199–212 (2009).
2. Burlaga, L.F.: J. Geophys. Res. **93**, 7217–7224 (1988).
3. de Koning, C.A., Pizzo, V.J., Biesecker, D.A.: Solar Phys. **256**, 167–181 (2009).
4. Hau, L.-N., Sonnerup, B.U.Ö.: J. Geophys. Res. **104**, 6899–6918 (1999)
5. Hasegawa, H., Sonnerup, B., Dunlop, M., Balogh, A., Haaland, S., Klecker, B., Paschmann, G., Lavraud, B., Dandouras, I., Rème, H.: Ann. Geophys. **22**, 1251–1266 (2004).

6. Hidalgo, M.A., Cid, C., Medina, J., Vinãs, A.F.: Solar Phys. **194**, 165–174 (2000).
7. Howard, T.A., Fry, C.D., Johnston, J.C., Webb, D.F.: Astrophys. J. **667**, 610–625 (2007).
8. Howard, T.A., Nandy, D., Koepke, A.C.: J. Geophys. Res. **113** doi:10.1029/2007JA012500 (2008).
9. Howard, T.A., Simnett, G.M.: J. Geophys. Res. **113**, doi:10.1029/2007JA0129209 (2008).
10. Howard, T.A., Tappin, S.J.: Solar Phys. **252**, 373–383 (2008).
11. Howard, T.A., Tappin, S.J.: Space Sci. Rev. **147**, 31–54 (2009).
12. Howard T.A., Webb, D.F., Tappin, S.J., Mizuno, D.R., Johnston, J.C.: **111**, doi:10.1029/2005JA011349 (2006) .
13. Hu, Q., Sonnerup, B.U.Ö.: Geophys. Res. Lett. **28**, 467–470 (2001).
14. Hu, Q., Smith, C.W., Ness, N.F., Skoug, R.M.: Geophys. Res. Lett. **30**, doi:10.1029/2002GL016653 (2003).
15. Leamon, R.J., Canfield, R.C., Jones, S.L., Lambkin, K., Lundberg, B.J., Pevtsov, A.A.: J. Geophys. Res. **109**, A05106, doi:10.1029/2003JA010324 (2004).
16. Lepping, R.P., Burlaga, L.F., Jones, J.A.: J. Geophys. Res. **95**, 11957–11965 (1990).
17. Liewer, P.C., De Jong, E.M., Hall, J.R., Howard, R.A., Thompson, W.T., Culhane, J.L., Bone, L., van Driel-Gesztelyi, L.: Solar Phys. **256**, 57–72 (2009).
18. Liu, J., Hu, Q., Howard, T.A., Yurchyshyn, V.B.: Astrophys. J. **659**, 758–772 (2007).
19. Lundquist, S.: Ark. Fys. **2**, 361–365 (1950).
20. Lynch, B.J., Gruesbeck, J.R., Zurbuchen, T.H., Antiochos, S.K.: J. Geophys. Res. **110**, A08107, doi:10.1029/2005JA011137 (2005).
21. Mierla, M., Inhester, B., Antunes, A., Boursier, Y., Byrne, J.P., Colaninno, R., Davila, J., de Koning, C.A., Gallagher, P.T., Gissot, S., Howard, R.A., Howard, T.A., Kramar, M., Lamy, P., Liewer, P.C., Maloney, S., Marquè, C., McAteer, R.T.J., Moran, T., Rodriguez, L., Srivastava, N., St. Cyr, O.C., Stenborg, G., Temmer, M., Thernisien, A. Vourlidas, A., West, M.J., Wood, B.E., Zhukov, A.N.: Ann. Geophys. **28**, 203–215 (2010).
22. Mierla, M., Inhester, B., Marquè, C., Rodriguez, L., Gissot, S., Zhukov, A.N., Berghmans, D., Davila, J.: Solar Phys. **259**, 123–141 (2009).
23. Moran, T.G., Davila, J.M., Thompson, W.T.: Astrophys. J. **712**, 453–458 (2010).
24. Riley, P.R., Linker, J.A., Lionello, R., Mikić, Odstrcil, D., Hidalgo, M.A., Cid, C., Hu, Q., Lepping, R.P., Lynch, B.J., Rees, A.: J. Atmos. Solar Terr. Phys. **66**, 1321–1331 (2004).
25. Sturrock, P.: In Geophysical and Laboratory Plasmas, Cambridge Univ. Press, New York (1994).
26. Tappin, S.J., Howard, T.A.: Space Sci. Rev. **147**, 55–87 (2009).
27. Temmer, M., Preiss, S., Veronig, A.M.: Solar Phys. **256**, 183–199 (2009).
28. Thernisien, A.F.R., Howard, R.A., Vourlidas, A.: Astrophys. J. **652**, 763–773 (2006).
29. Thernisien, A.F.R., Vourlidas, A., Howard, R.A.: Solar Phys. **256**, 111–130 (2009).
30. Wood, B.E., Howard, R.A., Thernisien, A., Plunkett, S.P., Socker, D.G.: Solar Phys. **259**, 163–178 (2009).

Chapter 6
Radio Astronomical Techniques

Now that we have discussed how we may extract physical properties of CMEs using the theory behind their detection in white light and knowledge of their associated phenomena, we move onto an alternative method for measuring ICMEs. When a dense structure moves through the heliosphere, it alters the signal from distant radio sources as it passes between the source and the observer. For ICMEs, this can be measured using low-frequency (of the order of a metre wavelength) sources. The best known such effect is interplanetary scintillation (IPS), which is the rapid flickering of relatively small sized radio sources caused by turbulence in the solar wind. IPS has been used to detect interplanetary transients (ICMEs and CIRs) since the 1960s (refer to Sect. 2.5).

Not only has IPS been detecting ICMEs since before the discovery of the CME, it has also provided observations of ICMEs throughout the 20 year time period between *Helios* and SMEI, where white light images of ICMEs were not available [19, 20]. Work continues to this day [14, 26] and IPS has been used not only to identify and track ICMEs, but we can also extract information on their density and structure.

A second propagation effect that may provide information about the passage of an ICME is Faraday rotation. This is the rotation of the direction of polarisation of the source, and could possibly be caused by the internal magnetic structure of the ICME. Hence by measuring Faraday rotation it is possible that one may identify the magnetic field vector within the ICME, which may allow the long-sought remote measurement of its magnetic field.

This chapter reviews the radio astronomical techniques that can be used in the study of ICMEs. We begin with a brief review of the theory of IPS and discuss some scientific breakthroughs using the technique. We then move onto a discussion on the potential for radio astronomy in the future using new instruments to measure Faraday rotation, which are currently under development.

6.1 Interplanetary Scintillation (IPS)

Interplanetary scintillation is the result of electromagnetic radiowave scattering arising from density irregularities in the interplanetary medium. It has been used since

T. Howard, *Coronal Mass Ejections: An Introduction*, Astrophysics and Space Science Library 376, DOI 10.1007/978-1-4419-8789-1_6,

the 1960s to study the solar wind [8], but can also be used to study interplanetary transients such as CMEs and CIRs. Here we review the IPS technique including the fundamental theory along with techniques used to extract information on the properties of ICMEs.

6.1.1 A Brief Review of IPS Theory

Radio scintillation got its start in the 1950s when Tony Hewish and others developed the theory for ionospheric scintillation [12, 13, 22]. The theory was applied to IPS around 15 years later by Little and Hewish [15] and since has been developed and discussed in many publications (e.g. Readhead [23], Marions [18] and Readhead et al. [24]). Budden and Uscinski [5–7] may be sought for some theoretical "heavy lifting" regarding IPS. The following theory has been adapted primarily from Tappin [27] and Alurkar [1].

The concept of IPS can be regarded as a plane wave from a distant radio source passing through a thin layer (or screen [15]) between the source and the observer, which contains density irregularities. By "thin", we mean that the thickness of the layer is much smaller than the distance of the layer to the observer. If the scattering is weak, then only a small component of the wave is diffracted resulting in a phase difference between the diffracted and non-diffracted waves at any distance from the layer. A diffraction pattern $F = F(x,y)$ is formed in a direction normal to the incident wave, which for an ideal point source of unit intensity may be defined as

$$F = \frac{I - \overline{I}}{\overline{I}}, \tag{6.1}$$

where $\overline{I}$ is the mean of the radio source intensity. We may then define the scintillation index m as

$$m = \sqrt{\overline{F^2}} = \frac{\sigma^2}{\overline{I}}, \tag{6.2}$$

where σ^2 is the variance of the source intensity. In real terms the scintillation index is the ratio of the root mean square (rms) scintillating flux density $S_{\rm rms}$ to the mean source flux density $\overline{S}$, or

$$m = \frac{S_{\rm rms}}{\overline{S}}, \tag{6.3}$$

where $S_{\rm rms}$ and $\overline{S}$ are in units of $\mathrm{Wm^{-2}Hz^{-1}}$, or more usually in Jy ($\mathrm{Wm^{-2}Hz^{-1}} \times 10^{-26}$). The scintillation index can be regarded as a measure of the turbulence in the thin layer. It varies between unity and zero.

> The classic analogy for IPS is viewing a light source through a bathroom window. Consider an old-fashioned bathroom window which consists of warped glass with no fogging. The idea of these was to let light into the bathroom from

the outside but distort the images for an observer looking through. This is for obvious reasons, but I digress. If one was to look at a light source through the window, what would it look like from the other side? If the source was small, one would observe a slightly spread signal with additional fainter patches surrounding. Now consider what would happen if the source was allowed to move slowly across the window. The pattern on the other side of the glass would be continually changing as the window refracted the light at different angles. If the source was large the distortion effects would not be observed. This is analogous to how structures in the interplanetary medium distort radio signals as they move through the sky from day to day, and demonstrates the importance of the size of the source-for IPS, the smaller the better.

A diagram of the situation regarding IPS scattering is shown in Fig. 6.1. A radio wave directed at the Earth passes through a thin layer containing density fluctuations. Let the layer have an electron density of N_e and the incident wave have an angular frequency of ω. The refractive index of the medium n is given by

$$n = \sqrt{\frac{1 - N_e e^2}{\varepsilon_0 m_e \omega^2}}, \tag{6.4}$$

where e, m_e and ε_0 denote the usual properties (Chap. 4). This can be approximated using the definition $c = \lambda\omega/(2\pi)$ (λ is the wavelength) by

$$\begin{aligned} n &\sim 1 - \frac{N_e e^2}{2\varepsilon_0 m_e \omega^2} \\ &= 1 - \frac{N_e e^2 \lambda^2}{4\pi^2 \varepsilon_0 m_e c^2} \\ &= 1 - \frac{N_e r_e \lambda^2}{2\pi}, \end{aligned} \tag{6.5}$$

where $r_e = e^2/4\pi\varepsilon_0 m_e c^2$ is the classical electron radius. The layer varies in density with time which results in corresponding changes to the refractive index. Following Alurkar [1] let the variations in density and refractive index be δN_e and δn. Equation (6.5) becomes

$$\delta n \sim -\frac{\delta N_e r_e \lambda^2}{2\pi}. \tag{6.6}$$

Now, as the wave emerges from the layer it will be refracted differently depending on the refractive index in the layer at the time of passage. This results in an emerging wave with a continually varying phase across its front. This is best

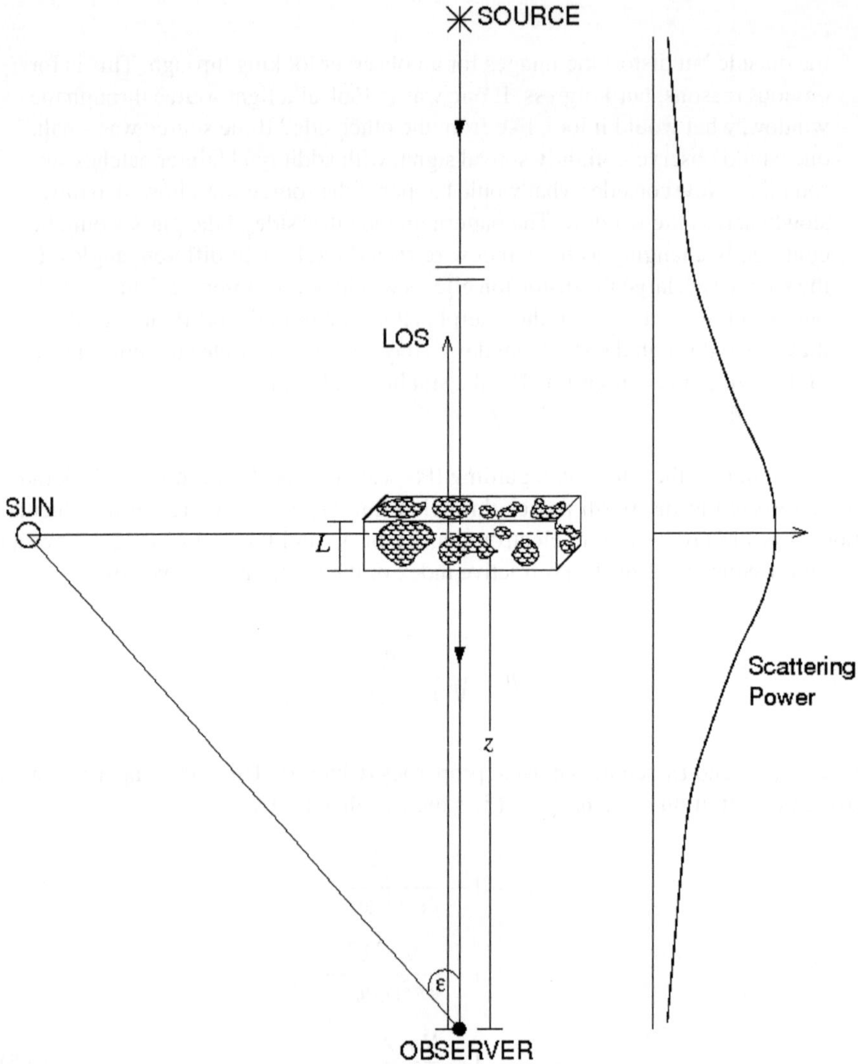

Fig. 6.1 Diagram of the geometry surrounding IPS observations. A radio wave from a distant source passes through a thin layer containing density variations before continuing through to the observer. The location of the Sun is noted as well as the thickness of the layer L and the distance from the observer to the layer z. The elongation ε is defined typically as the angle from the Sun-observer line to the layer. (Modified from Readhead et al. [24])

measured by comparing it with what the phase would have been if there were no scattering medium. So, the phase at any point x is defined [1] by

$$\phi(x) = \frac{-2\pi}{\lambda} \int \delta n \mathrm{d}z = -\lambda r_e \int \delta N_e \mathrm{d}z, \tag{6.7}$$

using (6.6). Recall that z is the direction from the observer to the source (and layer). Integrating along the LOS (z-axis) provides the rms of the phase:

$$\phi_{\mathrm{rms}} = \pi^{1/4}\sqrt{aL}\lambda r_e \sqrt{\overline{\delta N_e^2}}, \tag{6.8}$$

where a is the scale-size of the density changes and L is the thickness of the layer (Fig. 6.1).

Returning to the definition of the scintillation index, m can also be written in terms of the incident radio wave E, which is a function of x, y and z. In this form, we have

$$m \sim 2\frac{\sqrt{E_{\mathrm{rms}}^2}}{\overline{E}}, \tag{6.9}$$

or simply $m \sim 2\sqrt{E_{\mathrm{rms}}^2}$ when $\overline{E}$ is normalised to unity [1]. At the location of the source, a large distance from the regime containing the Sun, observer and scintillating layer, the mean of E *at the layer* can be approximated by

$$\overline{E^2} = \frac{1}{2}\phi_{\mathrm{rms}}^2, \tag{6.10}$$

and so

$$m \sim \sqrt{2}\phi_{\mathrm{rms}}. \tag{6.11}$$

We now have a relationship between the scintillation index m and the density fluctuations in the scintillating layer δN_e. The above derivation applies to a single LOS only through the layer to the source. In physical terms, by integrating along the collective of lines of sight (i.e. paths from a collective of radio sources) one can use this technique to measure quantities of the solar wind such as density and even speed. As an ICME can be regarded as a fluctuation in the solar wind, we can hence determine its properties as well.

6.1.2 How Can IPS Be Used to Determine ICME Properties?

6.1.2.1 Density

As shown in the previous section, by measuring the scintillation index m we can determine density fluctuations in the scintillating medium. In other words, we can determine the density of the part of the solar wind through which the radio source propagated to reach the observer. An ICME could be regarded as one such fluctuation. With a known value of m we can determine the rms of the phase of the emerging wavefront ϕ_{rms} using (6.11), which we can then apply to (6.8) to determine the mean of the density fluctuations $\overline{\delta N_e^2}$. The scale size of the fluctuation a and thickness of the layer L will need to be estimated via other means, probably including a careful analysis of the properties of the instrument.

While this technique is theoretically reasonable there are other methods by which one may determine the density using IPS, if one were to correlate the measurements with observations from other instruments. In reality, measuring the scintillation index m can be difficult, as while one can reasonably obtain a value for $S_{\rm rms}$, it is not so easy to obtain a value for $\overline{S}$. Instead, workers often rely on the relative scintillation index, denoted g. Using g has the advantage that it is independent of the intensity and only weakly dependent on the size of the radio source. The index is calculated and used for ICME density calculation using the procedure outlined below.

Procedure for Determining the Relative Scintillation Index, g

1. For every source, produce a value for each scintillation $S_{\rm rms}$ for each measurement. As we typically obtain a measurement a day, we can produce a plot of $S_{\rm rms}$ as the source moves around the sky throughout the year. This can be easily converted to a plot against elongation ε as the source will be at a different elongation each day. Figure 6.2 shows a $S_{\rm rms}$ vs ε plot for an example source: the quasar 3C 238 [23].
2. Fit a modeled curve to the distribution using an appropriate model, such as the RKH empirical model (Readhead-Kemp-Hewish [24]). Figure 6.2 shows this model superimposed on the data.
3. For each measurement, determine the difference between $S_{\rm rms}$ and the modeled fit at that elongation, $S_{\rm mod}(\varepsilon)$. The g index is defined as

$$g = \frac{S_{\rm rms}}{S_{\rm mod}(\varepsilon)}. \tag{6.12}$$

4. Conduct a study of "good" events where IPS transients are clearly associated with interplanetary shocks detected by in-situ spacecraft. The in-situ measurements provide the measurements of density and a relationship between g and N_e can be obtained. Such a study has been conducted (for example) by Tappin [28] and the results are shown in Fig. 6.3.
5. Use the relationship between g and N_e (in Tappin [28] the relationship is $g = (N_e(cm^{-3})/9)^{0.52}$) to convert the g index into density.

Hence, for each source, an estimation of the density of the transient can be determined. When performed collectively across every source, a total density can be determined, or at least a total density across the FOV of the instrument.

6.1.3 Structure

As the previous section has shown, there is a relationship between the g index and the density of the transient through which the source beam has passed. So it follows that by measuring the g value for every visible radio source in the sky (at your

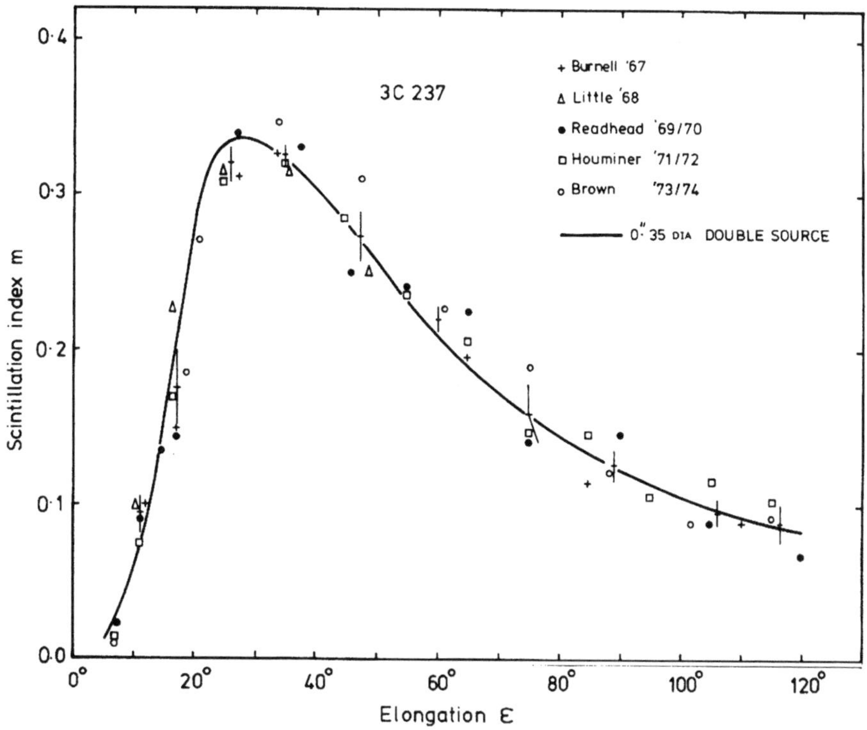

Fig. 6.2 Plot of scintillation S_{rms} (m) vs elongation ε for the radio source 3C 237 [23]. Each value was obtained at different times of the year, when the source was at different elongations. The superimposed curve represents the RKH modeled fit through the data [23]

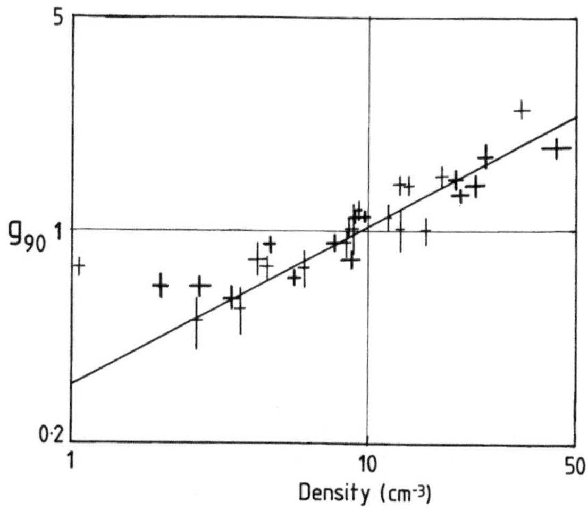

Fig. 6.3 Plot of the relative scintillation index g vs density N_e for a number of events where IPS transients were clearly associated with an in-situ ICME. Note the power-law relationship between the two (linear in this log–log plot), which can be used to relate g and N_e for future IPS measurements [28]. (Reprinted with permission from Elsevier)

frequency of interest), one can obtain a measurement of the density distribution in the heliosphere. Hence when an ICME travels past these sources its structure across the sky could be ascertained. An example of such a map of g is shown in Fig. 2.7 where each coloured square represents a different radio source. The shades of blue and red indicate the magnitude of g (intense red = large g, intense blue = small g). The loop-like structure of the ICME is clearly visible in red in this figure.

A sequence of IPS g maps is shown in Fig. 6.4, showing the passage of an ICME between 9–14 May 1980 [27]. Here different symbols represent different g values, but the same colour code applies (red for large values, blue for small values). Each source visible throughout a given day is assigned a g value and plotted on an appropriate sky map, which in this case is a Mollweide map (a mapping designed such that the latitude lines are parallel to the equator) with the Sun at the centre. As should be clear in the figure, some sources are not available every day, and the array used to produce the images (the 3.6 Hectare Array near Cambridge in the UK) is at high

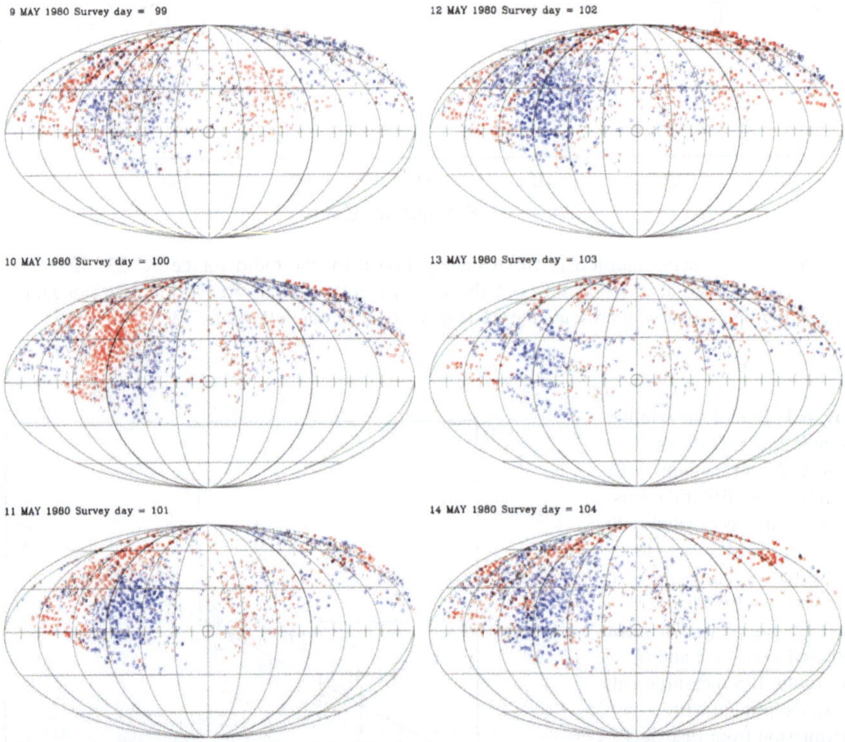

Fig. 6.4 Whole-day sky maps of g indices from 9–14 May 1980. Each symbol represents a different g value but brief large values are red (indicating large density change) and small values are blue. The grid is a Mollweide sky map in ecliptic coordinates with the circle in the centre representing the location of the Sun. These were obtained using the 3.6 Hectare Array which lies at a geographic latitude of around 52°N. The high-latitude location of this array explains why there are very few sources at locations south of the Sun (and nothing below 30°S). The images are arranged chronologically from top to bottom then from left to right [27]

northern latitudes, and so is unable to observe sources at southern latitudes in the sky. So, while structural measurements are possible using IPS they are limited to the field of view of the instrument and by the fact that only one sky map can be produced each day. A fast CME could travel an entire AU in this time. This limits the ability of IPS to track CMEs, but could be resolved with arrays at multiple sites worldwide.

6.1.3.1 Velocity (Speed and Direction)

The first method by which an ICME speed may be measured is the same as that used to measure ICME speed in white light images. Structural measurements discussed in the previous section could be used to identify a leading edge of the ICME, and appropriate geometry could be applied across a sequence of maps to identify its speed (see Tappin and Howard [26] for an example of how this may be achieved). There is, however, a technique by which small-scale velocities can be determined for a given radio source. Such techniques have been used to determine solar wind velocity in the heliosphere [2, 4, 9, 25] and can also be used for interplanetary transients [3, 14]. What is needed is a network of relatively closely-spaced (with separation of the order of 100 km) IPS arrays.

Let's begin with a speed measurement, for which we need two radio arrays. Each array obtains a measurement of scintillation for every source, and because of their spatial proximity a similar background signal will be observed at both locations. A moving density fluctuation, however, will be observed at slightly different times by each array as it passes by their respective LOS to the source. This is revealed in a cross-correlation analysis. If we assume that the LOS separation is small compared with the scale size, we can assume that the structure of the density irregularity does not change greatly as it moves from one LOS to the next. This enables the achievement of a good correlation if we apply an appropriate time delay between the arrays. Figure 6.5 shows one such correlation function for a network of IPS arrays established in the UK in the 1960s. They are called cross-correlograms and the time delay is clearly shown between the plots. This particular event had a speed of between 300 km/s and 490 km/s [9].

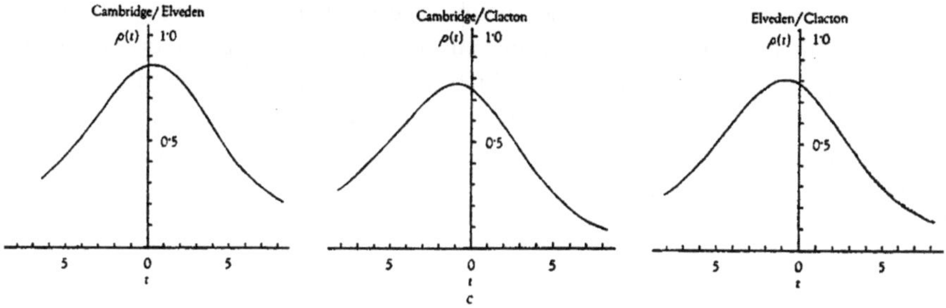

Fig. 6.5 Cross-correlograms from IPS observations on 9 May 1966, observed with a network of observatories in the UK. Note the time difference (x-axis) between each [9]. (Reprinted by permission from Macmillon Publishers Ltd: Nature [9], Copyright 1967)

So we have a time taken and a distance moved by the irregularity from one LOS to the next, therefore we have a speed. This is the bulk speed of the fluctuation projected into the sky plane. To determine the projected velocity we need a direction of travel, for which determination three arrays are required. Let's call them arrays A, B and C. Two vectors are formed combining A–B and A–C, resulting in two speed values by applying the technique discussed in the previous paragraph. These can each be resolved into x and y components (if the arrays have not already been situated to create two orthogonal vectors) which can then be combined to produce a velocity vector. This technique is limited to the scale size of the speed measurements, so is generally good for measuring small-scale motions in the solar wind plasma, but not so good at large-scale transient movement, such as ICMEs.

6.2 Faraday Rotation—The Future of ICME Study?

Using techniques discussed in other chapters in this book, the space physics community has been rapidly improving on the prediction of the arrival time of ICMEs at 1 AU and particularly (from a space weather forecasting perspective) at the Earth. I am confident that with the next generation of heliospheric imagers we will have a very good handle on the ICME's structure, trajectory and kinematic evolution and we will be able to predict the arrival time of most ICMEs early and with high degrees of accuracy.

Knowing the arrival time at the Earth, however, is not enough. We may know when an ICME is going to arrive at the Earth but still have no information on whether or not it will create a geomagnetic storm when it gets there. The geoeffectiveness of ICMEs is mostly dependent on the direction of its magnetic field as it must have a high rate of magnetic reconnection with the geomagnetic field when it arrives (Sects. 1.2.9 and 10.4). For now we cannot determine this property until it passes by an in-situ spacecraft with a magnetometer on-board.

For the moment, the best spacecraft for early ICME magnetic field detection is *ACE*, which effectively provides us with about an hour's warning before the ICME reaches the Earth. From a space weather forecasting perspective this is better than nothing, but in all honesty not much better. The ultimate goal in magnetic storm forecasting then becomes the quest not only for the accurate prediction of ICME arrival time, but also for a method by which we may measure the internal magnetic field of the ICME remotely, without the need for a network of in-situ spacecraft close to the Sun. Radio astronomy may provide the means to do this, using Faraday rotation. What we need are radio sources that are highly plane-polarised at low frequencies.

6.2.1 What is Faraday Rotation?

A plane polarised electromagnetic wave can regarded as the sum of two oppositely–directed circularly polarised waves. When such a wave propagates through an

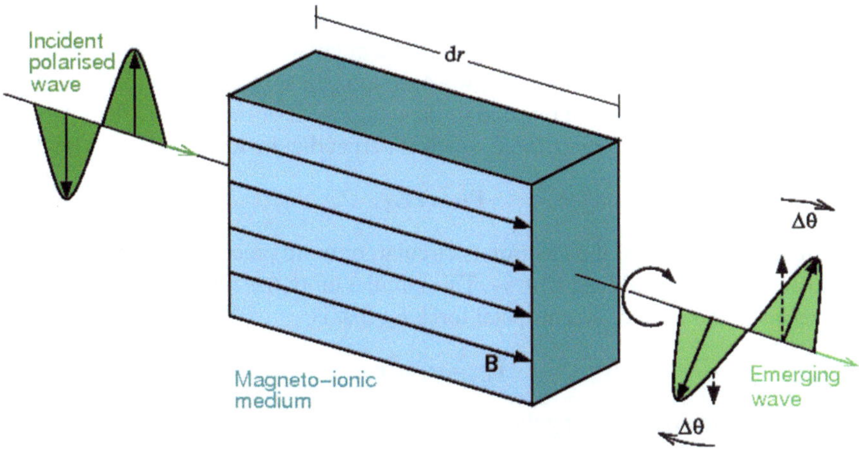

Fig. 6.6 Diagram showing Faraday rotation on an electromagnetic wave incident on a magnetic ionised (magneto-ionic) medium. In this case the internal magnetic field is directed along the direction of wave propagation and the medium has a thickness of dr. The emerging wave is rotated as a result of the passage, and in this case it rotates to the left when looking along the direction of the wave vector. The angle of rotation is governed by a number of properties, including the density and magnetic field strength of the medium and the frequency of the source wave

ionised medium with an internal magnetic field, the field causes one circular component to travel faster than the other, causing the resulting plane wave vector to rotate. This is called Faraday rotation, and by measuring the rotation of the polarised wave one can obtain information about the structure causing the rotation. Figure 6.6 shows a diagram describing the Faraday rotation of a polarised wave passing through a magnetised ionised medium.

The fundamental theory of Faraday rotation can be found in many texts dealing with the basics of astrophysics or electromagnetism. The following is, for the most part a large modification of Harwit [11], and includes a conversion from CGS to SI units. Also throughout this treatment, note that directions are measured relative to the internal magnetic field, so the notations $\perp$ and $\|$ indicate components that are perpendicular and parallel to the magnetic field, respectively.

Let's begin with an electron moving with velocity $\mathbf{v}$ through a magnetic field $\mathbf{B}$. It will experience a force $\mathbf{F}_L$ resulting from the electromagnetic force from the field on the charge,

$$\mathbf{F}_L = e\mathbf{v} \times \mathbf{B}. \tag{6.13}$$

This is called the Lorentz force and is considered to play an important role in CME evolution.[1] It is the component of the electron's velocity that is perpendicular to the magnetic field which is subjected to the Lorentz force, so

$$\mathbf{F}_L = \mathbf{v}_\perp \times \mathbf{B} = eB\mathbf{v}_\perp = eB\omega\mathbf{r}_\perp, \tag{6.14}$$

[1] There is more on the role played by the Lorentz force in CME evolution in Sect. 9.5.2.

where $B = |\mathbf{B}|$, $\mathbf{r}_\perp$ is the perpendicular component of the displacement vector of the electron and ω is the angular frequency of the polarised wave. Just as the electron is influenced by the frequency of the wave, so too is it subjected to a force $\mathbf{F}_E$ from its electric field $\mathbf{E}$. We are interested in the component of the electric field parallel to the direction of the Lorentz force which is perpendicular to the magnetic field, so

$$\mathbf{F}_E = e\mathbf{E}_\perp. \tag{6.15}$$

The resultant motion of the electron is circular, or more precisely a spiral when the circular motion is combined with $\mathbf{v}_\parallel$. This circular motion is represented in mechanical terms in the form of a centripetal force F_c, that is

$$\mathbf{F}_C = m_e\omega^2\mathbf{r}_\perp. \tag{6.16}$$

Hence, the equation balancing these forces is

$$e\mathbf{E}_\perp \pm eB\omega\mathrm{r}_\perp = -m_e\omega^2\mathbf{r}_\perp. \tag{6.17}$$

As has been already mentioned, the wave is plane polarised meaning that it can be represented as a combination of two circularly–polarised waves with opposite directions of rotation. The $\pm$ term in (6.17) indicates the direction of rotation ($+$ for left-handed, $-$ for right-handed when looking along $\mathbf{r}_\perp$). Rearranging to make $\mathbf{r}_\perp$ the subject we have

$$\mathbf{r}_\perp = -\frac{e}{m_e}\left(\frac{1}{\omega^2 \pm \frac{eB\omega}{m_e}}\right)\mathbf{E}_\perp. \tag{6.18}$$

Now, the displacement of the electron effectively establishes an electric dipole. When the collective of dipole fields is summed across a dense structure, the passing wave creates a polarisation field $\mathbf{P}$, defined by

$$\mathbf{P} = \frac{(\varepsilon - 1)\mathbf{E}}{4\pi}, \tag{6.19}$$

where ε is the relative permittivity of the medium, otherwise known as the dielectric constant. Assuming the wave is passing through a medium of density N_e we can rewrite $\mathbf{P}$ in terms of the total charge along the displacement vector $\mathbf{r}$,

$$\mathbf{P} = \frac{N_e e}{4\pi\varepsilon_0}\mathbf{r}. \tag{6.20}$$

Rearranging (6.19) and applying $\mathbf{P}$ from (6.20) we have the following for the dielectric constant:

$$\begin{aligned}
\varepsilon\mathbf{E} &= \frac{4\pi}{4\pi\varepsilon_0}(N_e e\mathbf{r}_\perp) + \mathbf{E} \\
&= \frac{N_e e}{\varepsilon_0}\left[\frac{-e}{m_e}\frac{1}{\left(\omega^2 \pm \frac{eB\omega}{m_e}\right)}\right]\mathbf{E} + \mathbf{E}
\end{aligned}$$

$$\Rightarrow \varepsilon = -\frac{N_e e^2}{\varepsilon_0 m_e \left(\omega^2 \pm \frac{eB\omega}{m_e}\right)} + 1$$

$$= 1 - \frac{N_e e^2}{\varepsilon_0 m_e \omega \left(\omega \pm \frac{eB}{m_e}\right)}$$

$$= 1 - \frac{N_e e^2}{\varepsilon_0 m_e \omega (\omega \pm \omega_c)}, \qquad (6.21)$$

where ω_c is the electron cyclotron frequency, defined by $\omega_c \equiv eB/m_e$.

Faraday rotation is the rotation of the polarisation angle brought about by the different speeds of propagation of the two circular components. If the magnetic field is parallel to the direction of propagation of the wave, then the left hand component propagates faster, while the right hand component is faster if the field is antiparallel. The difference in speed between the two components results in a phase difference which changes as the rotation continues. The rotation of the polarisation direction is at all times equal to half the phase lag between the two components [11]. This of course is compared with the polarisation direction of the incident wave.

The different speeds arise from the different refractive indices for each circularly polarised component. We denote the relative refractive index for the left- and right-handed components n_L and n_R. The difference between the squares of these two indices $\Delta n = n_L - n_r$ is related to the relative dielectric constants of each, which in turn is related to the angle of rotation of the polarisation vector. The relationship between the dielectric constants ε_L and ε_R is

$$n_L^2 - n_R^2 = \varepsilon_L - \varepsilon_R, \qquad (6.22)$$

and also we have

$$n_L^2 - n_R^2 = (n_L + n_R)(n_L - n_R) \sim (n_\omega + n_\omega)\Delta n = 2n_\omega \Delta n, \qquad (6.23)$$

where n_ω is the refractive index for the wave when it is not in a magneto-ionic medium. Because the Faraday rotation is relatively small, we can apply the approximation $n_L \sim n_R \sim n_\omega$. From (6.21),

$$n_\omega = \sqrt{1 - \frac{N_e e^2}{\varepsilon_0 m_e \omega (\omega \pm \omega_c)}}. \qquad (6.24)$$

We may now apply the mathematical approximation commonly known in the form $(1 \pm x)^n \sim (1 \pm nx)$, which applies when $x \ll 1$. Also, $\omega \gg \omega_c \Rightarrow \omega \pm \omega_c \sim \omega$. This allows us to rewrite (6.24) as

$$n_\omega \sim 1 - \frac{N_e e^2}{2\varepsilon_0 m_e \omega^2}. \qquad (6.25)$$

Now let's consider the right hand side of (6.22). Applying the derived expression for ε in (6.21) and adopting the convention for the handedness of polarisation rotation, we have

$$\begin{aligned}
\varepsilon_L - \varepsilon_R &= 1 - \frac{N_e e^2}{\varepsilon_0 m_e \omega\,(\omega + \omega_c)} - 1 + \frac{N_e e^2}{\varepsilon_0 m_e \omega\,(\omega - \omega_c)} \\
&= \frac{N_e e^2}{\varepsilon_0 m_e \omega}\left(\frac{1}{\omega + \omega_c} + \frac{1}{\omega - \omega_c}\right) \\
&= \frac{N_e e^2}{\varepsilon_0 m_e \omega}\left(\frac{2\omega_c}{\omega^2 - \omega_c^2}\right) \\
&\sim \frac{2N_e e^2 \omega_c}{\varepsilon_0 m_e \omega^3}. \qquad (6.26)
\end{aligned}$$

We can now combine (6.22) and (6.23) and make Δn the subject,

$$\Delta n = \frac{\varepsilon_L - \varepsilon_R}{2n_\omega}, \qquad (6.27)$$

which, when (6.25) and (6.26) are substituted in becomes

$$\Delta n = \frac{\frac{2N_e e^2 \omega_c}{\varepsilon_0 m_e \omega^3}}{2\left(1 - \frac{N_e e^2}{2\varepsilon_0 m_e \omega^2}\right)} = \frac{1}{2}\left(\frac{2N_e e^2 \omega_c}{\varepsilon_0 m_e \omega^3}\right), \qquad (6.28)$$

since $N_e e^2/\varepsilon_0 m_e \omega^2 \ll 1$. So, reapplying the definition of the electron cyclotron frequency ω_c we have

$$\Delta n = \frac{N_e e^3 B}{\varepsilon_0 m_e^2 \omega^3}. \qquad (6.29)$$

The distance lag per unit distance is the optical path length of the medium, $\mathrm{d}r\Delta n$ where $\mathrm{d}r$ is the distance traveled whilst undergoing Faraday rotation (the thickness of the medium). The phase lag is then $(2\pi/\lambda)\mathrm{d}r\Delta n$ and the polarisation vector is rotated by half the phase during this period. The rotation of the polarisation vector $\Delta\theta$ is therefore

$$\begin{aligned}
\Delta\theta &\sim \frac{1}{2}\left(\frac{2\pi}{\lambda}\right)\mathrm{d}r\Delta n \\
&= \frac{\omega \mathrm{d}r}{2c}\Delta n, \qquad (6.30)
\end{aligned}$$

since $2\pi/\lambda = \omega/c$. Finally, we substitute (6.29) and reveal

$$\Delta\theta \sim \frac{N_e e^3 B \mathrm{d}r}{2\varepsilon_0 m_e^2 c \omega^2}. \qquad (6.31)$$

This equation shows that the Faraday rotation angle is directly related to three variable parameters: density N_e, magnetic field B of the structure, and the frequency of the source ω.

6.2.2 Application to ICMEs

Equation (6.31) describes the Faraday rotation $\Delta\theta$ of a plane polarised electromagnetic wave propagating through a plasma with an embedded magnetic field. This description applies very well to an ICME moving through the interplanetary medium. Hence, a plane polarised electromagnetic wave moving through an ICME will undergo Faraday rotation with a magnitude of rotation defined by (6.31), provided a component of the magnetic field is parallel to the direction of propagation of the wave. The parameters in this equation can be applied to an ICME using the following parameters:

- N_e is the density of the ICME;
- B is the magnetic field component that is parallel to the LOS (from the observer through the ICME to the source);
- dr is the thickness of the ICME along the LOS;
- $\omega = 2\pi\nu$, where ν is the frequency of the source wave;
- ε_0, e, m_e and c are defined as usual.

So, if one can measure the Faraday rotation from a polarised source as it passes through an ICME, then one can determine the magnetic field of that ICME. Note once again that the magnetic field direction is only that which is *parallel to the LOS*, so for space weather interests one must either directly obtain a LOS with a significant component in the north–south direction, or make inferences of ICME structure to derive the required component.

What kind of angles of Faraday rotation might we expect from an ICME? This depends on the frequency of the source and the location of the ICME itself. Let us simulate an ICME using typical values for the parameters listed above (shown in Table 6.1). We will begin with the case where the LOS is exactly the north–south magnetic field ($\mathbf{B}_z$) component: when the ICME is at 1 AU. Figure 6.7 shows how the Faraday rotation angle $\Delta\theta$ changes as a function of source frequency for a

Table 6.1 Typical ICME properties at 1 AU to be used in our example to estimate Faraday rotation. A range of internal magnetic fields have been chosen

Parameter	Value
N_e	$10\ \text{cm}^{-3}$ ($10^7\ \text{m}^{-3}$)
B	[1, 5, 10, 20, 30, 40] nT
dr	0.1 AU (1.5×10^{10} m)
ε_0	$8.854 \times 10^{-12}\ \text{Fm}^{-1}$ ($\text{A}^2\text{s}^4\text{kg}^{-1}\text{m}^{-3}$)
e	1.6×10^{-19} C
m_e	9.109×10^{-31} kg
c	$3 \times 10^8\ \text{ms}^{-1}$

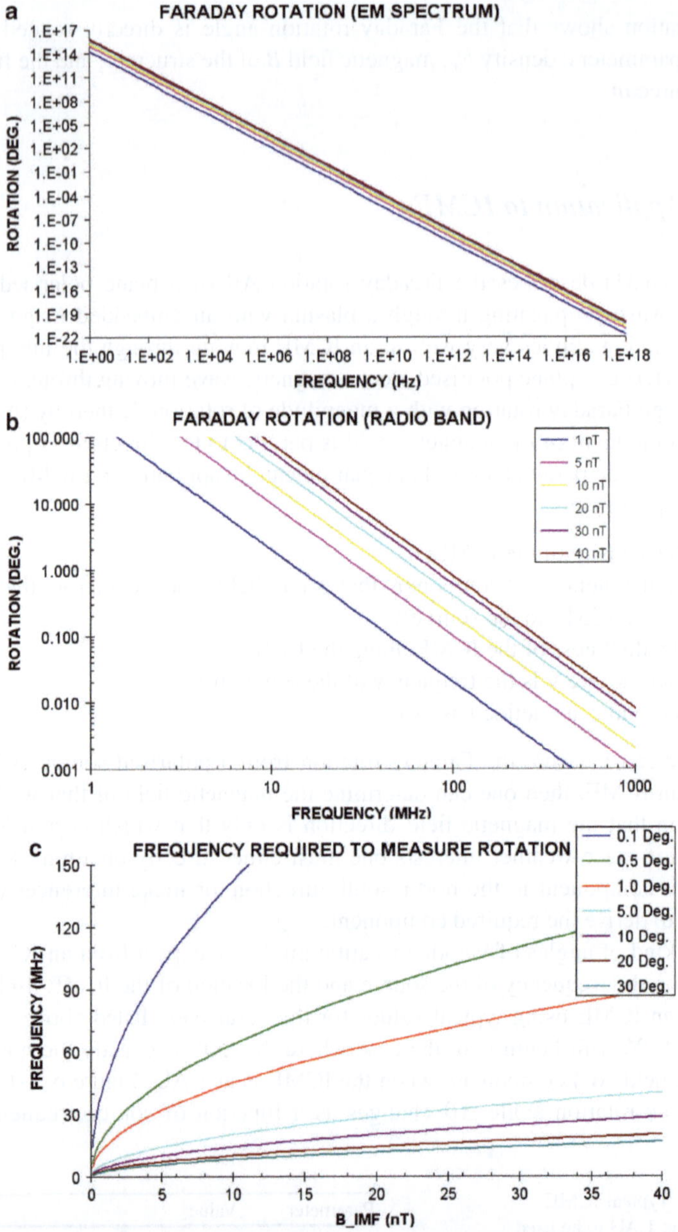

Fig. 6.7 Relationships involving Faraday rotation. (**a**) and (**b**) Log–log plots of rotation angle vs frequency of source wave, first (**a**) across the majority of the electromagnetic spectrum and then (**b**) across the frequency range where useful angles can be measured. Plots are shown for ICMEs at 1 AU with different magnetic field strengths (and all the other properties remaining constant), from 1 nT up to 40 nT. For a typical ICME at 1 AU the most useful Faraday rotation measurements occur when the source wave is in the radio band, around 10–100 MHz. (**c**) Plot of source frequency vs magnetic field strength causing a Faraday rotation angle of different values, ranging from 0.1–30°. If we assume that we can measure a rotation of around 1° then for an ICME at 1 AU of magnetic field strength 20 nT or larger we can measure the rotation if we looked at a source at around 80 MHz

typical ICME. Figure 6.7a shows the rotation across the electromagnetic spectrum, from 1 Hz (longwave radio) up to 10^{18} Hz (x-rays) for a range of internal magnetic field measurements. Useful rotation values can be regarded from around 0.1° to 100° which for our ICME lies in the radio frequency range, from around 1–100 MHz. This region of the spectrum is shown in Fig. 6.7b. As is implied in (6.31), the smaller the internal magnetic field, the smaller the Faraday rotation.

To get a feel for the frequencies at which we may consider to observe useful Faraday rotation, consider Fig. 6.7c. Here, plots of source frequency vs magnetic field are shown across a variety of rotation angles from 0.1° to 30°.[2] This plot shows that there are two things that can contribute to larger Faraday rotation:

1. Low observing frequency;
2. Strong internal magnetic field (i.e. a strong ICME).

Let's say for example that we can measure rotation to 1° accuracy with a reasonably high confidence level. Let's also say we have an ICME with an internal magnetic field along the LOS of 20 nT at 1 AU. We would then need a source frequency of 60 MHz or lower in order to measure the rotation at 1 AU. This raises a very important question: How many plane polarised radio sources are there at 60 MHz or even at 74 MHz, which is generally regarded as the lowest usable frequency at which to observe? The answer is we simply do not know, as no survey on polarised sources has been conducted at these frequencies. Efforts are moving in this direction (e.g. Pen et al. [21]) but we are must await the next generation of radio arrays which are currently under construction [17, 30].

6.2.2.1 Moving Closer to the Sun

The above discussion applies for Faraday rotation on ICMEs at 1 AU. Observing an ICME once it has impacted with the Earth is not particularly useful especially for space weather forecasting purposes. When we move closer to the Sun the Faraday rotation angles increase for any given ICME, as its magnetic field and density increase. Moving away from the Sun the radial component of both can be roughly approximated to vary as R^{-2} where R is the distance from the Sun. For magnetic field this is because each dimension varies at a rate proportional to R, and magnetic flux is proportional to the cross-sectional area through which the field flows (Sect. 9.2.1). The effects on Faraday rotation with ICMEs closer to the Sun are shown in Fig. 6.8, which contains plots of four separate ICMEs with a given magnetic field at 1 AU. For example, Fig. 6.8b is for an ICME with a magnetic field at 20 nT at 1 AU. This value increases at a rate of R^2 as we move closer to the Sun and we assume the density changes with a variation of $R^{2.45}$ in accordance with Tappin [29]. Each plot shows a different value of Faraday rotation, varying from 0.1° to 30°. For example, let us consider an ICME with a magnetic field strength of 20 nT

[2] Note that the frequency of the source does not vary, so this distribution shows a theoretical range of multiple radio sources.

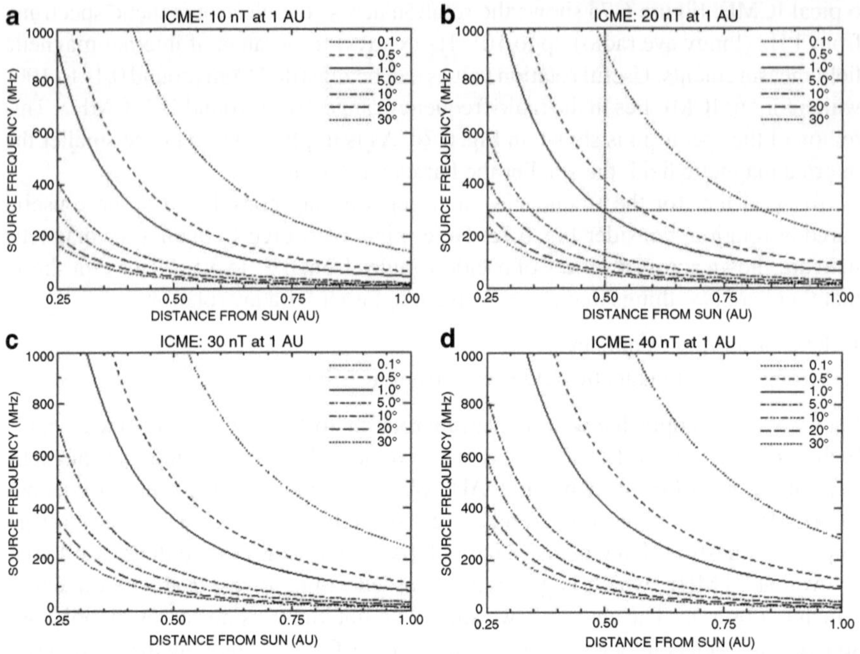

Fig. 6.8 Plots of source frequency vs distance from Sun for modeled ICMEs of magnetic field values of (**a**) 10 nT, (**b**) 20 nT, (**c**) 30 nT and (**d**) 40 nT at 1 AU. The density at 1 AU is constant at $10^7\ m^{-3}$ for each event but varies at a rate of $R^{2.45}$ where R is the distance from the Sun, as we move closer to the Sun. B is varied at a rate of B^2. Each plot shows a range of curves, each for a different angle of rotation from 0.1–30°

at 1 AU and let us assume that we can measure this ICME when it is 0.5 AU from the Sun. If we wished to measure the Faraday rotation with an accuracy of 1° then we would need to observe radio sources of a frequency of around 300 MHz. We could obtain the same accuracy with sources at higher frequency if we could measure the ICME when it was closer to the Sun, or it had a stronger magnetic field.

Recall that using this technique we are only able to measure the component of the magnetic field along the line of sight, and not the total magnetic field or the z component. As we move closer to the Sun, two important changes occur to the measured component of the field:

- The (x, y, z) coordinates become (R, T, N). This is the radial, tangential, normal coordinate system, where R is defined as the radial component from the Sun, T is the tangential component, or the direction of solar rotation relative to the observer, and N is the normal component which completes the right-hand orthogonal vector set. These coordinates become (x, y, z) when in the Sun-observer plane.
- The contribution of the z (or rather N) component of the magnetic relative to the total field becomes reduced.

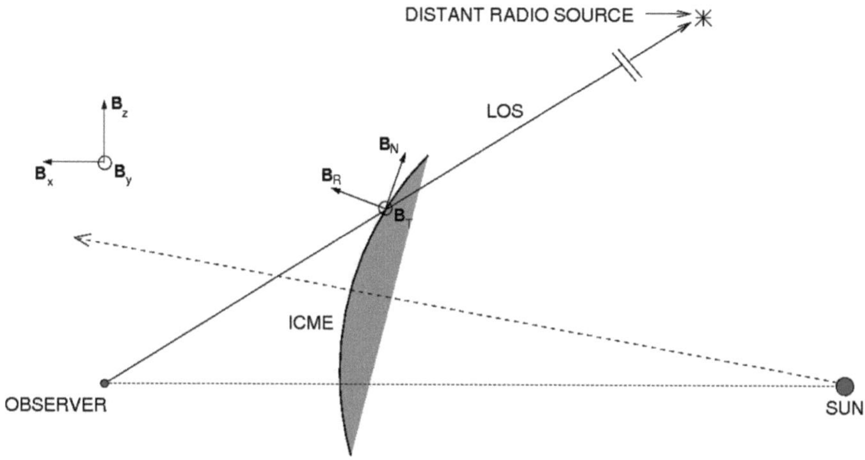

Fig. 6.9 Diagram showing the measurement of an ICME using a distant radio source and a line of sight (LOS) passing through the ICME. The location of the Sun and observer are indicated with the coordinates of each magnetic field component **B**. These are (R,T,N) coordinates which become (x,y,z) when in the Sun-Observer plane as shown

The former is illustrated in Fig. 6.9. Here the definition of the (R,T,N) coordinates are shown with respect to an ICME moving through the heliosphere and a LOS from the observer through the ICME to a distant radio source. The conversion to (x,y,z) is also shown. For such a conversion to occur it is necessary to apply an assumption of the magnetic structure of the ICME. If we assume the structure is large scale, then we may approximate $\mathbf{B}_N$ as $\mathbf{B}_z$.

The latter is a result of the variation of magnetic field in the interplanetary medium (Sect. 9.2.1). We can assume that the solar wind is roughly constant in this region, so the radial component of the area through which the magnetic field flows is assumed to be constant with R. The result is that $\mathbf{B}_N$ varies only with R while the total field continues to vary with R^2. As it is the N component we wish to measure it is important to consider the effects this has on Faraday rotation. These effects are shown in Fig. 6.10, which shows the same plots for the same ICMEs as in Fig. 6.8, but allowing for only a variation with R for the magnetic field. It is clear that the Faraday rotation angles are smaller for the same radio sources, meaning that lower frequencies are required for the same accuracy. Let us once again consider the example of the 20 nT ICME at 1 AU which has been detected at 0.5 AU from the Sun (Fig. 6.10b). Now we require sources at frequencies of 200 MHz rather than the 300 MHz from Fig. 6.8b. Hence, obtaining a measurement of the N component is somewhat more difficult than measuring the total field. Even if the z component did have a magnitude of 20 nT at 1 AU (which is not unreasonable for strong ICMEs), then the effects of observation of this component nearer the Sun are less significant.

The important thing to note is that theoretically it is possible to detect Faraday rotation for typical ICMEs at these radio frequencies, particularly if we are able to observe them when they are closer to the Sun. There are, however, some potential complications to consider. Leaving aside the technical complications and the

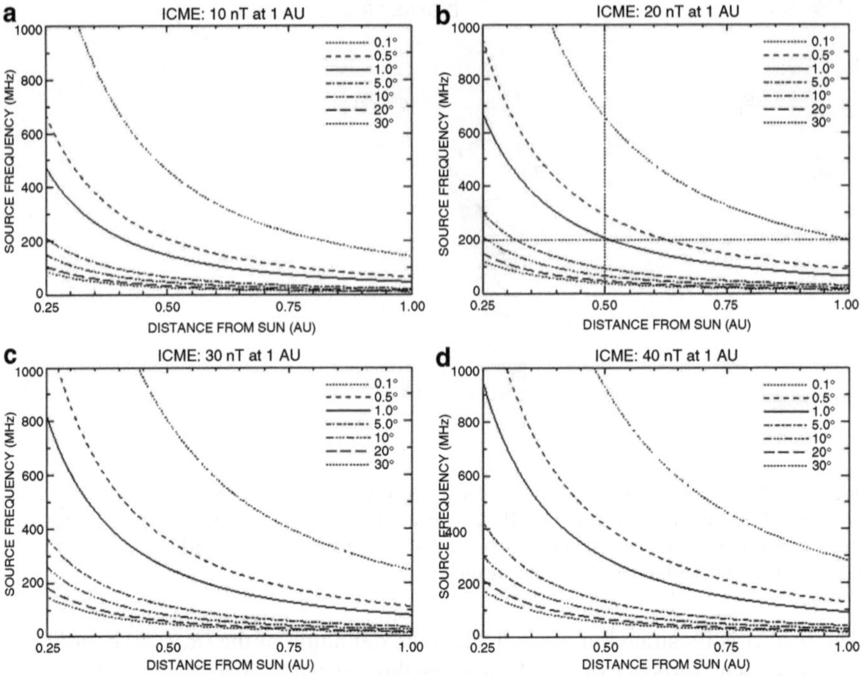

Fig. 6.10 The same plots as in Fig. 6.8, except we are now allowing for only a magnetic field variation with R rather than R^2

problems associated with ionospheric and other interference (although it should be noted that these are not by any stretch of the imagination insignificant at these frequencies), let's consider the theoretical limitations here. Firstly, we need to be able to measure polarisation to an accuracy of 1° (or ideally smaller), so we need sources that are strongly polarised. Secondly, the magnetic field causing the rotation is only in the component of the LOS, not the total. Thirdly, to be useful for space weather purposes, it is the north–south component of the magnetic field we need to measure, meaning we need a LOS with a significant north–south component. Here we encounter a paradox, that is the stronger the north–south component we can establish, the closer to the Earth we need to be. Also when measuring closer to the Sun we must measure the N component, so an assumption of ICME structure must be applied to convert this to a z component. Only experiment will determine how significant a component of the north–south vector we will need to be in order to reliably obtain a magnetic field measurement.

6.2.2.2 Measurements Very Close to the Sun

It should be clear from Fig. 6.8 that the closer to the Sun the ICME measurement can be made, the larger the angle of Faraday rotation and the higher the required frequencies for detection. Higher frequency sources are an advantage as there are known to be strongly polarised sources larger than 1 GHz (e.g. Grant et al. [10]). Unfortunately when we move very close to the Sun we have virtually no information on the z component of the magnetic field as very low latitude sources are needed to achieve these measurements. We can, however, infer the structure of the magnetic field if we apply some simple assumptions of magnetic field continuity.

Consider the situation illustrated in Fig. 6.11. We consider a number of polarised radio sources with their projected location near the Sun and an observer-directed (halo) CME passing between the observer and the sources. As the CME passes, a Faraday rotation will occur on the polarisation angle of the source signals. For the most part, this provides information on the radial component of the magnetic field only, but this is sufficient if we assume the magnetic field is continuous. For example, if the sources north of (above) the Sun show a magnetic field that is predominantly towards the observer and the sources south of (below) the Sun show a field away from the observer, then we may reasonably infer that the z component of the field must be directed southward. Such a technique has been investigated with a specific assumption by Liu et al. [16] and it is the technique by which investigators using the MWA (Murchison Widefield Array) plan to confirm that magnetic field measurements of ICMEs using Faraday rotation are possible.

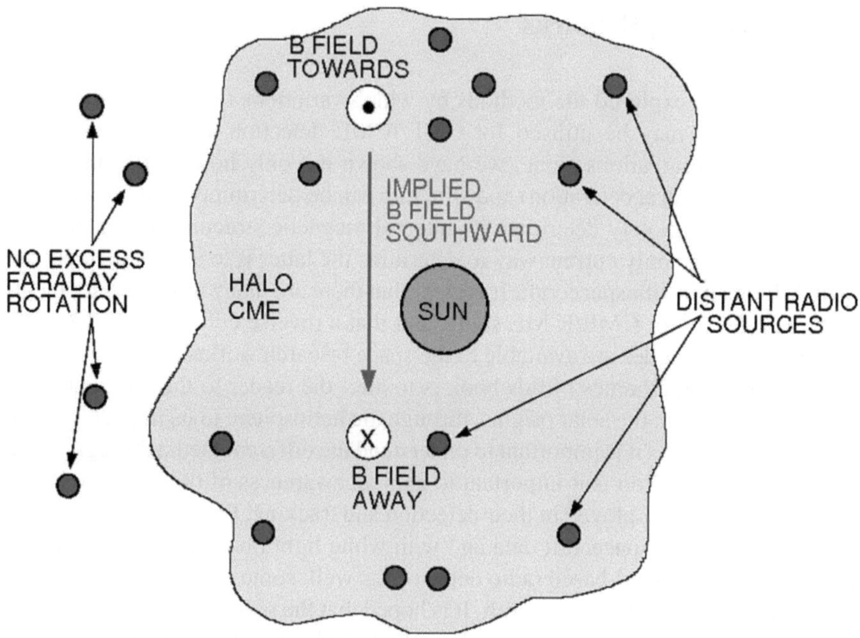

Fig. 6.11 A simple illustration of Faraday rotation of a CME when it is close to the Sun. Radio sources with their projected locations near the Sun are required. Outside the projected halo CME the sources experience no Faraday rotation *from the CME*. If the sources north of (*above*) the Sun provide Faraday rotation measurements that show a predominantly observer-directed magnetic field, and the sources south of (*below*) the Sun show a field directed away, then we may reasonably infer that the z component of the field is directed southward

The Liu et al. (2007) study [16] assumed the CME was magnetic cloud and used both the force free Lundquist solutions and Grad-Shafranov technique (discussed in Sect. 5.4) to approximate its magnetic field. They then determined the rate of change of Faraday rotation as the modeled ICME moved through the heliosphere. They found that the Faraday rotation could be measured and meaningful information about the CME geometry and magnetic field could be extracted.

This technique, unfortunately, also has some possible problems. Firstly, CMEs are often associated with radio bursts (e.g. Type II and Type IV bursts) (Sects. 2.3, 7.2.10 and 7.3.3) and it is possible that such bursts may interfere with the measurements of Faraday rotation from sources near the Sun. Secondly, the z component itself is not measured meaning that an assumption on the structure of the CME needs to be applied. Thirdly, for space weather prediction the magnetic structure of the CME is assumed not to change as it moves from the Sun to the Earth. This is not physically unreasonable as there is little that could disrupt this structure en route, but a combination of techniques is required in order to measure the CME as it moves from the Sun to the Earth. This would have to start with the near-Sun measurements and then move to higher latitude radio sources.

6.3 Concluding Remarks

This chapter has explored the methods by which variations of distant radio astronomical sources may be utilised for CME/ICME detection and tracking. Using ground-based observations alone, we have shown not only how CME kinematics (height-time, speed, acceleration) and structure can be determined, but also how we may be able to remotely determine its internal magnetic structure. Using conventional methods the only current way to determine the latter is to await the impact of the ICME on an in-situ spacecraft. It is clear that there are many more developments in the near future for CME/ICME study, and that a diverse collection of detection and analysis techniques are available to the space research worker.

One of the main themes of this book is to alert the reader to the big picture involving CMEs, from the solar origins, through the heliosphere to its impact with the Earth and beyond. As it is important to understand the different media through which the CME travels, so too is it important to gain an awareness of the variety of techniques that may be employed in their detection and tracking. Previous chapters have dealt primarily with spacecraft data and with white light observations, but innovations exist using ground-based radio detectors as well, some of which are currently impossible to achieve with spacecraft. It is hoped that the radio techniques discussed here along with the other data detection methods will equip the reader to appreciate the assortment of techniques by which we detect, track and measure CMEs.

References

1. Alurkar, S.K.: Solar and Interplanetary Disturbances, World Sci. Ser. Astron. Astrophys. **5**, World Scientific Publishing, Singapore (1997).
2. Armstrong, J.A., Coles, W.A.: J. Geophys. Res. **77**, 4602–4610 (1972).
3. Breen, A.R., Moran, P.J., Varley, C.A., Williams, P.J.S., Coles, W.A., Grall, R.R., Klinglesmith, M.T., Markkanen, J.: Adv. Space Res. **20**, 27–30 (1997).
4. Bourgois, B., Coles, W.A., Daigne, G., Silen, J., Turunen. T., Williams, P.J.: Astron. Astrophys. **144**, 452–462 (1985).
5. Budden, K.G., Uscinski, B.J.: Proc. R. Soc. Lond. **A316**, 315–339 (1970).
6. Budden, K.G., Uscinski, B.J.: Proc. R. Soc. Lond. **A321**, 15–40 (1971).
7. Budden, K.G., Uscinski, B.J.: Proc. R. Soc. Lond. **A330**, 65–77 (1972).
8. Coles, W.A.: Space Sci. Rev. **21**, 411–425 (1978).
9. Dennison, P.A., Hewish, A.: Nature **213**, 343–346 (1967).
10. Grant, J.K., Taylor, A.R., Stil, J.M., Ricci, R., O'Sullican, S.P., Landecker, T.L., Kothes, R., Ranson, R.R.: Proc. IAU **4**, 559–560, doi:10.1017/S1743921309031329 (2008).
11. Harwit, M.: Astrophysical Concepts, John Wiley & Sons, pp.210–213, New York (1973).
12. Hewish, A.: Proc. R. Soc. Lond. **A209**, 81–96 (1951).
13. Hewish, A.: Proc. R. Soc. Lond. **A214**, 494–514 (1952).
14. Jones, R.A., Breen, A.R., Fallows, R.A., Canals, A., Bisi, M.M., Lawrence, G.: J. Geophys. Res. **112**, doi:10.1029/2006JA011875 (2007).
15. Little, L.T., Hewish, A.: Mon. Not. R. Astron. Soc. **134**, 221–237 (1966).
16. Liu, Y., Manchester, W.B., IV, Kasper, J.C., Richardson, J.D., Belcher, J.W.: Astrophys. J. **665**, 1439–1447 (2007).
17. Lonsdale, C.J., Cappallo, R.J., Morales, M.F., Briggs, F.H., Benkevitch, L., Bowman, J.D., Bunton, J.D., Burns, S., Corey, B.E., deSouza, L., Doeleman, S.S., Derome, M., Deshpande, A., Gopalakrishna, M.R., Greenhill, L.J., Herne, D., Hewitt, J.N., Kamini, P.A., Kasper, J.C., Kincaid, B.B., Kocz, J., Kowald, E., Kratzenberg, E., Kumar, D., Lynch, M.J., Madhavi, S., Matejek, M., Mitchell, D., Morgan, E., Oberoi, D., Ord, S., Pathikulangara, J., Prabu, T., Rogers, A.E.E., Roshi, A., Salah, J.E., Sault, R.J., Udaya S.N., Srivani, K.S., Stevens, J., Tingay, S., Vaccarella, A., Waterson, M., Wayth, R.B., Webster, R.L., Whitney, A.R., Williams, A., Williams, C.: Proc. IEEE **97**, 1497–1506 (2009).
18. Marions, M.: Radio Sci. **10**, 115–119 (1975).
19. Manoharan, P.K.: Solar Phys. **235**, 345–368 (2006).
20. Manoharan, P.K., Tokumaru, M., Pick, M., Subramanian, P., Ipavich, F.M., Schenk, K., Kaiser, M.L., Lepping, R.P., Vourlidas, A.: Astrophys. J. **559**, 1180–1189 (2001).
21. Pen, U.-L., Chang, T.-C., Peterson, J.B., Roy, J., Gupta, Y., Hirata, C.M., Odegova, J., Sigurdson, K.: eprint arXiv:0807.1056 (2008).
22. Ratcliffe, J.A.: Rep. Prog. Phys. **19**, 188–267 (1956).
23. Readhead, A.C.S.: Mon. Not. R. Astron. Soc. **155**, 185–197 (1971).
24. Readhead, A.C.S., Kemp, M.C., Hewish, A.: Mon. Not. R. Astron. Soc. **185**, 207–255 (1978).
25. Rao, A.P., Balasubramanian, V., Coles, W.A.: Proc. Solar Wind Conf. p.94 (1995).
26. Tappin, S.J., Howard, T.A.: Solar Phys. **265**, 159 (2010).
27. Tappin, S.J.: PhD thesis, Univ. Cambridge (1984).
28. Tappin, S.J.: Planet. Space Sci. **34**, 93–97 (1986).
29. Tappin, S.J.: Solar Phys. **233**, 233–248 (2006).
30. Taylor, G., Henning, P., Gilfeather, F., Dickel, H.: The Long Wavelength array (LWA) Webpage, available via UNM. http://lwa.unm.edu/.

Chapter 7
Associated Phenomena

The previous three chapters have dealt with the CME and ICME as observed with white light imagers and using radio astronomical observations. However, CMEs are associated with many phenomena observed on and near the solar surface and in interplanetary space. These phenomena, while not the CMEs themselves, are probably related to their onset and evolution in some way. They certainly provide information on the response of the solar atmosphere and heliosphere to the CME and also on their likely 3-D location. Recall from Chap. 2 that solar flares, solar energetic particles and interplanetary shocks were known long before CMEs and are to this day invariably associated with them. Flares and other associated phenomena are detected by other means (e.g. visible light, UV and x-ray images, in-situ spacecraft, radio) and there are a large number of instruments currently in operation designed to detect them (see Chap. 3).

The relationship between these phenomena and CMEs is worthy of study, not just because of the additional analytical information they can provide, but also because they probably hold the key to allowing us to understand the CME launch mechanism (Chap. 8) and may tell us something about the physics of their evolution (Chap. 9). This chapter reviews the most common phenomena associated with CMEs and ICMEs, and the information about them they provide. It should be noted that no phenomenon has yet been discovered that has a 1–1 correspondence with CMEs (although some are more common than others), and so it is important to be aware of them all, their nature and their relationship to CMEs.

7.1 Magnetic Field Structure and Behaviour

Before we can discuss the associated phenomena it is important to review the physics describing the Sun in the photosphere, chromosphere and corona. This is discussed in further detail in Chap. 8 so only a brief review is presented here.

The lower solar atmosphere is governed by magnetic fields which extend from beneath the photosphere through the chromosphere, corona and out into the solar wind. Figure 7.1 shows diagrams illustrating a simplified version of the solar corona during solar minimum and maximum. Models that estimate the solar wind typically

T. Howard, *Coronal Mass Ejections: An Introduction*, Astrophysics and Space Science Library 376, DOI 10.1007/978-1-4419-8789-1_7,

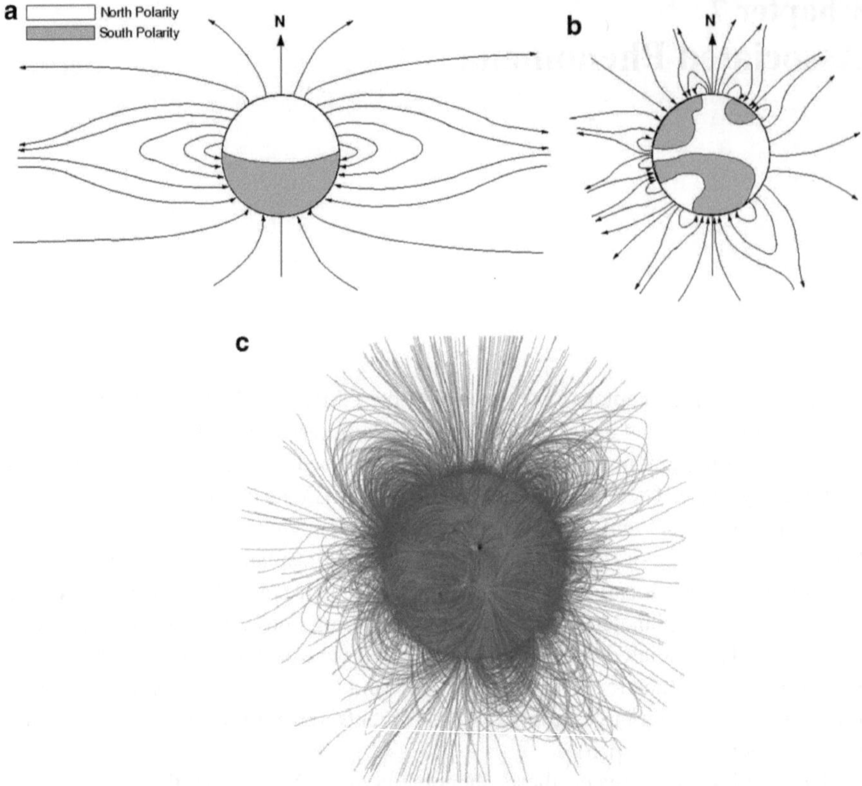

Fig. 7.1 Simplified diagram of the solar magnetic field during (**a**) solar minimum and (**b**) solar maximum. (**c**) Solar magnetic field reconstruction using the PFSS (Potential Field Source Surface) Model (Image courtesy of NASA [4])

derive the magnetic field structure from instruments that can measure the magnetic field in the photosphere by measuring the movement of plasma via Doppler shifting (e.g. *SOHO*/MDI, *SDO*/HMI). The fields are then extended into the corona. Magnetic structure is also identified with EUV imagers as shown (for example) in the *TRACE* image in 7.8a. The magnetic field lines are clearly visible even though the imager can only detect the plasma trapped along the field lines. The reason for this is not yet completely understood but it is termed selective heating [72].

The solar magnetic field is in a state of constant change due to movement on the Sun such as differential rotation, emerging flux from beneath the photosphere, magnetic shearing etc. These changes result in solar surface phenomena such as active region and sunspot formation, filament and arcade activity and magnetic helicity contribution. There are many texts which cover the magnetic field of the Sun and any will provide a reasonable review of these. One recommendation is *Solar Magnetic Fields* by Schüssler and Schmidt [83].

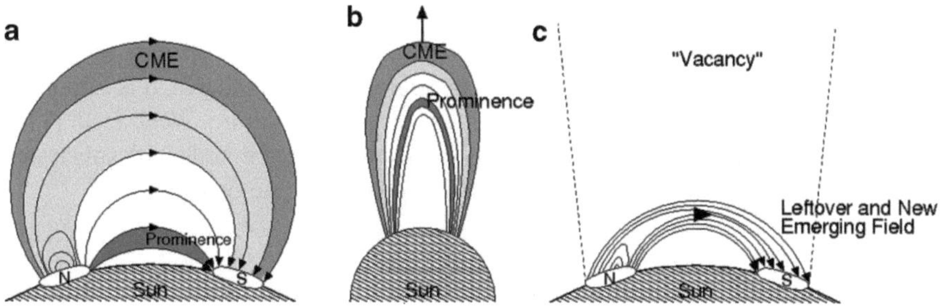

Fig. 7.2 Simplified 2-D depiction of the magnetic field on the Sun comprising the CME (**a**) before, (**b**) during and (**c**) after launch, showing the material comprising the CME and the associated prominence/filament below. When a CME erupts the overlying magnetic field is vacated, enabling the lower fields and structure to emerge and also erupt. This leads to a vacancy into which an emerging field can arise

The CME is believed to arise from large closed magnetic field regions that begin as stable structures anchored to the Sun. One can imagine something similar to the oversimplified depiction in Fig. 7.2, with smaller multiple magnetic field structures beneath a stable overlying structure. Importantly, the field and trapped plasma therein are bound to the lower solar atmosphere by the overlying magnetic structure and higher coronal plasma held in equilibrium.

When the CME erupts, the overlying magnetic structure is removed, thereby enabling the underlying field and plasma to erupt as well. For example, the filament that is commonly associated with a classic three-part CME (see Fig. 1.1) is lower in the erupting structure and is made up of plasma originating from the low solar atmosphere, that has been allowed to escape through the wake of overlying CME eruption. This leads to prominence/filament eruptions that are observed on the surface of the Sun in visible light (Hα) and EUV. Clearly this eruption leads to a disturbance in the lower magnetic field and as it is reconfigured during and following the eruption other phenomena are observed on the solar surface. The exact relationship between many of these phenomena and the CME itself is not yet completely understood.

7.2 CME-Associated Phenomena

The first thing one notices about CMEs is that even by solar standards they are very large and fast. They therefore contain a great deal of mechanical energy. An average CME contains of the order of 10^{25} J of kinetic energy alone, and that is before we consider the gravitational potential and magnetic energy components. Given that CMEs are confined to magnetic fields and those fields originate from the solar surface, it is not unreasonable to expect that the lower solar atmosphere will respond to the launch of the CME. This response may or may not take the form of an explosive flare, but some movement of the lower coronal and/or photospheric plasma is commonly observed around the onset time of a CME. The most minor effects observed are reorientation of plasma loops or transportation of plasma around an active region on the Sun. While mediocre by solar eruptive standards, it is important to recognize that such transportation requires energy expenditure. Webb et al. [112] has suggested that of the order of 10^{24} J of work may be performed on plasma transportation and magnetic field convection in the lower solar atmosphere.

7.2.1 Active Regions

Generally speaking, an active region is an area on the Sun where solar activity is enhanced. The term "activity" is very general, and may involve the appearance of sunspots, structured and/or dynamic magnetic fields, plasma dynamics or enhanced electromagnetic emission. Today, active regions are defined when at least two ground solar observatories independently label the same region on the Sun as "active", and it must have sunspots. The region is then assigned a number. The present numbering system began in January 1972 and 10,000 were labeled by June 2002. From then the numbering system was reset and as of 1 September 2010 a further 1,100 have been labeled. Figure 7.3 shows images of an active region observed

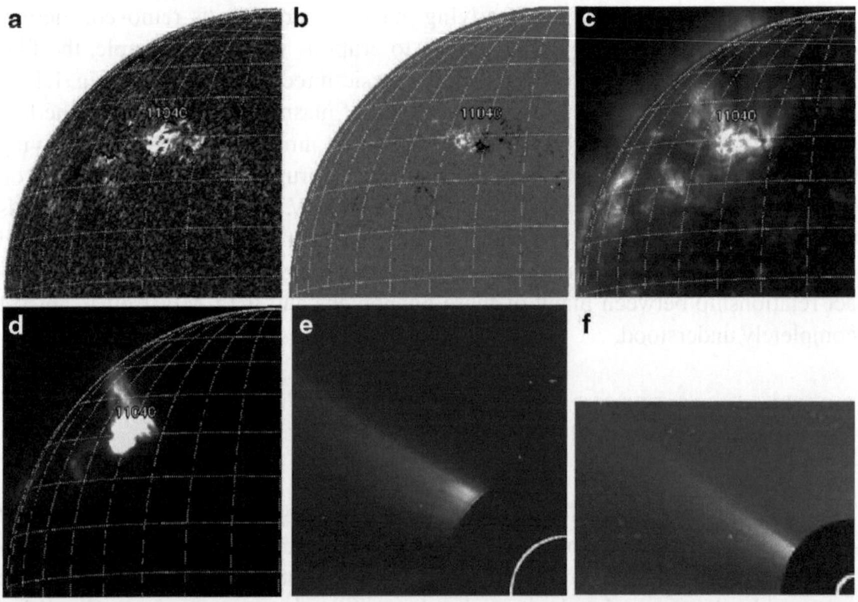

Fig. 7.3 Active region AR11040 (1040) observed in January 2010. It is shown here on 10 January observed by (**a**) Hα (Kanzelhoehe), (**b**) magnetogram (*SOHO*/MDI, and (**c**) EUV (*SOHO*/EIT 195 Å). (**d**) The x-ray (*Hinode*/XRT) image was obtained on 9 January and the *SOHO*/LASCO (**e**) C2 and (**f**) C3 coronagraph images were obtained on 8 January, when the active region was near the northeast limb. Data supplied courtesy of Solarmonitor.org (Trinity College, Dublin)

from different observatories. The University of Hawaii maintains an archive of active region maps, which can be found at http://www.solar.ifa.hawaii.edu/ARMaps/armaps.html.

In the coronal area above active regions a well-defined structure is often observed. This is known as a helmet streamer (Fig. 7.3e) and is a dense bright dynamic plasma region believed to contain a current sheet where magnetic fields of opposite polarity meet. At solar minimum there is a so-called streamer belt on the neutral line at low latitudes (Fig. 7.1a). Their structure extends out to several solar radii in the high corona, and they are probably the source of at least some of the slow solar wind [23]. Narrow streamers exist over active regions, but the larger, longer-lived ones (helmet streamers) overlie prominences/filaments, active regions and other types of complete solar activity.

The relationship between active regions and CMEs remains a subject of some controversy. Narrower CMEs (of widths say within 30° position angle) are often aligned with streamers, leading many to believe that the streamers themselves, and therefore the active regions, cause these CMEs. This may or may not be the case, but narrow CMEs are regularly observed traversing streamers. Figure 7.4 shows such a CME observed by *SMM* and a simplified illustration of the associated magnetic activity.

It should be obvious that, while they may be the source of many narrow CMEs, active regions alone cannot accommodate for larger CMEs or those that occur outside the streamer region. Here the association with these types of CMEs is much more tenuous. While active regions, like solar flares, are probably associated with

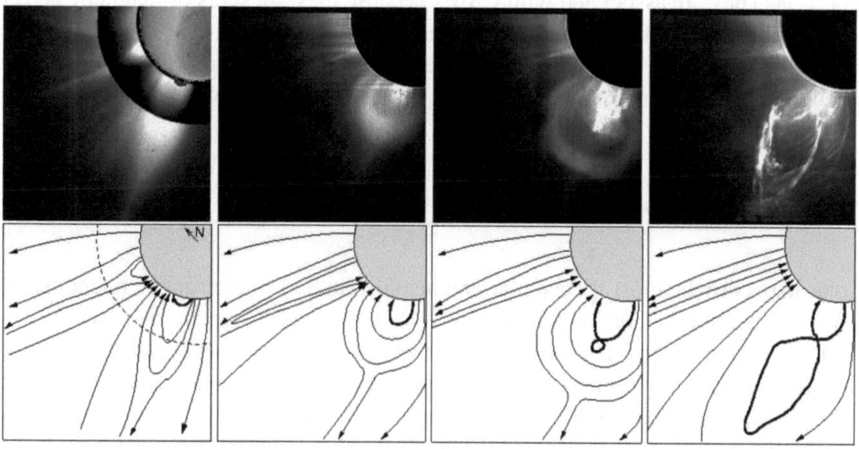

Fig. 7.4 *Top row*: A CME erupting through a helmet streamer as observed by *SMM* (outer corona) and the Mauna Loa prominence monitor (inner corona on the left image) on 18 August 1980. Images available courtesy of HAO (http://mlso.hao.ucar.edu/smm/smmcp_events/1980aug18.html). *Bottom row*: Simplified illustration of the behaviour of the magnetic structure that may be associated with this CME. The magnetic field polarities are estimated. The heavy field line represents the prominence

one CME footpoint the overall CME extends well beyond the area governed by the active region. It should also be noted that CMEs are often observed without an associated active region.

7.2.2 *Sunspots*

Sunspots are often associated with active regions and, as mentioned in the previous section, their appearance can lead to the judgment of an active region. They are dark areas on the Sun typically around the size of the Earth that corotate across the solar disk. Figures 1.3 and 7.6 show images of sunspots related to CMEs as they appear on the whole solar disk. Sunspots are used as indicators of solar activity as they are known to increase and decrease in number with the arrival of solar maximum and minimum. Figure 7.5b shows a plot of sunspot number for each month since 1750. When the numbering of solar cycles began (we are currently in Cycle 24). They are also known to migrate towards lower latitudes as the solar cycle increases, from latitudes of around 30° near solar minimum to near the equator around solar maximum. When plotted as a function of latitude against time an easily-recognisable pattern emerges. An example of these so-called butterfly diagrams is shown in Fig. 7.5c.

Sunspots arise from magnetic fields emerging through the photosphere from beneath the solar surface and are a tracer of strong magnetic activity visible on the solar surface. They appear dark because the material comprising them is cooler than the surrounding photosphere. Close-up a sunspot has a distinct appearance, with the dark region called the "umbra" surrounded by a dynamic "penumbra", shown in Fig. 7.5a. They are associated with CMEs because they are signs of magnetic activity near the solar surface. Several proposed launch mechanisms require a magnetic instability in order to launch the CME (see Chap. 8), and so signs of emerging magnetic fields from beneath the photosphere may be the source of such an instability.

7.2.3 *Solar Flares*

Solar flares are the most spectacular and heavily studied solar eruptive phenomena. Their description has taken many forms over the decades, but the most basic definition is that a flare is a sudden enhancement of electromagnetic radiation from a localised region on the Sun. This enhancement is typically broadband, and has been observed in the radio, visible, UV and x-ray bands. Their observational legacy dates back to the mid-nineteenth century (Chap. 2), and along with CMEs they have been associated with particle acceleration [19], plasma heating and transport [95], magnetic field reconfiguration [10,29,62] and magnetic reconnection [97,98]. Countless textbooks and papers have been published on solar flares and it would not do them justice to present a detailed review here. Instead the reader is referred to Sturrock [97] for a general review and to Strong et al. [95] for a review of more recent developments.

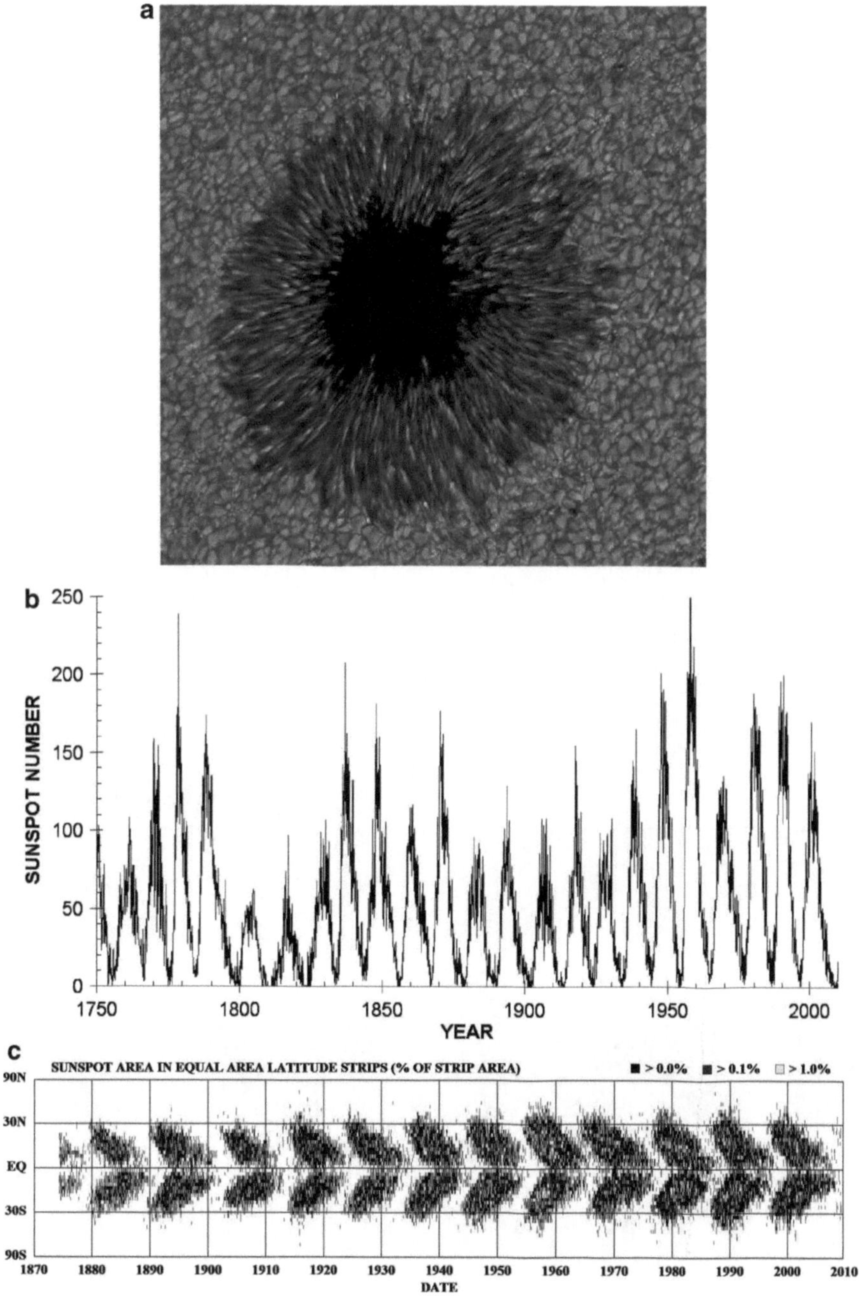

Fig. 7.5 (**a**) An image of a sunspot taken from the Big Bear Solar Observatory on 27 August 1999 [14]. (Reproduced with kind permission of Springer Science and Business Media). (**b**) and (**c**) trends across the solar cycle for sunspot activity. (**b**) Sunspot number per month as a function of time from 1750 to 2010. (**c**) Sunspot location (heliographic latitude) and area against time across 12 solar cycles (12–23) from 1875 to 2010 (Available courtesy of NASA)

Solar flares often (but not always) occur in active regions and arise from solar magnetic activity (probably reconnection [2]). Their emission output exceeds 10^{23} J [16] and they can reach intensities up to 50% brighter than the solar disk. Flares are classified using both visible light (Hα) observatories and orbiting x-ray telescopes. The Hα classification involves a letter or number (S and 1–4) indicating its area, followed by a letter (F, N, B) indicating its intensity. So, for example the solar flare associated with the famous "Bastille Day" event on 14 July 2000 (Fig. 7.6) had a Hα classification of 3B (bright, area between 12.5–24.7$^{\circ 2}$) and a GOES x-ray class of X5.7 (Peak x-ray flux of 5.7×10^{-4} W/m^2). Table 7.1 summarises these classifications. Because the emission energy allocated to the flare is not uniform

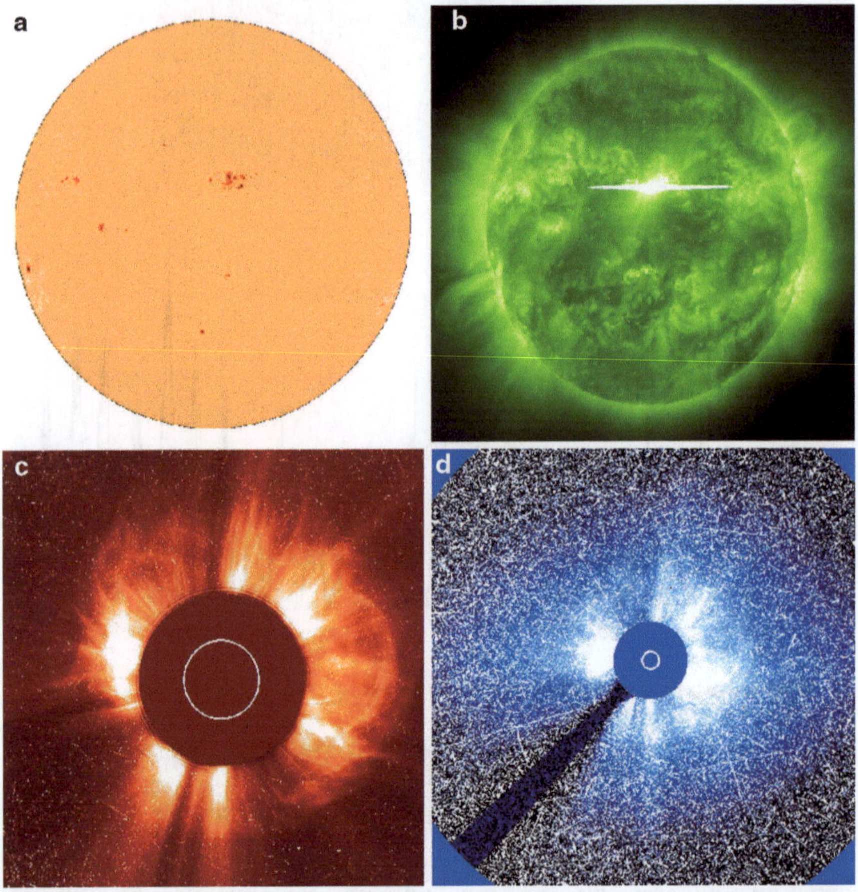

Fig. 7.6 Images of the famous "Bastille Day" flare on 14 July 2000. The solar disk during the flare is shown in (**a**) Dopplergram (*SOHO*/MDI (taken 20:48 UT), (**b**) EUV (*SOHO*/EIT), and white light coronagraphs (**c**) LASCO/C2 (at 10:54 UT) and (**d**) LASCO C3 (at 11:42 UT). The "snowstorm" appearance in C3 is due to energetic particles impacting the camera (Images courtesy of NASA)

Table 7.1 Classification of flares via (left columns) Hα and (right columns) x-ray. The last column multiplies by "Class" meaning the peak flux corresponding to the x-ray letter

Hα				X-ray			
Class	Area deg.^{2}	Class	Brightness	Class	Peak flux W/m^2 (1–8 Å)	Class	Peak flux W/m^2 (1–8 Å)
S	<2.0	F	Faint	A	$10^{-8} - 10^{-7}$	0–2	0–2 × Class
1	2.0–5.1	N	Normal	B	$10^{-7} - 10^{-6}$	2–4	2–4 × Class
2	5.2–12.5	B	Bright	C	$10^{-6} - 10^{-5}$	4–6	4–6 × Class
3	12.5–24.7			M	$10^{-5} - 10^{-4}$	6–8	6–8 × Class
4	>24.7			X	$>10^{-4}$	8–9.9·	8–9.9· × Class

across the electromagnetic spectrum, there is no clear relationship between Hα flare classification and those in the x-rays. It is not uncommon for a flare to be recorded in x-rays without a Hα counterpart, and vice versa. Figure 7.6 shows a sequence of images associated with the Bastille Day CME, including the flare and MDI image. A similar sequence appears in Fig. 1.3 for the Halloween event.

The association between flares and CMEs has been a subject of debate over the years but it generally believed that the two are associated by an as-yet unidentified physical mechanism. While it is possible that CMEs may give rise to the launch of a solar flare it has been established that the inverse is not the case (recall The Solar Flare Myth from Sect. 2.4). It is commonplace for CMEs to launch without an associated flare and vice versa, but the fastest CMEs have a tendency to be associated with bright flares [35]. Finally it is important to realise that the energy contained within a solar flare is at least an order of magnitude less than that in a CME [9, 16, 112].

7.2.4 Disappearing Filaments and Erupting Prominences

When one discusses filaments and prominences on the Sun they are in fact referring to the same phenomena. They are both structured magnetic fields embodying plasma suspended above the solar photosphere. They typically comprise material that is cooler than the solar photosphere, so appear dark when observed on the solar disk. These are called filaments. When, however, the same structure is observed on the solar limb they appear bright against the background corona. These are called prominences.

The event shown in Fig. 7.7 was observed to erupt in the early hours of 18 February 2003 (accompanied by a CME). after which time a large part the filament had disappeared and that remaining had faded (Fig. 7.7d). This is imaginatively called a disappearing filament and indicates that the filament has either collapsed or erupted. Either way it is a signature that a magnetic destabilisation has taken place.

Figure 7.7c shows the eruption of a prominence observed by *SOHO*/EIT towards the northwest. This is from the same region as the filament in Fig. 7.7b but the EIT

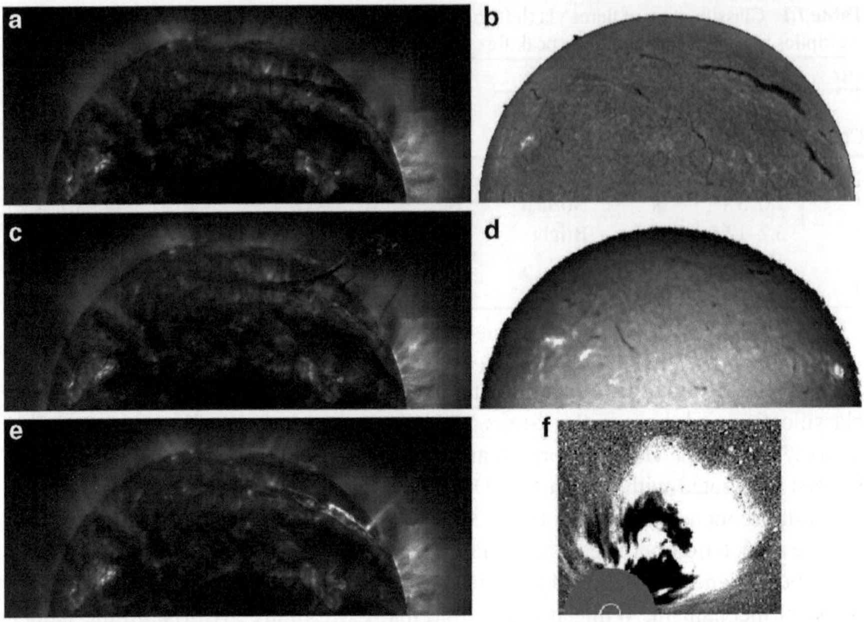

Fig. 7.7 Images of the prominence observed in February 2003. This appears as (**a**) a prominence in EUV (here observed with *SOHO*/EIT at 00:12 UT on 18 February) and (**b**) a filament in Hα (at 10:51 UT on 17 Feb [32]). (**c**) The prominence is observed to erupt around 02:00 UT on 18 Feb, and afterwards (**d**) a large portion of the Hα filament has disappeared and that remaining has faded (image from the Meudon spectroheliograph at 08:14 on 19 Feb) and (**e**) a post-eruptive arcade forms in its place (image on 03:36 UT on 18 Feb). (**f**) This event was associated with a coronal mass ejection, seen here in LASCO/C3 on 05:54 UT on 18 Feb [32] (Figures reproduced by permission of the AAS)

images show the timing of actual eruption. This is the erupting prominence and in this case is clearly the same signature as a disappearing filament. Finally, consider the CME shown in Fig. 7.7f which has a projected onset time close to the launch time of the prominence. The CME and erupting prominence (disappearing filament) are clearly related, and it is likely that they are different parts of the same erupting magnetic structure.

The relationship between prominences/filaments (hereafter referred to as simply prominences) and CMEs is more straightforward than with active regions and flares. Erupting prominences (disappearing filaments) have long been associated with CMEs [63, 113] and geomagnetic activity [43]. The magnetic field comprising the prominence material is confined to near the solar surface by pressure forces from the overlying magnetic structure that extends well into the corona. When the overlying structure is destabilised (i.e. it erupts and becomes the CME) a vacancy is created above the prominence and it too erupts. So, while the prominence is not the CME itself, it does erupt and propagate in the wake of the CME. The underlying filament inside the classic three-part CME (Fig. 1.1) is almost certainly prominence material from lower in the solar atmosphere. A review of prominences can be found in Gopalswamy et al. [26].

As with all CME-associated phenomena there is not a 1–1 relationship between erupting prominences and CMEs. Many CMEs are observed without a prominence counterpart probably because there is not always a region of dense material suspended below the high coronal structure. Prominence eruptions are not always associated with CMEs either. The magnetic structure comprising the prominence material may collapse for other reasons, or there is an eruption of the overlying magnetic field that is invisible to coronagraphs. The latter possibility is discussed further in Sect. 7.4.2

7.2.5 *Post-Eruptive Arcades (Post-Flare Loops)*

Figure 7.7e shows a bright highly-structured arcade which followed the launch of the CME above. This is called a post-eruptive arcade (PEA) or a post-flare loop, and is known to occasionally form as part of the magnetic reconfiguration following the launch of a CME. Some have attributed their formation to be a consequence of magnetic reconnection above [46] and they are sometimes connected with two ribbon flares. Current theories suggest that the arcade forms as a result of the underside of the magnetic reconnection process following the launch of a CME compressing the fields in an arcade toward the Sun. Studies of PEAs include Hudson et al. [40], Sterling et al. [90] and Tripathi et al. [107], and Fig. 7.8a shows a PEA following the Bastille Day CME on 14 July 2000. PEAs have been likened to a cylindrical magnetic flux rope, not unlike those comprising a magnetic cloud (Sect. 2.3.1.1).

PEAs are clearly not part of the CME itself and are not always present following the launch of the CME, but they are probably aligned with the CME structure. Indeed, a remarkable coincidence has been identified between the locations of PEAs and those of their associated CME footpoints. Tripathi et al. [107] surveyed 236 PEAs, their associated CMEs and the underlying photospheric field. They found a clear CME association for 210 (92%) of the PEAs and that none were at high latitudes (above 60°). Their timing and location compared very well with those of the associated CME. Figure 7.8b shows their description of the magnetic field configuration responsible for PEAs.

7.2.6 *Coronal Dimming*

Consider the sequence of images shown in Fig. 7.9. These are EIT 195 Å images and their difference images (with each subtracted from a single base image on 07:14 UT) for the event observed on 6 July 2006 [53]. Soon after the flare begins to erupt (but before it reaches peak intensity) on 08:36 UT, a dark region appears on the solar disk southwest of the flare, which grows further to the southwest. A second dark region appears to the northeast of the flare site which grows and appears to migrate eastward. This phenomenon, once termed a transient coronal hole [80,93,94] is now called coronal dimming [91].

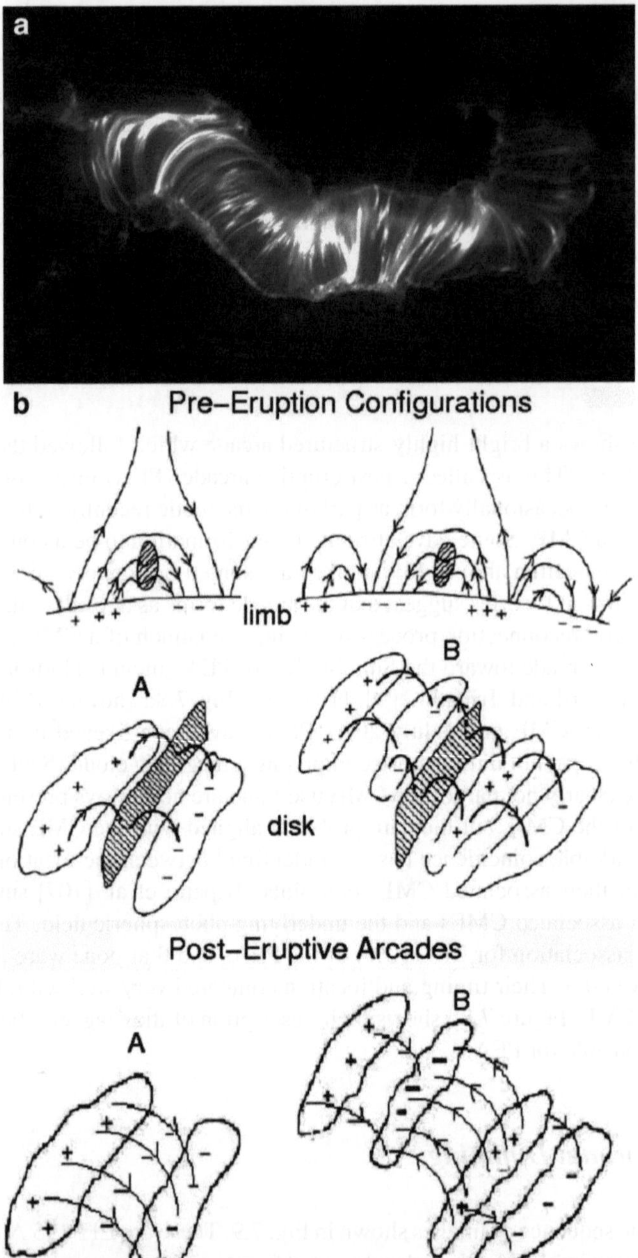

Fig. 7.8 (**a**) Post-eruptive arcade observed by *TRACE* following the launch of the Bastille Day CME on 14 July 2000 [18]. (Reproduced with kind permission of Springer Science and Business Media). (**b**) Diagram by Tripathi et al. [107] of the magnetic structure prior to and following the launch of a CME and prominence. This is shown for two scenarios: Single bipolar (A) and in-between bipolar regions (B). The dark shaded regions are the prominence material. (Reproduced with permission © ESO)

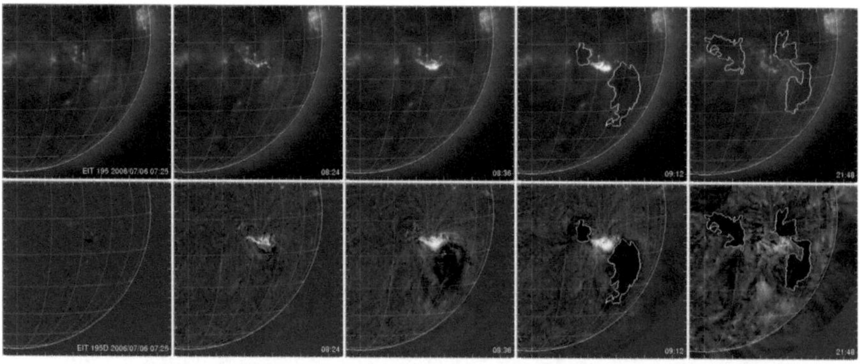

Fig. 7.9 Coronal dimming event observed around the launch of the CME and flare on 6 July 2006 [53]. The *top row* shows the EIT 195 Å images at 07:25, 08:24, 08:36, 09:12 and 21:48 UT. The *bottom row* shows the same images subtracted from a single base image at 08:36 UT. The last two images on the right have the dark regions highlighted by the white border [53]. (Reproduced by permission of the AAS)

Like PEAs, coronal dimmings generally appear following the launch of a CME and they can last for several hours (note that for the event in Fig. 7.9 the dimming regions persisted well into the next day). They were originally identified in x-rays [81, 91] but have also been observed in EUV [103, 105] and Hα [42]. Following detailed spectral analysis [30, 33] it is now widely accepted that coronal dimmings are due to a reduction of density rather than a change in temperature.

Coronal dimmings may therefore be regarded as the removal of coronal (and possibly chromospheric and photospheric) mass following the launch of a CME. Generally, they are the largest CME-related phenomena by area, and their structure appears to match reasonably well with that of the CME [75]. They may therefore be the best indicator of the CME projection onto the solar disk available.

The nature of coronal dimming remains largely unknown as is its exact relationship with a CME. Some believe [30] that at least some of the CME material arises from the dimming while others [33] believe that, like prominences, dimmings are secondary phenomena triggered by the launch of CMEs. It does seem possible that the dimming represents a coronal hole in every respect, including as a source of a fast-flowing plasma stream along open field lines [53].

7.2.7 EUV (EIT) Waves and Moreton Waves

Early in the *SOHO* years, large waves were observed with EIT (195 Å) moving across the solar disk and beginning around the onset time of CMEs [15, 104]. The waves as observed in EIT crossed the solar disk in less than an hour, always associated with CMEs, and immediately identified as a possible candidate for a CME-associated Moreton wave [104]. Figure 7.10 shows an EIT sequence of an

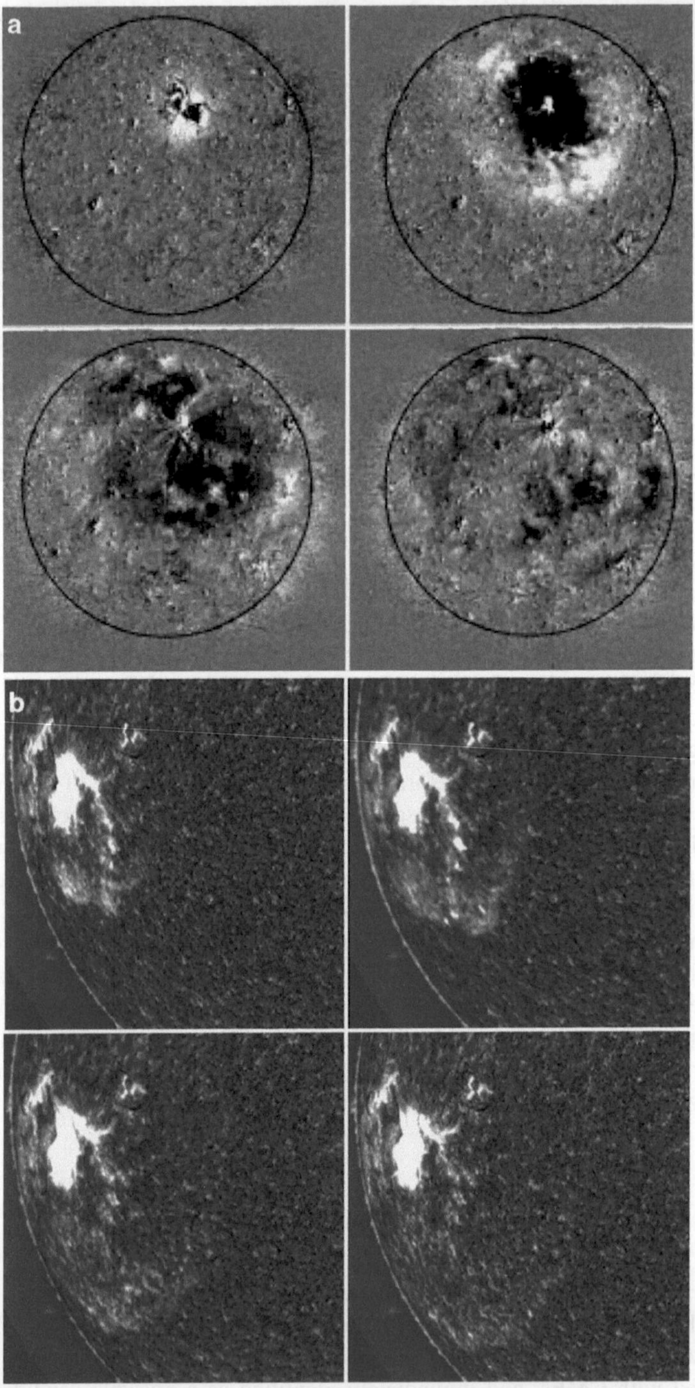

Fig. 7.10 (**a**) Running difference EIT sequence showing the wave associated with the event on 12 May 1997 [105]. (**b**) Moreton wave observed in Hα by the ground ISOON solar observatory on 6 December 2006. (Modified from Gilbert et al. [22])

EUV wave (also known as an EIT wave, coronal wave or solar tsunami) reported by Thompson et al. [105]. A review of the observations of EUV waves in these early years is given by Thompson [102].

According to the theory of Uchida [108], a solar eruption may trigger an impulse that will propagate across the solar atmosphere, observed as a travelling front manifest as an increase in solar emission. When this occurs in the photosphere/chromosphere it is sometimes observed in Hα as a Moreton wave [60,61,88]. The impulse was initially believed to be provided by the flare, but Thompson et al. [104] suggest that a CME could provide the impulse. They also suggest that the EIT wave is the coronal manifestation of a Moreton wave. Figure 7.10b provides an image sequence for a Moreton wave observed on 6 December 2006 [22]. EIT waves are uncommon (they have been associated with less than 1% of all observed CMEs) and Moreton waves are even rarer. Studies comparing EIT with coincident Moreton waves include Thompson et al. [103] and Warmuth et al. [109, 110], while Narukage et al. [64] have compared a Moreton wave with a wave observed in soft x-rays.

While it seems physically reasonable for Moreton waves to be manifest in the corona as EIT waves there are a number of observations that appear to contradict this theory. Most significantly, Moreton waves are much faster than EIT waves and do not always coincide with onset time or source location. EIT waves have an average speed of 270 km/s [45] whereas Moreton waves have a speeds averaging 1,000 km/s [88]. This means that they are rarely co-spatial, even though it can be shown following the theory of Uchida [108] that they should be. One theory [12] is that the EIT waves are produced by the opening of the closed magnetic field lines created by the overhead erupting CME while the Moreton waves (and soft x-ray waves) propagate as an impulse through the solar plasma. Some evidence for this has been provided theoretically and by comparing Hα with soft x-ray waves [12,64]. If true, then EIT waves and Moreton waves, while occasionally observed coincidentally, are not propagated by the same physical mechanism and are therefore not the same phenomenon. It may even be the case that an EIT wave is not a wave at all, but the opening of a coronal hole as a result of the CME launch (i.e. a form of coronal dimming) or as a propagating series of loop reconnections.

7.2.8 X-ray Sigmoids

Active regions when observed in soft x-rays consist of bright loops which sometimes form an "S" or sigmoid shape [1, 82]. They evolve from existing arcades into an arcade of loops or a diffuse cloud. Canfield et al. [11] provides a statistical survey of these structures and Fig. 7.11 shows an x-ray sigmoid image observed by the *Hinode* XRT camera.

Soft x-ray sigmoids have been associated with CMEs, often occurring up to several days before their launch. The work presenting these findings is an extension of earlier work, which connected long-decay x-ray flares (so-called Long-Duration Events) with arcades [44] and CMEs [85, 111]. The first of these was associated with a halo CME on 7 April 1997 and reported by Sterling and Hudson [91]. They found

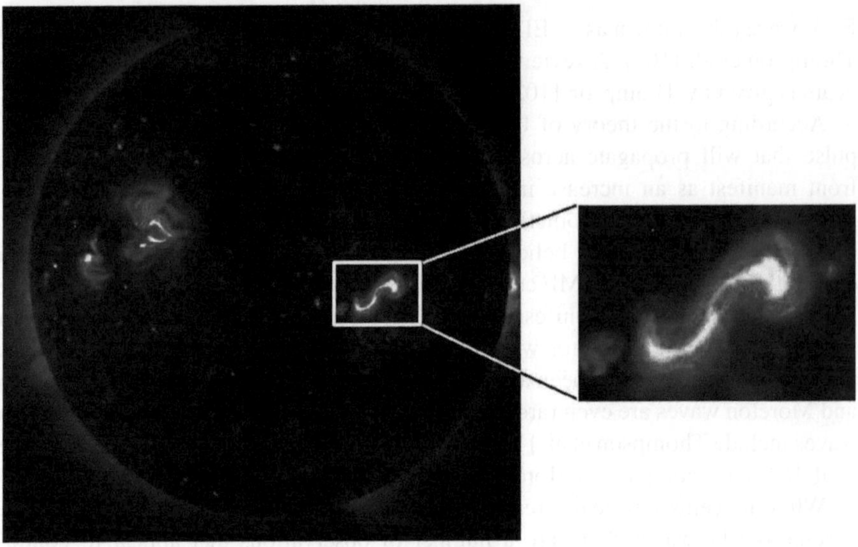

Fig. 7.11 *Hinode*/XRT image of the Sun on 12 February 2007, shortly before the launch of a CME. The "S" shaped sigmoid structure is visible in the *bottom right* and enlarged in the second box (Modified from McKenzie and Canfield [54])

that the sigmoid structure existed in the associated active region for 2 days prior to the launch of the CME and afterward (and after the launch of the flare), the sigmoid reconfigured itself to form a cusp-like structure. The sigmoid does not appear as well defined in EUV images [92, 115]. A review of these early observations appears in Sterling [90].

More recent observations by *Hinode*/XRT has revealed greater details about the structure of the x-ray sigmoids. It has been shown that the sigmoid is in fact a combination of two separate "J" structures that are intertwined [55, 56]. One model by Titov and Demoulin [106] is believed to describe this formation and eruption.

While it is not known exactly the relationship between the sigmoids and CMEs and it is not possible to predict the their eruption, they appear to be a possible precursor to CME activity. Physically they are probably indicators of the CME onset and launch mechanism (Chap. 8). Statistical Studies of this phenomenon have shown that the likelihood of a flare and CME occurring increases with the appearance of an x-ray sigmoid.

7.2.9 Solar Energetic Particles

As discussed in Sect. 2.3, energetic particle signatures associated with flares and CMEs were known long before the discovery of the CME itself, with observations dating back to the 1940s and 1950s [19, 57, 74]. It was suggested in the 1960s that these energetic particles were accelerated in two stages, firstly by energy supplied by a flare and secondly to higher energies via an associated heliospheric shock [114]. Energetic particles have since been almost as heavily studied as solar flares and we visit them only very briefly here.

Put simply, solar eruptive events such are flares and CMEs impart some of their energy to solar particles (typically protons and electrons, but energetic neutrons and heavier ions exist as well), which are accelerated to relativistic speeds. These solar energetic particles (SEPs) move through the heliosphere along magnetic field lines, where they can be detected by in-situ spacecraft, at the Earth or by a solar imager (see, for example, the snow storm in Fig. 7.6d). Hence, by studying the properties and behaviour of SEPs we can learn a great deal about the nature of CMEs and flares, about the magnetic nature of the heliosphere, and about particle acceleration processes in general.

The initial acceleration process for SEPs remains largely unknown. Several theoretical processes have been investigated throughout the years and may be divided into three categories: electric field acceleration [58], shock acceleration [21] or stochastic (time varying electromagnetic field) acceleration [96]. Once above the low corona they travel along heliospheric field lines (the so-called Parker spiral[1]) and those field lines intersecting the Earth originate from near the western limb of the Sun. While particles arriving from here will do so directly, other particles can arrive from different sources on the Sun, the heliosphere or even interstellar space (the latter, of course, are not called solar energetic particles). Variations in the directions of these particles are measured by their pitch angle and anisotropy. Energetic particles from outside the heliosphere are called cosmic rays. The energetic particle population of both SEPs and cosmic rays change with the solar cycle [47].

The reader is encouraged to read Lin [50] and references therein [50] for a review of energetic electrons, Simnett (1991) and references therein [87] for a review on energetic protons, and *Physics of the Space Environment* [24], or *Heliophysical Processes* [25] for reviews on SEPs in general.

7.2.10 Type III Radio Bursts

Solar energetic electron beams (1–100 keV) caused by flares produce regions of increased electron population density, produced because faster electrons catch up with the slower electrons ahead. This creates an instability, giving rise to Langmuir waves in the corona [51]. The waves are electromagnetic, originate at frequencies of the order of 100 MHz, and over the course of a few hours drift downwards in frequency as the disturbance moves outward and the surrounding plasma density decreases. These are called Type III radio bursts and the appearance of one on a

[1] The Parker spiral is named after Eugene Parker who in 1958 [67] first showed that the interplanetary magnetic field assumes the shape of an Archimedean spiral due to its corotation with the Sun. Parker also predicted the existence of the solar wind, and it and the IMF structure were confirmed by spacecraft measurements in the space age.

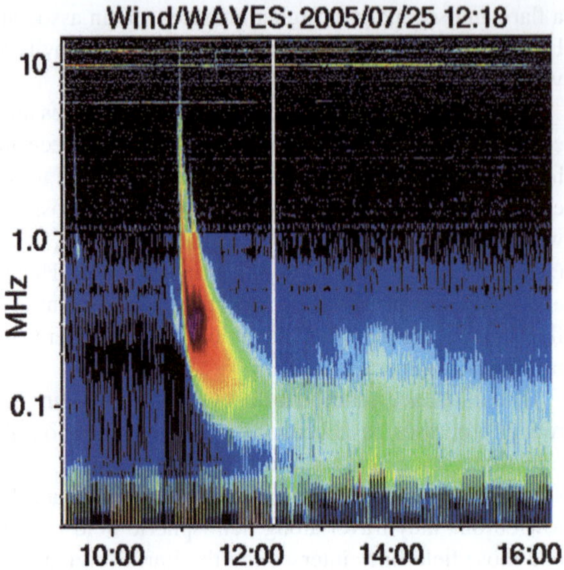

Fig. 7.12 Dynamic spectrum (with the log of frequency) of a Type III radio burst observed with the *WIND*/Waves radio telescope on 25 July 2005. The burst lasted for over 2 h and spanned several orders of frequency magnitude [27]. (Reproduced by permission of the AAS)

dynamic spectrum is given in Fig. 7.12. Their physics and those of solar energetic electrons are invariably connected. The history of the discovery of Type III radio bursts is briefly discussed in Sect. 2.3.

As Type III bursts are generated primarily by solar flares their association with CMEs is the same as those of flares. That is, their detection indicates the presence of solar eruptive activity, but not necessarily the eruption of a CME. Also like flares, if they are associated with a CME they contain very little information about the CME itself. Complete radio spectra including Type III can provide information on the associated CME, if the Type III burst is accompanied by a Type II burst (Sect. 7.3.3). Hence with a single instrument, one may identify properties related to flares and CMEs simultaneously.

7.3 ICME-Associated Phenomena

As the CME propagates through the heliosphere, it becomes an ICME and is tracked by heliospheric imagers, via IPS, and identified near 1 AU by spacecraft that monitor the ICME in-situ as it passes by them. There are a number of characteristic signatures of ICMEs that can be identified directly, or secondary phenomena caused by the ICME are often detected. In this section we review the most common ICME associations as measured by in-situ spacecraft and radio telescopes. Many of these are also discussed in Chap. 2.

7.3.1 Interplanetary Shocks

Many CMEs move through the heliosphere at speeds much greater than the sound speed of the surrounding interplanetary medium. They are therefore supersonic and so give rise to collisionless shocks that form ahead of the CME structure. It is not clear whether heliospheric imagers observe the CME itself or solar wind plasma that has built up at the CME shock, or both.

Along with plasma signatures such as HAEs (Sect. 2.3.1) there are a number of well-known signatures associated with interplanetary shocks measured with in-situ data.

The first is a consequence of the built-up magnetic field and plasma at the shock front. A forward shock can easily be identified by a sudden increase in magnetic field strength, solar wind plasma density, solar wind speed and temperature. In a similar manner, a reverse shock can be identified by a sudden decrease in magnetic field and density but an *increase* in solar wind speed and temperature. The solar wind speed increase arises because one samples the downstream region prior to observing the upstream region for such shocks, and reverse shocks are propagating sunward in the solar wind rest frame. Figure 7.13 shows these signatures for an interplanetary shock observed by *ACE* and *WIND* for the ICME on 7 April 2000. Reverse shocks are very rarely observed for in-situ ICMEs.

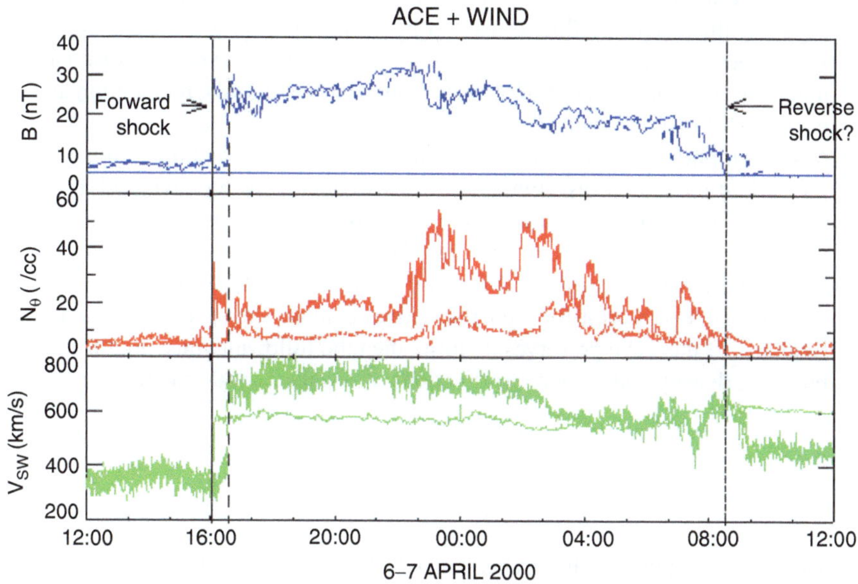

Fig. 7.13 Forward interplanetary shock signature observed at on 7 April 2000 by *ACE* (*solid traces*) and *WIND* (*dashed traces*). The shock is clearly visible as a sudden increase in magnetic field (*B*, *blue*), solar wind plasma density (*N*, *red*) and solar wind speed (*V*, *green*). Later, a possible reverse shock is seen, characterized by a sharp decrease in magnetic field and density but an *increase* in solar wind speed. Each is indicated with an arrow and label

In many, but not all cases a shock can also be identified by its energetic particle signature. Low-energy (10's to 100's of keV) particles streaming towards the Sun will encounter a "solid" shock front and will be reflected back away from the Sun. Likewise antisunward-travelling particles will be reflected toward the Sun. Hence a shock can be identified by a sudden increase in anti-sunward low energy particles immediately before its arrival [101] and antisunward-travelling particles afterwards. Figures 7.14 and 7.15 show the particle signature for a shock combination observed by *Ulysses* at 3.15 AU on 27 May 1991. Figure 7.14 shows the pitch angle plots for various energy bands, while Fig. 7.15 shows the direction of flow of the particle beams during this event, including during the times of the forward and reverse shocks. While the anisotropies of the electrons and ions are not identical, it is clear that a direction change in the flow of particles occurs at both shocks [101].

Finally, to be theoretically classified as a shock it must satisfy the Rankine-Hugoniot relations [41, 73]. These are briefly reviewed below.

Rankine-Hugoniot Relations These relations describe the conservation laws of a fluid across a shock front and can be found in any textbook on fluid dynamics that discusses compressible flow. The following describes a highly simplified ideal case and has been modified from Fetter and Walecka [17].

The relations apply when the hydrodynamic equations describing wave propagation through a fluid break down when the wave becomes discontinuous. At this stage the flow ceases to be isentropic (i.e. the entropy of the system begins to change), and a shock wave is formed. Despite the discontinuity, the fundamental laws of physics (conservation of mass, momentum and energy) still must be apply.

For simplicity, consider a one-dimensional shock propagating with speed u in a fluid initially at rest (Fig. 7.16). We will denote the region ahead of the shock to the right of the figure Region 1 with density and speed ρ_1 and v_1 and the region behind the shock to the left Region 2 with density and speed ρ_2 and v_2. Likewise the pressure p, permittivity ε and entropy s will be denoted with the same subscript for their respective regions. The speed v_1 is zero in Region 1. If the wave preserves its form, then the following must hold for the pressure, density and velocity as functions of distance x and time t:

$$\begin{aligned} p(x,t) &= p(x-ut) \\ \rho(x,t) &= \rho(x-ut) \\ v(x,t) &= v(x-ut). \end{aligned} \tag{7.1}$$

We assume that the conditions become steady far ahead and behind the shock ($x \to \pm\infty$) and that, once formed, the shock will propagate with a constant speed u. The shock is assumed to be thin, and the transition between Regions 1 and 2 occurs on a surface layer involving irreversible processes of heat conduction and viscosity. We will not consider the theory of the plasma dynamics within the shock (Fetter and Walecka [17] recommend Zel'dovich and Raizer [116] for more information), but instead consider the plasma in Regions 1 and 2.

The conservation of mass in one dimension

$$\frac{\partial \rho}{\partial t} + v\frac{\partial \rho}{\partial x} + \rho\frac{\partial v}{\partial x} = 0 \tag{7.2}$$

must hold, and so assuming the time derivatives are equivalent to spatial derivatives $-u\partial/\partial x$ this equation becomes

$$-u\frac{\partial \rho}{\partial x} + \frac{\partial}{\partial x}(v\rho) = \frac{\partial}{\partial x}(v\rho - u\rho) = 0. \tag{7.3}$$

We can now integrate with respect to x across the shock front, applying the conservation across the location of the shock,

$$-u\rho_2 + v_2\rho_2 = -u\rho_1 + v_1\rho_1. \tag{7.4}$$

We may apply $v_1 = 0$ if we wish, but will leave it as non-zero for generality.

The conservation of momentum must also hold. This may be written as

$$\frac{\partial(\rho v_i)}{\partial t} + \sum_{j=1}^{3}\frac{\partial T_{ij}}{\partial x_j} = \rho f_i, \tag{7.5}$$

where T is the stress tensor for non-viscous fluids, given by

$$T_{ij} \equiv p\delta_{ij} + \rho v_i v_j, \tag{7.6}$$

with $\delta_{ij} = 1$ if $i = j$ and 0 if $i \neq j$ (i.e. the Kronecker delta). For $\mathbf{f} = 0$ and assuming once again that $\partial/\partial t \equiv -u\partial/\partial x$ we have

$$-u\frac{\partial(\rho v)}{\partial x} + \frac{\partial}{\partial x}(p + \rho v^2) = 0. \tag{7.7}$$

Integrating and applying the continuity across the shock we arrive at

$$-u\rho_2 v_2 + p_2 + \rho_2 v_2^2 = -u\rho_1 v_1 + p_1 + \rho_1 v_1^2. \tag{7.8}$$

Finally we have the conservation of energy:

$$\frac{\partial}{\partial t}\left(\frac{1}{2}\rho v^2 + \rho\varepsilon\right) + \nabla\cdot\left[\left(\frac{1}{2}\rho v^2 + \rho\varepsilon + p\right)\mathbf{v}\right] = \rho\mathbf{f}\cdot\mathbf{v}, \tag{7.9}$$

which becomes

$$-u\frac{\partial}{\partial x}\left(\frac{1}{2}\rho v^2+\rho\varepsilon\right)+\frac{\partial}{\partial x}\left[\left(\frac{1}{2}\rho v^2+\rho\varepsilon+p\right)v\right]=0, \tag{7.10}$$

and then, integrated with the continuity condition,

$$-u\rho_2\left(\frac{1}{2}v_2^2+\varepsilon_2\right)+\rho_2 v_2\left(\frac{1}{2}v_2^2+\varepsilon_2+\frac{p_2}{\rho_2}\right)$$
$$=-u\rho_1\left(\frac{1}{2}v_1^2+\varepsilon_1\right)+\rho_1 v_1\left(\frac{1}{2}v_1^2+\varepsilon_1+\frac{p_1}{\rho_1}\right). \tag{7.11}$$

We can now simplify these conservation expressions by considering them relative to the reference frame of the travelling shock front. Here the v_i of the surrounding plasma ($i=1$ or 2 depending on the Region) become $v_{si}=v_i-u$, where the subscript s denotes the shock reference frame. Substituting into (7.4), (7.8) and (7.11) respectively we have

$$\rho_2 v_{s2}=\rho_1 v_{s1}, \tag{7.12}$$
$$p_2+\rho_2 v_{s2}^2=p_1+\rho_1 v_{s1}^2, \tag{7.13}$$
$$\varepsilon_2+\frac{1}{2}v_{s2}^2+\frac{p_2}{\rho_2}=\varepsilon_1+\frac{1}{2}v_{s1}^2+\frac{p_1}{\rho_1}. \tag{7.14}$$

These are one form of the Rankine-Hugoniot relations and describe the conservation of mass, momentum and energy across a shock front, from the shock reference frame.

Interplanetary shocks are the oldest recognised signatures of CMEs and they remain a strong indicator of the existence of one at the location of the in-situ spacecraft. Unfortunately, not all CMEs produce shocks and so many will not be recognised if one searches for shocks alone.

7.3.2 Magnetic Clouds

Magnetic clouds have been discussed in some detail in Chap. 2 (Sects. 2.3.1.1 and 2.6.1) and Chap. 5 (Sect. 5.4), so we only briefly review them here.

Magnetic clouds are large, highly structured magnetic field configurations that are observed in the interplanetary medium. Their signatures are characterised in-situ by spacecraft as:

1. Low plasma temperatures;
2. High magnetic field strength;
3. Smoothly rotating magnetic field vector.

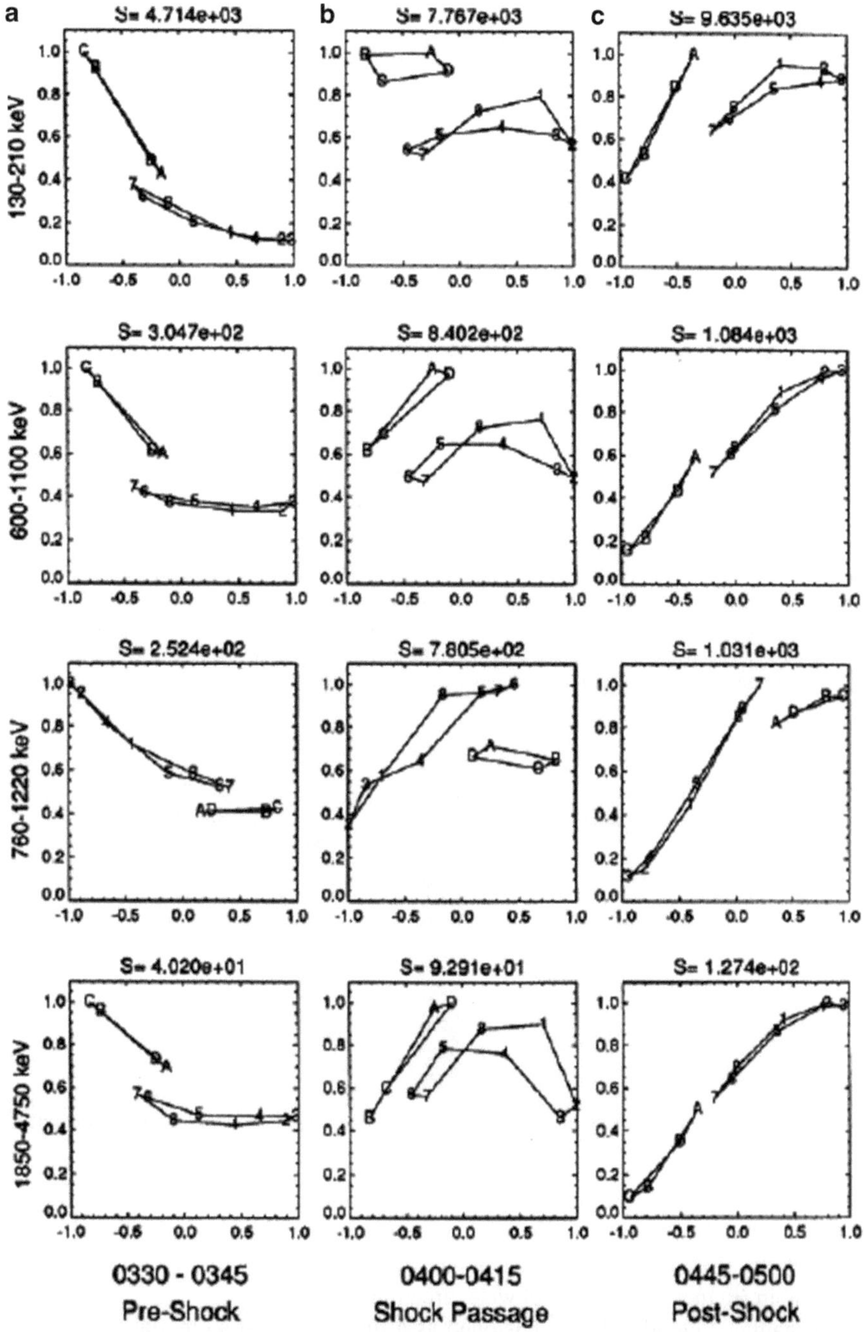

Fig. 7.14 Energetic particle signatures for the event observed by *Ulysses* on 25–29 May 1991. Pitch angle plots of particle flux for three 15 min periods (**a**) before, (**b**) during and (**c**) after the passage of the forward shock. They indicate a strongly field-aligned beam in directions parallel before and anti-parallel after. This indicates a change of direction of energetic particle flow at the shock front [101]. (Reproduced with permission © ESO)

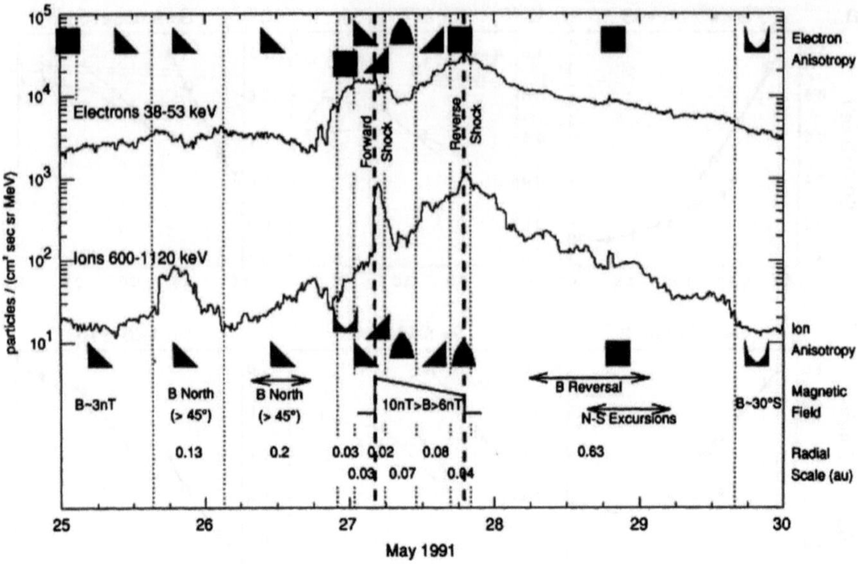

Fig. 7.15 Energetic particle signatures for the event observed by *Ulysses* on 25–29 May 1991. Spin averaged particle fluxes in the 38–53 keV LEMS30 electron and 600–1,120 keV LEMS120 ion channels. The symbols indicate the direction of propagation of the particles and the timing of the forward and reverse shocks [101]. (Reproduced with permission © ESO)

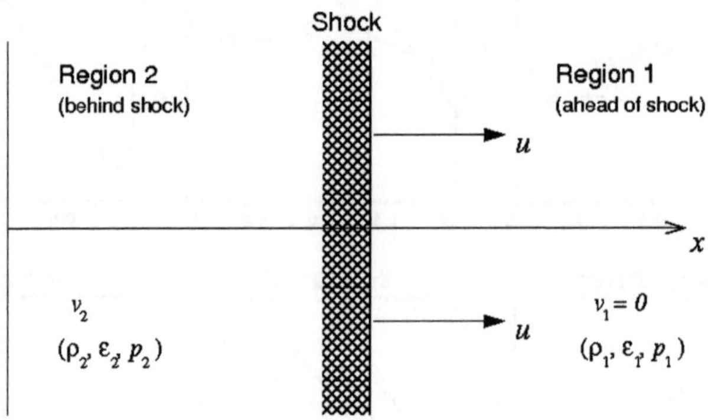

Fig. 7.16 Propagation of a one-dimensional shock front moving through a fluid initially at rest

Figure 7.17 shows a signature of one such magnetic cloud observed on 12–13 August 2000 [48]. Notice the smooth variation in the polar coordinates of the B field Θ, Φ. This indicates that the magnetic field vector is likely to be smoothly rotating in (in this case) the clockwise direction.

A magnetic cloud is best envisioned as a flux rope with a circular (or perhaps elliptical) cross-section spiralling within. The structure of the post-eruptive arcade shown in Fig. 7.8 may be the best way to envision this structure, but it must not be confused with a PEA which never escapes from the low corona.

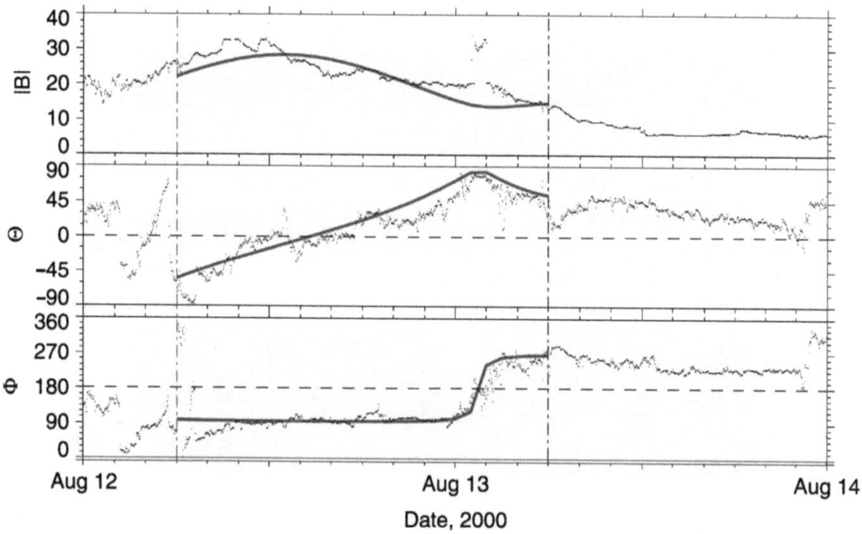

Fig. 7.17 Magnetic field plot of the magnetic cloud observed in August 2000 with the *WIND* spacecraft. The *top plot* represents the total magnetic field while the second and third plots are the spherical coordinates (Θ, Φ) of the measured field. The *signature of the cloud* is represented by the smooth rotation of the magnetic field [48]. The grey overlayed curve is the force-free solution determined by Leamon et al. [48]

As shown in Fig. 1.4, it cannot be known which part of the cloud impacts the spacecraft, and so deductions must be made about its structure based on the single-track measurements. A number of such deductive methods exist, but all are based generally on the idea of a magnetic flux rope with a helical structure. As discussed in Sect. 5.4, some assume the field is force free [5, 37, 49, 52], while others assume a non-force free field [31, 38, 39]. Riley et al. [78] discusses many of the popular techniques used to estimate magnetic cloud structure.

7.3.3 *Type II Radio Bursts*

Just as Langmuir waves are generated by energetic electrons arising from the eruption of a flare (Sect. 7.2.10), so they can be produced by electrons accelerated by an interplanetary shock. This is believed to give rise to a separate type of radio emission called a Type II burst. A history of the discovery of Type II is briefly reviewed in Sect. 2.3.

Like Type III, Type II bursts are long duration and decrease in frequency with time. They are different to Type III in that they are longer-lived and have a more gradual decrease in frequency. Figure 7.18 shows a Type II dynamic spectrum observed with two spacecraft, along with another with a Type II and coincident Type III. The rate of the downward drift in frequency for Type II is consistent with

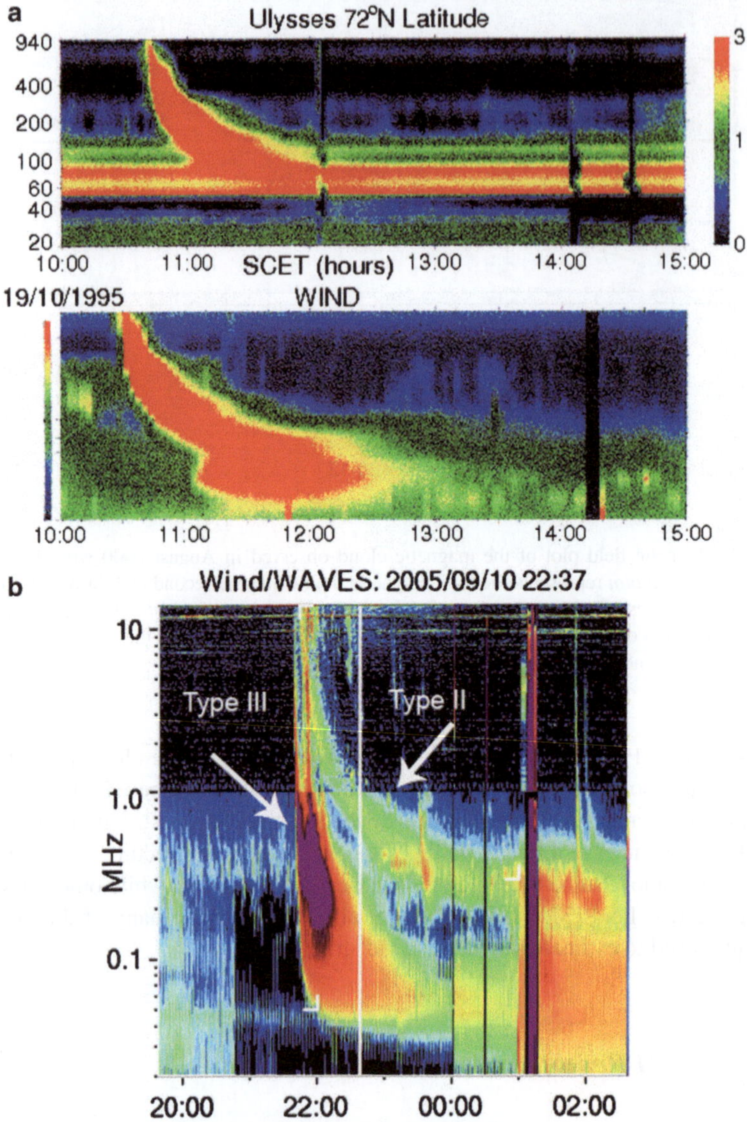

Fig. 7.18 (**a**) Type II burst observed with *Ulysses* and *WIND* [66]. (**b**) Dynamic spectrum of a Type II and accompanying Type III radio burst observed with the *WIND*/Waves radio telescope on 10 September 2005 [27]. (Reproduced by permission of the AAS)

a shock moving through the corona and solar wind [65] and they have been directly associated with interplanetary shocks observed with in-situ spacecraft [8,76]. These, along with other evidence [3, 6, 7, 76], have led to the general acceptance that Type II bursts are produced by electrons accelerated by CME-induced shocks.

Armed with this knowledge, we can use information about the Type II bursts to provide information about the ICME shock front. Furthermore, these properties can be tracked as the ICME propagates through the heliosphere, so Type II bursts therefore provide an additional means of tracking ICMEs through the heliosphere. A number of CME evolution models, for example, use Type II information to provide boundary conditions for the CME disturbance (e.g. Fry et al. [20]).

7.4 Interplanetary Transients Without a CME Counterpart

The previous sections deal with phenomena observed in the heliosphere that can be associated with a coronagraph CME near the Sun. However there are transients that have been observed by in-situ spacecraft and heliospheric imagers but do not have a CME counterpart. These take two forms: Corotating interaction regions and "invisible" CMEs.

7.4.1 Corotating Interaction Regions

Corotating interaction regions (CIRs) are not caused by CMEs and are not related to CMEs. In fact, many would not regard a corotating interaction region as a transient disturbance at all as they corotate with the Sun and are often observed to recur with subsequent solar rotations. They arise from the interaction between solar wind streams of different speeds. If a fast solar wind stream, typically from a coronal hole, is at low solar latitudes it will rotate with the Sun causing an interaction with the surrounding slow solar wind streams. The interaction regions have a high pressure, so are bound by forward and reverse waves that eventually develop into interplanetary shocks at large distances from the Sun. The shocks are expected to form beyond 2 AU, but they are commonly observed at 1 AU by in-situ spacecraft [13, 28, 59] and have recently been observed within 1 AU by heliospheric imagers [79, 100]. Because it is bound by solar wind streams the structure of a CIR looks like the so-called Parker spiral. Figure 7.19a shows a simple diagram of a CIR [100].

More recently, CIRs have also been tracked by heliospheric imagers. They have been observed in *STEREO*/HI-B by Sheeley et al. [84], HI-A by Rouillard et al. [79] and by both *STEREO* HIs and SMEI by Tappin and Howard [100]. When one considers the location of the *STEREO* spacecraft relative to the spiral structure of the CIR one will realise that each will observe different parts of the CIR at different times. *STEREO-A* will observe the trailing outer part of the CIR while *STEREO-B* observes the inner part of its structure as the spiral moves into its field of view. It is unlikely that very much material would have accumulated this close to the Sun, and so *STEREO-B* more likely observes waves in the solar wind forming ahead of

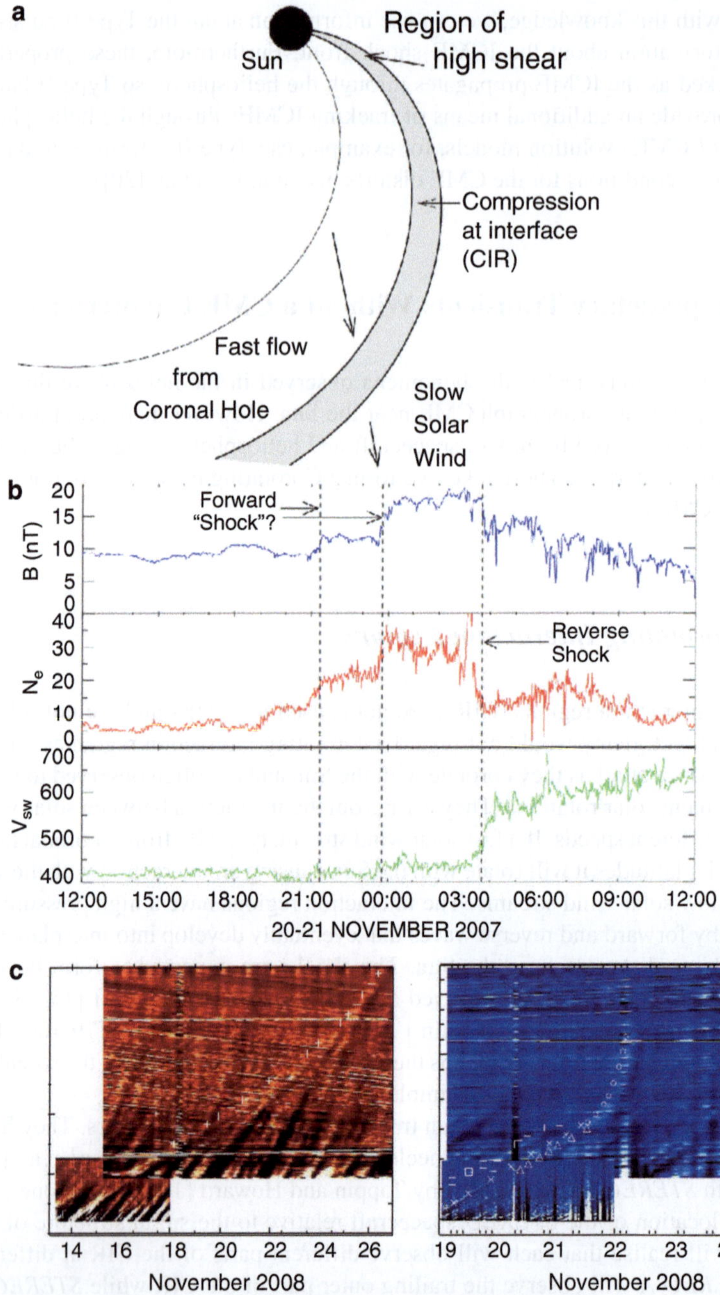

Fig. 7.19 (**a**) Schematic diagram showing the basic structure of a CIR in the solar equatorial plane [100]. A low-latitude fast solar wind stream corotates with the Sun and interacts with the surrounding slow solar wind creating a compression region at the interface. (**b**) A CIR as observed by the *STEREO-A* in-situ instrument on 20–21 November 2007 [36]. (**c**) A CIR as observed by heliospheric imagers. These are effectively polar plots of elongation-time for (*left*) HI-2A and (*right*) HI-2B. (**a**) and (**c**). (Reproduced by permission of the AAS)

the CIR. *STEREO-A* would be more likely to observe shocked material, but it would have a very slow transit time across its field of view (over 10 days). This is what has been observed – one example is shown in Fig. 7.19c.

CIRs are sometimes mistaken for ICMEs when observed in white light, but they do have in-situ signatures that distinguish them from ICMEs. These include the nature of the shocks: For an ICME the forward shock is stronger than the reverse (and a reverse shock is rarely seen at all) but for a CIR the reverse shock is stronger than the forward; and low-energy particles: For an ICME they are directed anti-sunward before a forward shock and sunward after the reverse shock but for a CIR they are sunward both before the forward and after the reverse shock (Fig. 7.19b). CIRs are also identified by the absence of an associated coronagraph CME and the existence of a low latitude coronal hole [77]. These two criteria must simultaneously be satisfied, as the former does not imply that no CME as present (this is discussed in the next section).

Further information on CIRs can be sought from Smith and Wolfe [89], Pizzo [68–70], Tappin et al. [99] and Pizzo and Gosling [71].

7.4.2 Erupting Magnetic Structures ("Invisible" CMEs)

Recent work by Simnett and colleagues has revealed a class of ICME without a CME counterpart. Originally identified using SMEI data [86] they have also been identified using in-situ data [35] and PEAs [107]. Here an unmistakable signature of a CME in heliospheric imagers or in-situ data are traced back to the Sun and there is no coronagraph CME signature there, or within several days of the projected onset time. Statistical studies of these have shown them to occur around 5–15% of the time [34, 35, 86, 107].

Figure 7.20 provides a clue as to what is likely to be the reason for the absent coronagraph CME. Here are SMEI and LASCO images of a CME that was observed on 28–29 July 2003. The SMEI event is clearly visible (by SMEI standards) but its LASCO counterpart is a very faint, very slow CME moving northwards. This CME was so faint that it was not identified in any of the online CME catalogues.[2] Now look closely at the dark band to the left of the faint CME in both LASCO images. These are running differences, so the dark band represents the location of the CME in the prior image (which has been subtracted away). This may therefore be just a component (probably the western flank) of a much larger CME, the majority of which is invisible to the coronagraph.

Recall from Chap. 4 that coronagraphs observe light that has been scattered off the free electrons in the plasma cloud comprising the CME. They also have an intensity detection range, so do not observe features that are too faint. Hence, if

[2] Please note the absence of this CME from the catalogs as a warning to those who intend to rely solely on those catalogues: They do not list *every* CME.

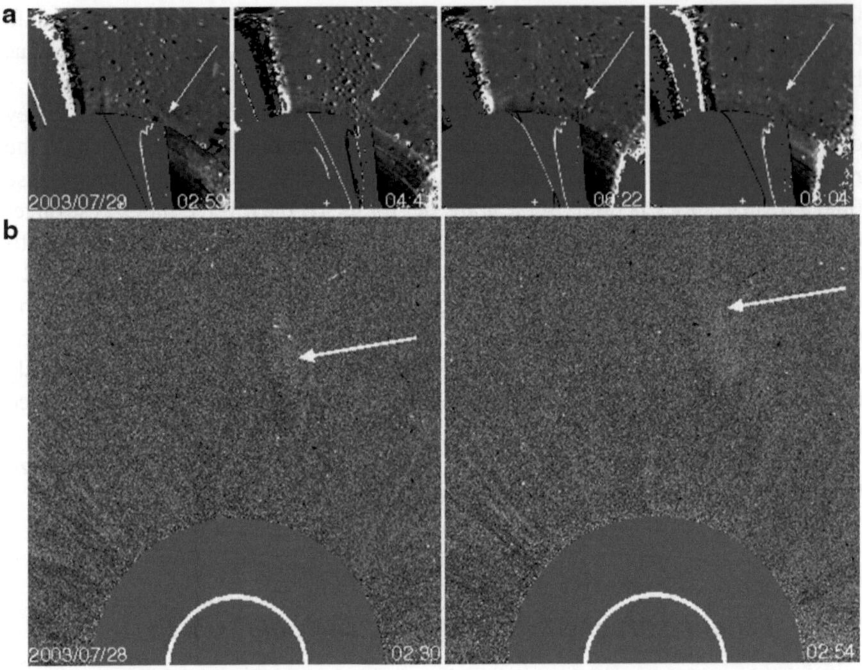

Fig. 7.20 (**a**) SMEI and (**b**) LASCO C2 running difference images of a CME that occurred on 28–29 July 2003 [34]. The SMEI ICME is indicated with the arrow and the associated LASCO signature is arrowed in (*b*). This is the only LASCO CME associated with this CME and it is very slow and faint. It seems likely that this is the flank of a faster CME, the reaming structure of which is invisible to the coronagraph

a CME does not contain sufficient plasma density for its scattered light to achieve an intensity above the detection threshold of the coronagraph, then it will not be observed by the instrument. This is probably what is occurring with the invisible CMEs. They become detected in the heliosphere because they are large and fast enough to accumulate solar wind material ahead of them (snow plough), which builds up in density to a point where their scattered light intensity is above the detection threshold. Their particle and magnetic field signatures will also be observed by in-situ spacecraft in the same manner as a regular ICME would.

If true, then these erupting magnetic structures (EMS [86]) may be occurring regularly in the solar corona and passing undetected by coronagraphs. Along with their associated ICME signatures they may also account for solar eruptions that occur without an associated CME. For example it is commonplace for flares to erupt without an associated CME. Perhaps there are EMS erupting overhead but they do not contain sufficient plasma to be detected by coronagraphs. Further information on EMS is available from Howard and Simnett [34].

7.5 Discussion

This chapter reviews the many phenomena that are known to be related to CMEs and ICMEs. Figure 7.21 illustrates and attempts in a very simple manner to estimate their relationship with CMEs. It is important to realise that the exact relationship between each of these and the CME itself remains largely unknown and that there is no single phenomena that has a 1–1 relationship with CMEs. Many of these occur without an identified CME counterpart and many CMEs erupt without an obvious association on the solar disk or in the interplanetary medium. It should also be noted that each of these phenomena represent a field of research in their own right, and many of these (e.g. solar flares, SEPs) have been studied more intensely than CMEs. The reader should regard this chapter as a brief review of each and should investigate more dedicated literature on these fields for more information.

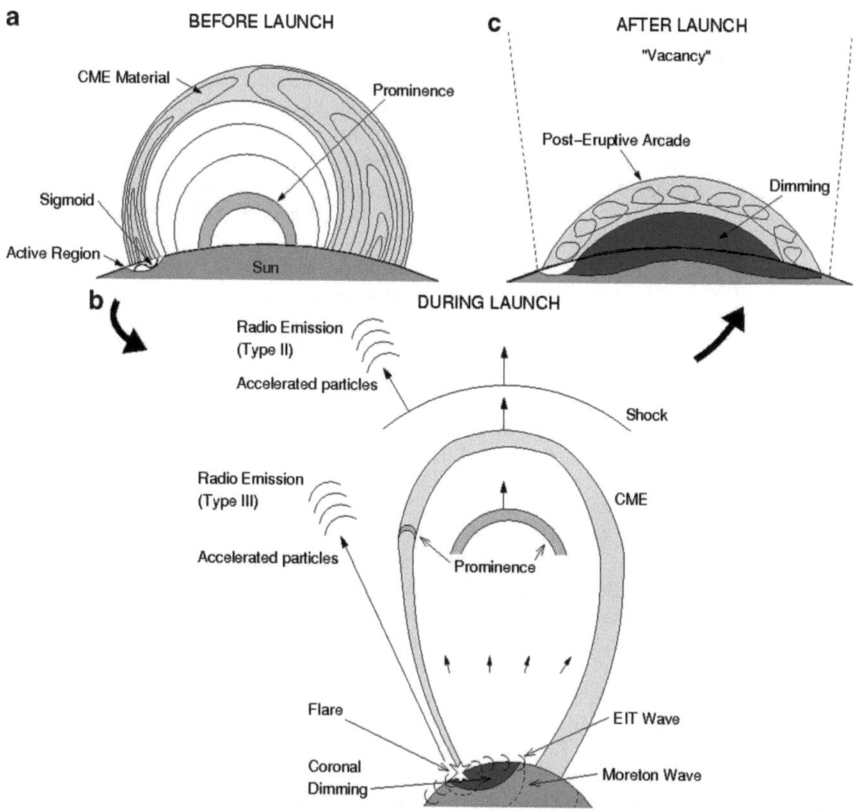

Fig. 7.21 Illustration of many of the solar surface eruptions and an estimate of their physical relationship with (**a**) before, (**b**) during and (**c**) after the launch of the CME

One final note, an absence of solar surface eruptions can be regarded as a good indicator that an observed CME is back-sided (i.e. is directed away from the Earth on the other side of the Sun), but this is no guarantee. It is entirely possible that the CME simply does not have any detectable surface associations. Such CMEs are typically small and have low energies, but not always. Such CMEs are now known as stealth CMEs.

References

1. Acton, L., Tsuneta, S., Ogawara, Y., Bentley, R., Bruner, M., Canfield, R., Culhane, L., Doschek, G., Hiei, E., Hirayama, T.: Science **258**, 618–625 (1992).
2. Antiochos, S.K., DeVore, C.R.: In Burch, J.L., Carovillano, R.L., Antiochos, S.K. (eds.), Sun-Earth Plasma Connections, AGU, Washington DC, pp.113–120 (1999).
3. Bale, S.D., Reiner, M.J., Bougeret, J.-L., Kaiser, M.L., Krucker, S., Larson, D.E., Lin, R.P.: Geophys. Res. Lett **26**, 1573–1576 (1999).
4. Bridgman, T., Kucera, T., Kaiser, M., Howard, R., DeRosa, M.L., Michels, D.: In: Grand Tour of the Coronal Loops Model, available via NASA/GSFC Scientific Visualization Studio. http://www.nasaimages.org/luna/servlet/detail/NSVS~3~3~13647~113647: Grand-Tour-of-the-Coronal-Loops-Mod. Cited March 2006.
5. Burlaga, L.F.: J. Geophys. Res. **93**, 7217–7224 (1988).
6. Cairns, I.H.: J. Geophys. Res. **91**, 2975–2988 (1986).
7. Cane, H.V., Sheeley, N.R., Jr., Howard, R.A.: J. Geophys. Res. **92**, 9869–9874 (1987).
8. Cane, H.V., Stone, R.G., Fainberg, J., Steinberg, J.L., Hoang, S.: Solar Phys. **78**, 187–198 (1982).
9. Canfield, R.C., Cheng, C.-C., Dere, K.P., Dulk, G.A., McLean, D.J., Robinson, R.D., Jr., Schmahl, E.J., Schoolman, S.A.: In Sturrock, P.A (ed.), Solar Flares: A Monograph From Skylab Workshop II, Colo. Assoc. Uni. Press, Boulder, p.451 (1980).
10. Canfield, R.C., de La Beaujardiere, Leka, K.D.: Phil Trans. R. Soc. Lond. **336**, 381–388 (1991).
11. Canfield, R.C., Hudson, H.S., McKenzie, D.E.: Geophys. Res. Lett. **26**, 627–630 (1999).
12. Chen, P.F., Ding, M.D., Fang, C.: Space Sci. Rev. **121**, 201–211 (2005).
13. Choi, Y., Moon, Y.-J., Choi, S., Baek, J.-H., Kim, S.S., Cho, K.-S., Choe, G.S.: Solar Phys. **254**, 311–323 (2008).
14. Denker, C., Yang, G., Wang, H.: Solar Phys. **202**, 63–70 (1999).
15. Dere, K.P., Brueckner, G.E., Howard, R.A., Koomen, M.J., Korendyke, C.M., Kreplin, R.W., Michels, D.J., Moses, J.D., Moulton, N.E., Socker, D G., St. Cyr, O.C., Delaboudiniére, J.-P., Artzner, G.E., Brunaud, J., Gabriel, A.H., Hochedez, J.F., Millier, F., Song, X.Y., Chauvineau, J.P., Marioge, J.P., Defise, J.M., Jamar, C., Rochus, P., Catura, R.C., Lemen, J.R., Gurman, J.B., Neupert, W., Clette, F., Cugnon, P., van Dessel, E.L., Lamy, P.L., Llebaria, A., Schwenn, R., Simnett, G.M.: Solar Phys. **175**, 601–612 (1997).
16. Emslie, A.G., Kucharek, H., Dennis, B.R., Gopalswamy, N., Holman, G.D., Share, G.H., Vourlidas, A., Forbes, T.G., Gallagher, P.T., Mason, G.M., Metcalfe, T.R., Mewaldt, R.A., Murphy, R.J., Schwartz, R.A., Zurbuchen, T.H.: J. Geophys. Res. **109**, doi:10.1029/2004JA010571 (2004).
17. Fetter, A.L., Walecka, J.D.: Theoretical Mechanics of Particles and Continua, McGraw-Hill, New York, Ch. 9 (1980).
18. Fletcher, L., Hudson, H.: Solar Phys. **204**, 69–89 (2001).
19. Forbush, S.E., Gill, P.S., Vallarta, M.S.: Rev. Mod. Phys. **21**, 44–48 (1949).
20. Fry, C.D., Dryer, M., Smith, Z., Sun, W., Deehr, C.S., Akasofu, S.-I.: J. Geophys. Res. **108**, 1070, doi:10.1029/2002JA009474 (2003).

21. Giacalone, J., Jóta, J.: Space Sci. Rev. **124**, 277–288 (2006).
22. Gilbert, H.R., Daou, A.G., Young, D., Tripathi, D., Alexander, D.: Astrophys. J. **685**, 629–645 (2008).
23. Golub, L., Pasachoff, J.M.: The Solar Corona, Cambridge Univ. Press (1997).
24. Gombosi, T.I. (ed.): Physics of the Space Environment, Cambridge Univ. Press, New York, Ch. 13 (1998).
25. Gopalswamy, N., Hasan, S.S., Ambastha, A.: Heliosphysical Processes, Springer-Verlag, Berlin, pp. 73–282 (2010).
26. Gopalswamy, N., Shimojo, M., Lu, W., Yashiro, S., Shibasaki, K., Howard, R.A.: Astrophys. J. **586**, 562–578 (2003).
27. Gopalswamy, N., Yashiro, S., Xie, H., Akiyama, S., Aguilar-Rodriguez, E., Kaiser, M.L., Howard, R.A., Bougeret, J.-L., Astrophys. J. **674**, 560–569 (2008).
28. Gosling, J.T., Pizzo, V.J.: Space Sci. Rev. **89**. 21–52 (1999).
29. Hagyard, M.J.: Solar Phys. **115**, 107–124 (1988).
30. Harrison, R.A., Bryans, P., Simnett, G.M., Lyons, M.: Astron. Astrophys. **400**, 1071–1083 (2003).
31. Hasegawa, H., Sonnerup, B., Dunlop, M., Balogh, A., Haaland, S., Klecker, B., Paschmann, G., Lavraud, B., Dandouras, I., Rème, H.: Ann. Geophys. **22**, 1251–1266 (2004).
32. Howard, T.A., Fry, C.D., Johnston, J.C., Webb, D.F.: Astrophys. J. **667**, 610–625 (2007).
33. Howard, T.A., Harrison, R.A.: Solar Phys. **219**, 315–342 (2004).
34. Howard, T.A., Simnett, G.M.: J. Geophys. Res. **113**, A08102, doi:10.1029/2007JA012920 (2008).
35. Howard, T.A., Tappin, S.J.: Geophys. Res. Lett. **32**, L14106, doi:10.1029/2005GL023056 (2005).
36. Howard, T.A., Tappin, S.J.: Space Sci. Rev. **147**, 89–110 (2009).
37. Hidalgo, M.A., Cid, C., Medina, J., Vinãs, A.F.: Solar Phys. **194**, 165–174 (2000).
38. Hu, Q., Sonnerup, B.U.Ö.: Geophys. Res. Lett. **28**, 467–470 (2001).
39. Hu, Q., Smith, C.W., Ness, N.F., Skoug, R.M.: Geophys. Res. Lett. **30**, doi:10.1029/2002GL016653 (2003).
40. Hudson, H.S., Lemen, J.R., St. Cyr., O.C., Sterling, A.C., Webb, D.F., Geophys. Res. Lett. **25**, 2481–2484 (1998).
41. Hugoniot, H.: J. de l'Ecole Polytechnique **57**, 3 (1887).
42. Jiang, Y., Ji, H., Wang, H., Chen, H.: Astrophys. J. **597**, L161–L164 (2003).
43. Joselyn, J.A., McIntosh, P.S.: J. Geophys. Res. **86**, 4555–4564 (1981).
44. Kahler, S.: Astrophys. J. **214**, 891–897 (1977).
45. Klassen, A., Aurass, H., Mann, G., Thompson, B.J.: Astron. Astrophys. Supp. **141**, 357–369 (2000).
46. Kopp, R.A., Pneuman, G.W.: Solar Phys. **50**, 85–98 (1976).
47. Lario, D., Simnett, G.M.: In Pap, J.M., Fox, P. (eds.), Solar Variability and its Effects on Climate, AGU, Washington DC, pp.195–216 (2004).
48. Leamon, R.J., Canfield, R.C., Jones, S.L., Lambkin, K., Lundberg, B.J., Pevtsov, A.A., J. Geophys. Res. **109**, A05106, doi:10.1029/2003JA010324 (2004).
49. Lepping, R.P., Burlaga, L.F., Jones, J.A.: J. Geophys. Res. **95**, 11957–11965 (1990).
50. Lin, R.P.: Solar Phys. **100**, 537–561 (1985).
51. Lin, R.P., Potter, D.W., Gurnett, D.A., Scarf, F.L.: Astrophys. J. **251**, 364–373 (1981).
52. Lynch, B.J., Gruesbeck, J.R., Zurbuchen, T.H., Antiochos, S.K.: J. Geophys. Res. **110**, A08107, doi:10.1029/2005JA011137 (2005).
53. McIntosh, S.W., Leamon, R.J., Davey, A.R., Wills-Davey, M.J.: Astrophys. J. **660**, 1653–1659 (2007).
54. McKenzie, D.E., Canfield, R.C.: In Press Release, Montana State University, New Telescope Making It Easier to Predict Solar Activity, Webpage available via MSU. http://solar.physics.montana.edu/press/XRT_Sigmoid.html. Cited 30 May 2007.
55. McKenzie, D.E., Canfield, R.C.: Bull. American Astron. Soc. **38**, 180 (2007).
56. McKenzie, D.E., Canfield, R.C.: Astron. Astrophys. **481**, L65–L68 (2008).

57. Meyer, P., Parker, E.N., Simpson, J.A.: Phys. Rev. **104**, 768–783 (1956).
58. Miller, J.A.: Space Sci. Rev. **86**, 79–105 (1998).
59. Möbius, E., Morris, D., Lee, M.A., Mason, G.M., Klecker, B., Mazur, J.E., Popecki, M.A., Galvin, A.B., Kistler, M.: Proc. Int. Conf. Cosmic Rays, 293 (2001).
60. Moreton, G.F.: Astron. J. **65**, 494 (1960).
61. Moreton, G.F.: Sky and Telescope **21**, 145 (1961).
62. Moreton, G.F., Severney, A.B: Solar Phys. **3**, 282–297 (1968).
63. Munro, R.H., Gosling, J.T., Hildner, E., MacQueen, R.M., Poland, A.I., Ross, C.L.: Solar Phys. **61**, 201–215 (1979).
64. Narukage, N., Hudson, H.S., Morimoto, T., Akiyama, S., Kitai, R., Kurokawa, H., Shibata, K.: Astrophys. J. **572**, L109–L112 (2002).
65. Nelson, G.J., Melrose, D.B.: In Solar Radiophysics: Studies of Emission from the Sun at Metre Wavelengths, Cambridge Univ. Press, New York, pp.333–359 (1985).
66. Ogilvie, K.W., Desch, M.D.: Adv. Space Res. **20**, 559–568 (1997).
67. Parker, E.N.: Astrophys. J. **128**, 664 (1958).
68. Pizzo, V.J.: J. Geophys. Res. **83**, 5563–5572 (1978).
69. Pizzo, V.J.: J. Geophys. Res. **85**, 727–743 (1980).
70. Pizzo, V.J.: J. Geophys. Res. **87**, 4374–4394 (1982).
71. Pizzo, V.J., Gosling, J.T.: Geophys. Res. Lett. **21**, 2063–2066 (1994).
72. Plowman, J.E., Kankelborg, C.C., Longcope, D.W.: Astrophys. J. **706**, 108–112 (2009).
73. Rankine, W.J.M.: Phil. Trans. R. Soc. **160**, 277–288 (1870).
74. Reid, G.C., Leinbach, H.: J. Geophys. Res. **64**, 1801–1805 (1959).
75. Reinard, A.A., Biesecker, D.A.: Astrophys. J. **674**, 576–585 (2008).
76. Reiner, M.J., Kaiser, M.L., Fainberg, J., Bougeret, J.-L., Stone, R.G.: Geophys. Res. Lett. **25**, 2493–2496 (1998).
77. Richardson, I.G., Webb, D.F., Zhang, J., Berdichevsky, D.B., Biesecker, D.A., Kasper, J.C., Kataoka, R., Steinberg, J.T., Thompson, B.J., Wu, C-C., Zhukov, A.N.: J. Geophys. Res. **111**, A07S09, doi:10.1029/2005JA011476 (2006).
78. Riley, P.R., Linker, J.A., Lionello, R., Mikić, Odstrcil, D., Hidalgo, M.A., Cid, C., Hu, Q., Lepping, R.P., Lynch, B.J., Rees, A.: J. Atmos. Solar Terr. Phys. **66**, 1321–1331 (2004).
79. Rouillard, A.P., Davies, J.A., Forsyth, R.J., Rees, A., Davis, C.J., Harrison, R.A., Lockwood, M., Bewsher, D., Crothers, S.R., Eyles, C.J., Hapgood, M., Perry, C.H.: Geophys. Res. Lett. **35**, L10110 (2008).
80. Rust, D.M.: Space Sci. Rev. **34**, 21–36 (1983).
81. Rust, D.M., Hildner, E.: Solar Phys. **48**, 381–387 (1976).
82. Rust, D.M., Kumar, A.: Astrophys. J. **464**, L199 (1996).
83. Schüssler, M., Schmidt, W. (eds.): Physics of the Solar Corona, Univ. Cambridge Press, Cambridge (1994).
84. Sheeley, N.R., Jr., Herbst, A.D., Palatchi, C.A., Wang, Y.-M., Howard, R.A., Moses, J.D., Vourlidas, A., Newmark, J.S., Socker, D.G., Plunkett, S.P., Korendyke, C.M., Burlaga, L.F., Davila, J.M., Thompson, W.T., St. Cyr, O.C., Harrison, R.A., Davis, C.J., Eyles, C.J., Halain, J.-P., Wang, D., Rich, N.B., Battams, K., Esfandiari, E., Stenborg, G.: Astrophys. J. **675**, 853–862 (2008).
85. Sheeley, N.R., Jr., Howard, R.A., Koomen, M.J., Michels, D.J.: Astrophys. J. **272**, 349–354 (1983).
86. Simnett, G.M.: In Fleck, B., Zurbuchen, T.H. (eds.), Proc. Solar Wind 11/SOHO 16 Connecting Sun and Heliosphere, p.767, ESA SP-592, ESTEC, Noordwijk, Netherlands (2005).
87. Simnett, G.M.: Mem. S. A. It. **62**, 359–388 (1991).
88. Smith, S.F., Harvey, K.L: In Macris, C.J. (ed.), Physics of the Solar Corona, D. Reidel Pub. Co., 156–167 (1971).
89. Smith, E.J., Wolfe, J.H.: Geophys. Res. Lett. **3**, 137–140 (1976).
90. Sterling, A.C.: J. Atmos. Solar Terr. Phys. **62**, 1427–1435 (2000).
91. Sterling, A.C., Hudson, H.S.: Astrophys. J. **491**, L55–L58 (1997).
92. Sterling, A.C., Hudson, H.S., Thompson, B.J., Zarro, D.M.: Astrophys. J. **532**, 628–647 (2000).

93. Stewart, R.T., Howard, R.A., Hansen, S.F., Gergely, T., Kundo, M.: Solar Phys. **36**, 219–231 (1974a).
94. Stewart, R.T., McCabe, M.K., Koomen, M.J., Hansen, R.T., Dulk, G.A.: Solar Phys. **36**, 203–217 (1974b).
95. Strong, K.T., Saba, J.L.R., Haisch, B.M., Schmelz, J.T. (eds.): The Many Faces of the Sun: A Summary of Results from NASA's Solar Maximum Mission, Springer-Verlag, New York (1999).
96. Sturrock, P.A.: Phys. Rev. **141**, 186–191 (1966).
97. Sturrock, P.A.: Solar Flares, Colorado Assoc. Univ. Press, Boulder, CO. (1980).
98. Sweet, P.A.: In Lehnert, B. (ed.), Electromagnetic Phenomena in Cosmic Physics, Cambridge Univ. Press, New York (1958).
99. Tappin, S.J., Hewish, A., Gapper, G.R.: Planet. Space Sci. **32**, 1273–1281 (1984).
100. Tappin, S.J., Howard, T.A.: Astrophys. J. **702**, 862–870 (2009).
101. Tappin, S.J., Roelof, E.C., Lanzerotti, L.J.: Astron. Astrophys. **292**, 311–329 (1994).
102. Thompson, B.J.: In Burch, J.L., Carovillano, R.L., Antiochos, S.K. (eds.), Sun-Earth Plasma Connections, AGU, Washington DC, p.42 (1999a).
103. Thompson, B.J., Cliver, E.W., Nitta, N., Delannèe, C., Delaboudiniére, J.-P.: Geophys. Res. Lett. **27**, 1431–1434 (2000).
104. Thompson, B.J., Gurman, J.B., Neupert, W.M., Newmark, J.S., Delaboudiniére, J.-P., St. Cyr, O.C., Stezelberger, S., Dere, K.P., Howard, R.A., Michels, D.J.: Astrophys. J. **517**, L151–L154 (1999b).
105. Thompson, B.J., Plunkett, S.P., Gurman, J.B., Newmark, J.S., St. Cyr, O.C., Michels, D.J., Delaboudiniére, J.-P.: Geophys. Res. Lett. **25**, 2461–2464 (1998).
106. Titov, V.S., Dèmoulin, P.: Astron. Astrophys. **351**, 707–720 (1999).
107. Tripathi, D., Bothmer, V., Cremades, H.: Astron. Astrophys. **422**, 337–349 (2004).
108. Uchida, Y.: Solar Phys. **4**, 30–44 (1968).
109. Warmuth, A., Vršnak, B., Aurass, H., Hanslmeier, A: Astrophys. J. **560**, L105–L109 (2001).
110. Warmuth, A., Vršnak, B., Aurass, H., Hanslmeier: Proc. SOLSPA: The Second Solar Cycle and Space Weather Euroconference (ESA SP-477), Vico Equense, Italy (2002).
111. Webb, D.: Proc. IAU Colloq. 133, Eruptive Solar Flares, pp.234–247 (1992).
112. Webb, D.F., Cheng, C.-C., Dulk, G.A., Edberg, S.J., Martin, S.F., McKenna-Lawlor, S., McLean, D.J.: In Sturrock, P.A (ed.), Solar Flares: A Monograph From Skylab Workshop II, Colo. Assoc. Uni. Press, Boulder, p.471 (1980).
113. Webb, D.F., Krieger, A.S., Rust, D.M.: Solar Phys. **48**, 159–186 (1976).
114. Wild, J.P., Smerd, S.F., Weiss, A.A.: Ann. Rev. Astron. Astrophys. **1**, 291–366 (1963).
115. Zarro, D.M., Sterling, A.C., Thompson, B.J., Hudson, H.S., Nitta, N.: Astrophys. J. **520**, L139–L142 (1999).
116. Zel'dovich, Y.B., Raizer, Y.P.: Physics of Shock Waves and High-Temperature Hydrodynamic Phenomena **1**, Academic, New York (1966).

Chapter 8
CME Onset and Initial Acceleration

To date, we have been unable to observe the onset mechanism of CMEs directly. While we review the many phenomena that are associated with CME onset and early evolution in Chap. 7, none of these are likely to be the onset mechanism of the CME itself. Some of them, however, are probably connected with it in some way, or caused by the same mechanism that launches the CME. It would appear that the Sun reveals its most important secrets reluctantly and it has succeeded in eluding the community on this crucial mechanism.

The sparsity of observational data means that CME onset and early evolution can only be described with models based on physical speculation, beginning with what we do know about the physics of the solar corona. We know, for example, that the plasma β in the corona is low, meaning it is a region dominated by magnetic activity. So it is reasonable to conclude that CME onset is a magnetic phenomenon. Many models have emerged describing the launch, formation and early acceleration of the CME but have been met with mixed success when comparing with coronagraph CME observations. The strength of models is that boundary conditions can be adjusted until they match the observations. This is also their weakness, as often the conditions are adjusted without any real physical justification. Any ideal model of an onset mechanism must not only accurately describe the CME as it appears in coronagraphs, but also the related associated phenomena described in Chap. 7. No model yet exists that is capable of achieving this objective but it is likely that different types of CME are best described by different models. Indeed, it is possible that most, if not all of the models discussed in this chapter may be appropriate to describe some CMEs under certain conditions.

For the purposes of onset and early acceleration description, CMEs may be divided into two types: low speed, low acceleration; and high speed, high acceleration. The first category involves CMEs that appear to take a long time to develop and move away from the Sun at a relatively gradual rate. The second type has a high rate of acceleration and appears to be explosively released from the Sun. The former type are often associated with streamers while the second are often accompanied by energetic phenomena such as solar flares. These two categories are probably separated also by the physics surrounding their early evolution, where the slow variety

T. Howard, *Coronal Mass Ejections: An Introduction*, Astrophysics and Space Science Library 376, DOI 10.1007/978-1-4419-8789-1_8,

may be regarded as "drifting" into the solar wind (recall from Sect. 1.2.8 that the natural state of the corona is toward expansion), while the other involves a rapid delivery of large quantities of energy.

8.1 Origins

Before we move onto the models describing the CME onset mechanism itself, it is helpful to review the state of the corona before the onset of the CME and some fundamental physics to explore what is reasonable for the eruption to occur. In this section we mostly regard CMEs of the explosive variety, addressing the more passive slow variety when describing the different mechanism later.

There are two sources of importance for the launch of an explosive CME–the triggering mechanism and the energy source for its early evolution. These may not necessarily be physically related or even spatially localised to each other. The general idea is that a magnetic structure, held in equillibrium in the corona (and lower in the solar atmosphere) by a delicate balance of gravity, magnetic and hydrodynamic pressure, is disrupted by some mechanism (the CME onset), which disturbs the equillibrium causing parts or all of the structure to erupt. During this process the erupting field must have access to very large amounts of energy in a short period of time, as it must accelerate a mass of order 10^{13} kg to a speed of order 1,000 km/s over a distance of the order of a solar radius (the early acceleration).

The disruption (onset) may take many forms. It could be some emerging structure from beneath the solar surface (herniation) or magnetic reconnection occurring anywhere in the field structure, or even a small reconfiguration of the local field. For the purposes of the early evolution of the CME the disruption source itself is largely irrelevant: whatever its form the Sun will eventually find one.

It is generally accepted that the CME launch is a coronal phenomenon, i.e. the CME is initiated in the corona. This means that the energy responsible for its early acceleration must arise from this regions. The energy associated with coronal electric currents is called the free magnetic energy [26]. We know that the plasma β in the corona is typically low, and so gas pressure alone cannot be solely responsible. This would mean that the CME's energy is predominantly provided by the magnetic field, and only the field associated with the corona is available to drive it. This energy builds up in the corona over time due to increasingly arising new magnetic fields and eventually, following a disruption (onset), is explosively released. This enables the emergence of new fields into a less energetic and complex region. Hence, the state of the coronal field before the launch of the CME governs the launch and behaviour of the CME. The following quotation is from Forbes et al. [18] and describes the general assumptions applied when regarding CME launch:

> Most CME initiation models today are based on the premise that CMEs and flares derive their energy from the coronal magnetic field. The currents that build up in the corona as a result of flux emergence and surface flows slowly evolve to a state where a stable equilibrium is no longer possible. Once this happens, the field erupts. If the eruption is sufficiently

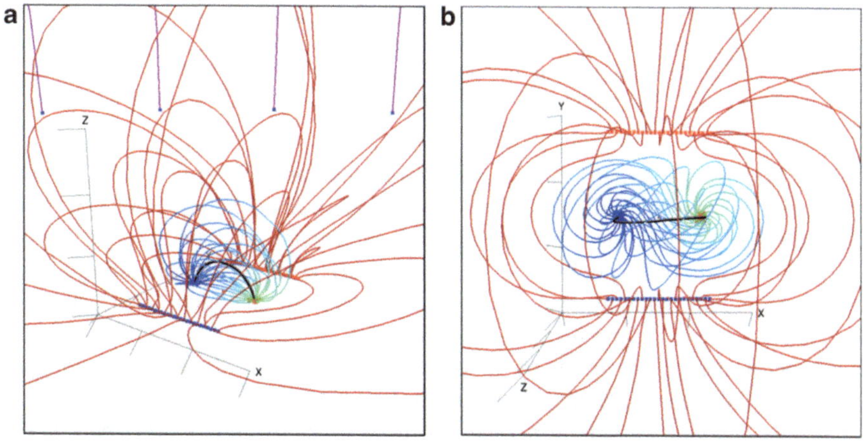

Fig. 8.1 A 3-D view of an ideal scenario involving a CME pre-launch magnetic field configuration. This consists of two structures: An underlying (core) field (*blue* and *green*) and an overlying straddling field (*red*) which acts to hold down the core. This is shown from two perspectives: (**a**) from an arbitrary 3-D direction, and (**b**) top–down, after a single twist has been introduced to the system. From Rachmeler et al. [50] and reproduced by permission of the AAS

> strong and the overlying fields not too constraining, plasma is ejected into interplanetary space. If strong magnetic fields exist in the erupted region, then bright, flare-like emission occurs. (p. 254 [18])

It is important to note that the images of the CME magnetic structure depicted in Chap. 7 are much too overly simplified. A better way of picturing the pre-launch CME is to consider it as two separate 3-D magnetic field structures as shown in Fig. 8.1. In this view the structure that will become the CME is the blue–green (known as a core) field which is held down by an overlying (red) field that straddles the core. When the CME launches it must either push the straddling field aside or stretch it out to infinity.

The important factor is the angle between the magnetic field in the core and its axis. When the two are aligned the tension force dominates causing it to shrink and reach compact equillibrium. When they are at a significant angle, magnetic pressure destabilises the core and so expansion becomes energetically favourable. Recall from Sect. 1.2.8 that CMEs erupt because they remove built-up magnetic energy from the Sun. This, however, raises a perplexing paradox. If the CME (the blue core field) were to stretch the overlying magnetic field (red) as it erupts and evolves, then the total energy of the system would *increase*, not decrease. Furthermore, as the CME continuously expands, the energy would increase further. Aly [1] and Sturrock [55] have theoretically demonstrated that the maximum energy state of any force-free magnetic field is the fully open one. So why would a CME spontaneously erupt into a more energetic state? There are, several means by which we may escape this paradox. Quoting Forbes et al. [18] again:

> First, the magnetic field may not be simply connected and contain knotted field lines. Second, it may contain field lines that are completely disconnected from the surface. Third,

> an ideal-MHD eruption can still extend field lines as long as it does not open them all the way to infinity. Fourth, an ideal-MHD eruption may be possible if it only opens a portion of the closed field lines. Fifth, small deviations from a perfectly force-free initial state might make a difference. And finally, a non-ideal process, specifically magnetic reconnection, invalidates the constraint. (p. 255 [18])

Another likely alternative arises from more recent modelling in 3-D, which has demonstrated that the straddling field can also be pushed aside to make way for the erupting magnetic structure beneath. This can occur with even the smallest disruption to the straddling field [50]. There may also be an interaction during launch between the two fields by way of magnetic reconnection. Which, if any, of these the Sun undergoes during the launch of the CME remains unknown, but there are a number of theoretical means by which we may launch a CME without stretching the associated magnetic fields out to infinity.

In this chapter, we briefly review the more popular models describing CME launch. Most of them require the occurrence of some destabilizing event, such as herniation or magnetic reconnection, and a mechanism by which great amounts of energy can be transferred to the erupting CME structure. We discuss the requirements of the model, the physics of their evolution and, where appropriate, their comparison with observations. For further reading on these mechanisms, good places to start are the review papers by Low [41] and Forbes et al. [18]. Lin et al. [32] provide a more recent review. We may generally divide the current theories into two classes: Those that require magnetic reconnection and those that do not, although some of these mechanisms may take place with or without magnetic reconnection.

8.2 Models Not Requiring Magnetic Reconnection

8.2.1 Magnetic Buoyancy

Magnetic buoyancy is probably the mechanism that is best for describing the evolution of the slow, slow-accelerating CMEs. It is based on the equillibrium between two natural effects [40]:

1. The expansion of the heated coronal plasma into interplanetary space;
2. The resistance of the coronal magnetic field against being opened.

The thermal energy in the corona is such that it overcomes the resistance from magnetic forces and gravity and is hence in a state of constant expansion. This is manifest in the form of the solar wind. According to Low [40], the Lorentz force exerted by the magnetic field enhances the expansion of the coronal plasma, but this is resisted if the magnetic field is curved towards the Sun. The corona is thus a combination of these two types of magnetic fields, resulting in regions in different states of plasma flow.

Magnetic buoyancy is therefore a consequence of the coronal plasma naturally expanding into the heliosphere. Parts of the corona are held back by magnetic pressure from closed fields within the corona, but will be released when these field structures are disrupted or re-oriented. One way to achieve this is via new magnetic flux emerging into the corona from the photosphere. This model has been advocated in a number of reports [39–41, 56, 61, 62] but some workers (e.g. [4]) argue that this would require an association between CMEs and large masses of material moving toward the Sun. This has not been observed in coronagraphs.

8.2.2 Flux Injection (Toroidal Instability)

The flux injection model [6] (also called toroidal instability) assumes that the CME is comprised of a magnetic flux rope. It is initially at equilibrium, and erupts as a result of poloidal magnetic flux being injected into the rope. Figure 8.2 gives a diagram of the conditions for this model.

The loop contains a toroidal current I_t and is encircled by a poloidal magnetic field B_p surrounded by a solar magnetic field B_s. Once launched, the total current **J** and field **B** give rise to a Lorentz force **J**×**B** driving the loop away from the Sun. The forces governing the balance of the flux rope F are given by

$$F = \frac{\ell I_t^2}{c^2 R} = \left[\ln\left(\frac{8R}{a}\right) + \frac{1}{2}\beta_p - \frac{B_t^2}{2B_{pa}^2} + 2\left(\frac{R}{a}\right) - 1 + \frac{\zeta}{2} \right] + F_g + F_d, \tag{8.1}$$

where B_t is the average toroidal magnetic field inside the loop, R and a are the major and minor radii of the loop, F_g and F_d are the forces of gravity and drag, B_{pa} is the poloidal field at a, (i.e. $B_{pa} = 2I_t/ca$), and ζ is the internal inductance. $\beta_p = 8\pi(p - p_a)/B_{pa}^2$ (p_a is the average pressure inside the loop and p is the coronal pressure). R is related to the distance of the transient in the z direction Z by

$$R = \left(Z^2 + s_0^2\right)/2Z, \tag{8.2}$$

where s_0 is the separation of the footpoints, given by $s_0 = 2R\sin\theta_0$. ℓ is the length of the flux rope, $\ell = 2r(\pi - \theta_0)$.

While this may be adequate to describe the evolution of the CME (the topic of Chap. 9), it does little to describe the launch mechanism itself. We can surmise that the ejection is caused by magnetic flux being injected from the photosphere causing an instability, but it is difficult to reconcile how the instability and initial acceleration can be accomplished without a kink instability and/or processes involving magnetic reconnection. It may be possible if a great amount of twist can be achieved in the flux rope over a small enough timescale. This must be accomplished in such a way such that the flux rope does not kink, because if it did then it would be more appropriately described by the kink instability model (Sect. 8.2.3). Once the initial acceleration is complete, however, the flux rope comprising the CME can then be driven by a Lorentz force. This means that the energy required to launch the CME arises from the photosphere. Flux Injection has been criticised because the energetic surge from the photosphere required for the eruption does not reconcile with photospheric ob-

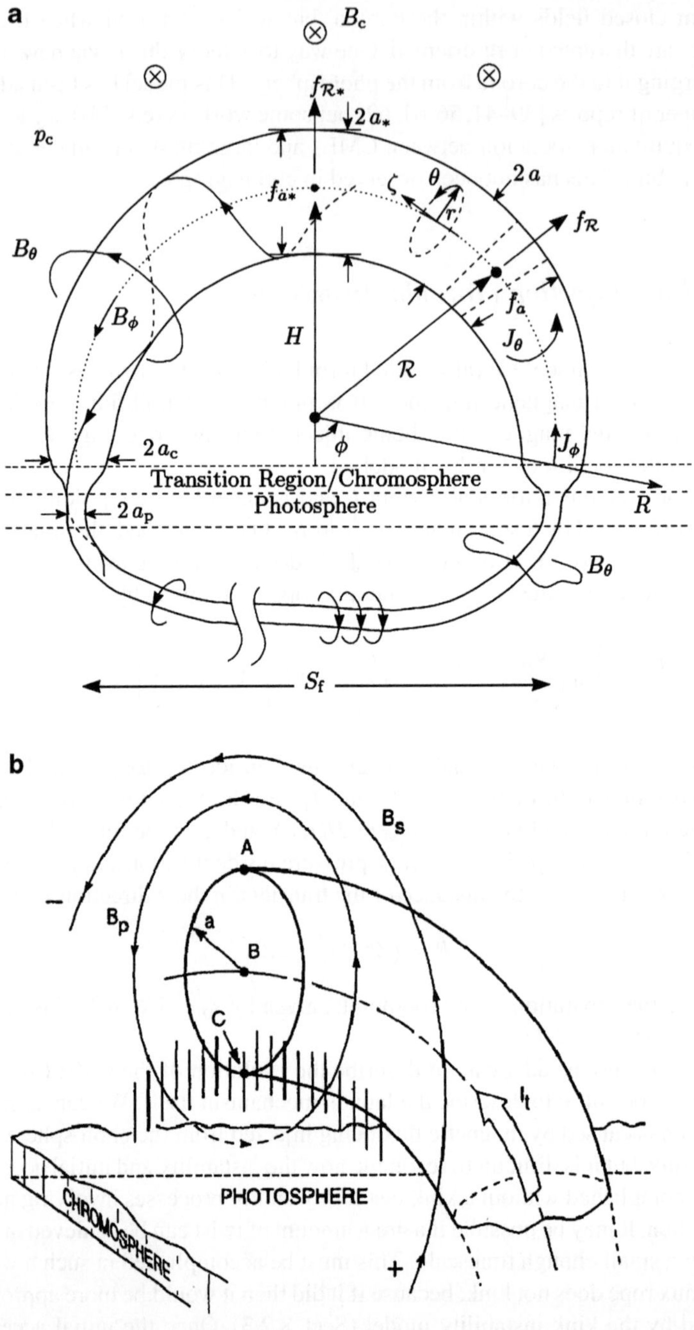

Fig. 8.2 (**a**) The conditions surrounding the flux injection model, modified from Chen [7]. The subscripts t and p refer to toroidal and poloidal respectively [52] (Reproduced by permission of the AAS). (**b**) The geometry of the flux rope itself [7]

servations, which show no such burst of energy during CMEs and flares. It has also been criticised as the flux rope contains no twist, which has often been observed and is considered essential for the physical evolution of CMEs [16, 51]. Chen and co-workers have replied to these criticisms in a number of publications [8–10, 28, 29].

Most recently, an attempt has been made to use photospheric observations to identify the energy available from the photosphere for launching the CME. Schuck [52] determined this energy for a single CME observed on 12 September 2000 (considered an ideal event for a flux rope model) using *SOHO*/MDI data . They could not provide sufficient energy required for the CME launch. It remains unresolved whether the flux injection model is appropriate to describe the launch of some types of CME.

8.2.3 Kink Instability

The Kink Instability and Mass Loading (Sect. 8.2.4) models attempt to describe not only the launch of the CME, but also the process by which magnetic energy may be stored in the corona over long periods. To put it simply, Kink Instability continually twists magnetic flux ropes until they erupt through the overlying magnetic field. Publications discussing this model include Török and Kliem [58, 59], Fan and Gibson [13, 14] and Rachmeler et al. [50]. This model may erupt CMEs with or without magnetic reconnection, so under some conditions this may be regarded as part of Sect. 8.3 as well.

Kink Instability involves a flux rope system in the corona with footpoints fixed in the photosphere. The footpoints turn, creating a twist (kink) in the flux rope which increases in tension until a critical value is achieved, resulting in an instability [22]. Simulations [13, 14, 57] have shown that following "supercritical twist", two current sheets form. The first is helical in structure and wraps around the rising kinked flux rope, while the second is a vertical sheet below the flux rope. The formation of the vertical current sheet enables the eruption, which then may or may not occur via magnetic reconnection (Rachmeler et al. [50] show a case where an eruption may occur without reconnection, where in 3-D the overlying strapping field is pushed aside to make way for the erupting core CME). Figure 8.3 shows a sequence with a flux rope eruption under the Kink Instability from Fan [12]. Here also the overlying red strapping field is pushed aside.

The Kink Instability model can describe many phenomena observed in CME-associated solar phenomena, including soft x-ray sigmoids [31] and the twisted structure that often appears in CMEs and prominences (see Fig. 7.4). It also enables the build-up of significant quantities of mass and results in an energy release from the system – simulations by Török and Kliem [59] have shown that up to 25% of the energy may be released. It has, however, generally been regarded as a plausible explanation only for confined events [20], as some simulations have shown that a current sheet does not always form. Also, the amount of twist required to produce an explosive eruption is considered by some to be nonphysical.

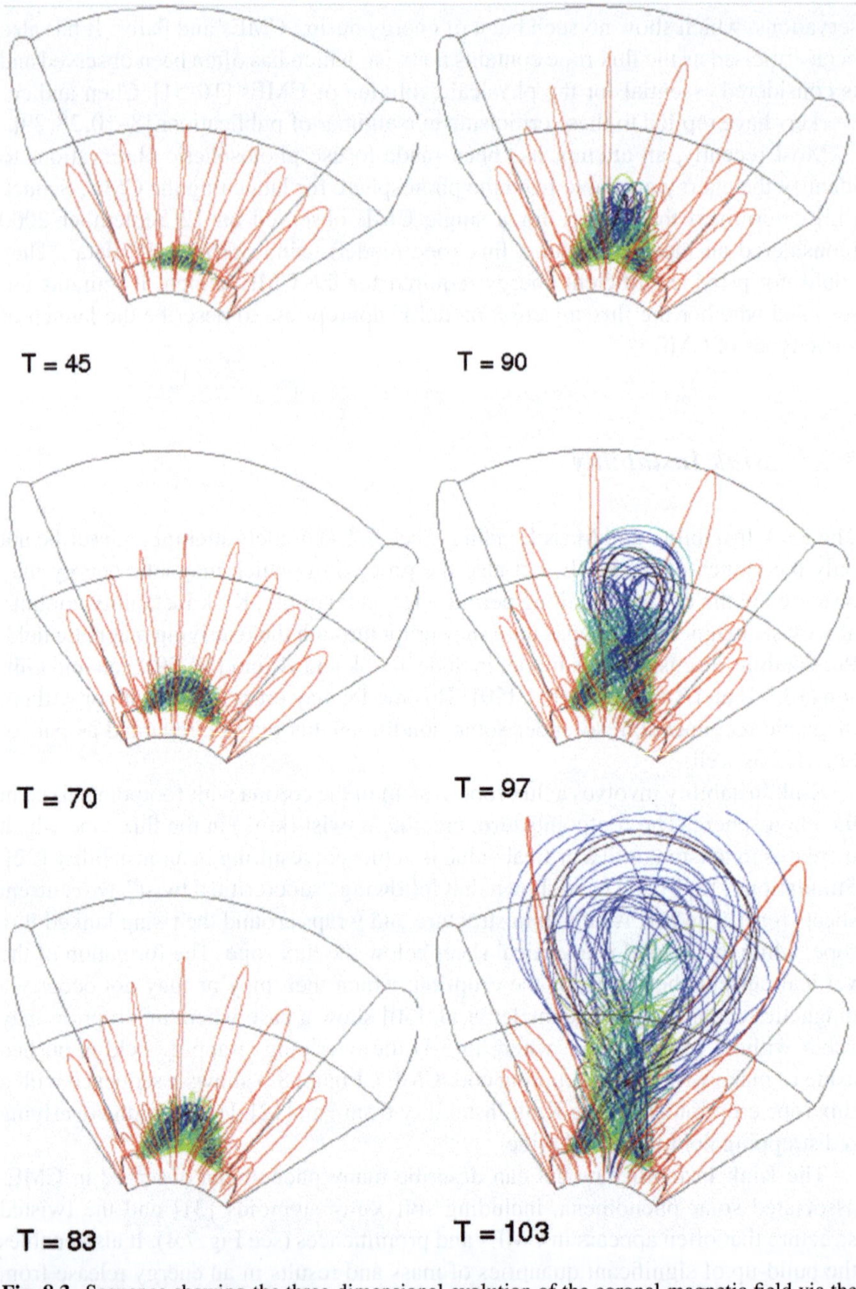

Fig. 8.3 Sequence showing the three-dimensional evolution of the coronal magnetic field via the Kink Instabilility Model [12]. The heavy blue/green lines represent the kinked flux rope, which erupts through the overlying magnetic structure (*red*), which is pushed aside. (Reproduced by permission of the AAS)

8.2.4 Mass Loading (Unloading)

Mass Loading is based on the premise that the mass structures within prominences are sufficient to sustain equilibrium between the gravitational binding energy and the magnetic forces expanding the prominence. Provided the structure is sufficiently stable, this allows the build-up of mass and magnetic energy to large quantities over time. When the mass structure is disturbed such as via a reorientation of the magnetic field or a re-distribution of the mass, the equilibrium is broken and an eruption ensues. By this method, sufficient quantities of mass and magnetic energy may be accumulated in the corona before the eruption. One means by which the mass may accumulate is via dips in the prominence structure, where the accumulation occurs under gravity. In this case a disruption may lead to the accumulated mass spilling over across the rest of the prominence. This is a possible [25, 30] but not required scenario [24]. The theoretical means by which free magnetic energy may be stored in the corona has been established by Klimchuk and Sturrock [26] and the mass loading due to prominences is discussed by Fong et al. [15] and Zhang and Low [63].

The magnetic field of a prominence may assume two forms: normal or inverse. This depends on whether the prominence has its field pointing in the same or opposite direction. Fong et al. [15] demonstrate that for the inverse case the prominence sheet takes the role of anchoring a magnetic flux rope that, if not for its anchor, would rise into the atmosphere. So when the mass is lost from the prominence an eruption occurs. Zhang and Low [63] demonstrate that the normal prominences are subject to a Lorentz force that is always radially outward, and so needs less azimuthal flux to store enough energy to be energetically beneficial for launch. They conclude that prominences with normal configuration are more likely to erupt. Figure 8.4 shows modeled fields of the two types of prominences.

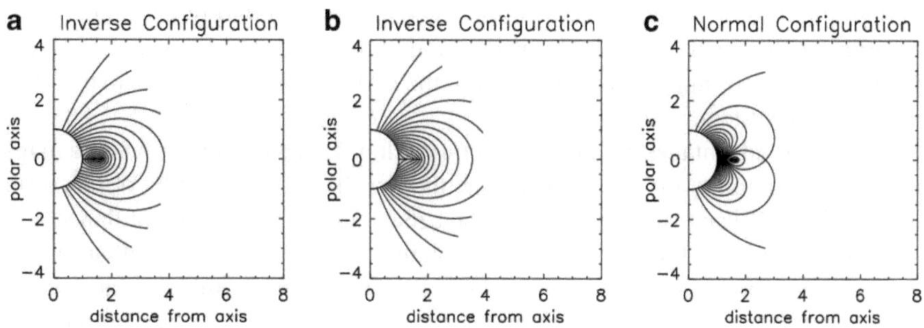

Fig. 8.4 Modeled magnetic field configurations for the two types of prominences, governed by the Mass Loading model, with the prominence sheet in the equatorial plane [15]. (**a**) An inverse prominence with the base normal flux distribution in the form of a dipole potential field. The resulting Lorentz force arises from the tension force being outward near the base but inward near the rim. (**b**) Inverse prominence with the dipole field increased so that the Lorentz force is radially inward along the prominence sheet. (**c**) Normal prominence with the dipole field reduced to allow closed field lines to form over the current. (Reproduced by permission of the AAS)

8.3 Models Requiring Magnetic Reconnection

8.3.1 Tether Cutting (Tether Release)

Sturrock [54] demonstrated that a simple loss of equilibrium alone cannot be responsible for the onset of large magnetic structures such as CMEs. He therefore stated that the most likely eruption mechanism must therefore be some type of instability. Reconnection was offered as one such instability, which was reinforced by Moore and Roumeliotis [49] for solar flares using observations and theory.[1] These and other reports have led many to the belief that magnetic reconnection may be the mechanism that triggers CMEs, but more importantly, it may also be responsible for providing energy for their early acceleration. Reconnection is an attractive concept as it provides the means for the rapid transfer of energy from magnetic fields to the surrounding plasma and it provides an instability required to disrupt and disconnect the magnetic fields comprising the CME. An argument advocating magnetic reconnection as a launching mechanism for CMEs is given by Antiochos and DeVore [3].

The Tether Cutting process works by reducing the strapping tension force in the overlying field by magnetic reconnection beneath. It begins with a core field which becomes strongly sheared as a result of changing pressure forces changing the angle between the field and the flux rope axis. The sheared field runs almost parallel to the neutral line [46]. This is embedded in a surrounding envelope of less-sheared field, which runs over the top of the neutral line (the strapping overlying field, Moore and Roumeliotis [49] refer to it as the envelope). The core field rises initially due to the imbalance between the pressure from the sheared field against the tension from the overlying field, and the surrounding field collapses into the vacancy beneath. A current sheet is created there, which drives magnetic reconnection across the sheared field lines underneath the core. This reduces the strapping field by breaking it in places, allowing the core field to penetrate. The erupting core further stretches remaining field lines, which excites further reconnection providing more energy to the eruption. Figure 8.5 shows a basic diagram of this process by Sturrock [54]. By this process it does not cost energy to erupt the CME. In fact, it reduces overall the energy of the system.

Among the advantages of magnetic reconnection, Tether Cutting provides a magnetic configuration to the pre-launch stage that has been confirmed by observation [11,21,23,47,48] and can also describe the behaviour of certain types of flares (e.g. two-ribbon flares) as they evolve. The problem with Tether Cutting is that historically much of the work has been conducted with solar flares in mind, and while it may be appropriate for flares, it cannot accommodate for the energy required to launch a CME. According to Antiochus et al. [4], Tether Cutting cannot provide sufficient energy to the launching CME to overcome the Aly-Sturrock (open field)

[1] It is noteworthy that Moore and Roumeliotis [49] regarded the flare as the source of the CME, and so the two were indistinguishable to them at the time.

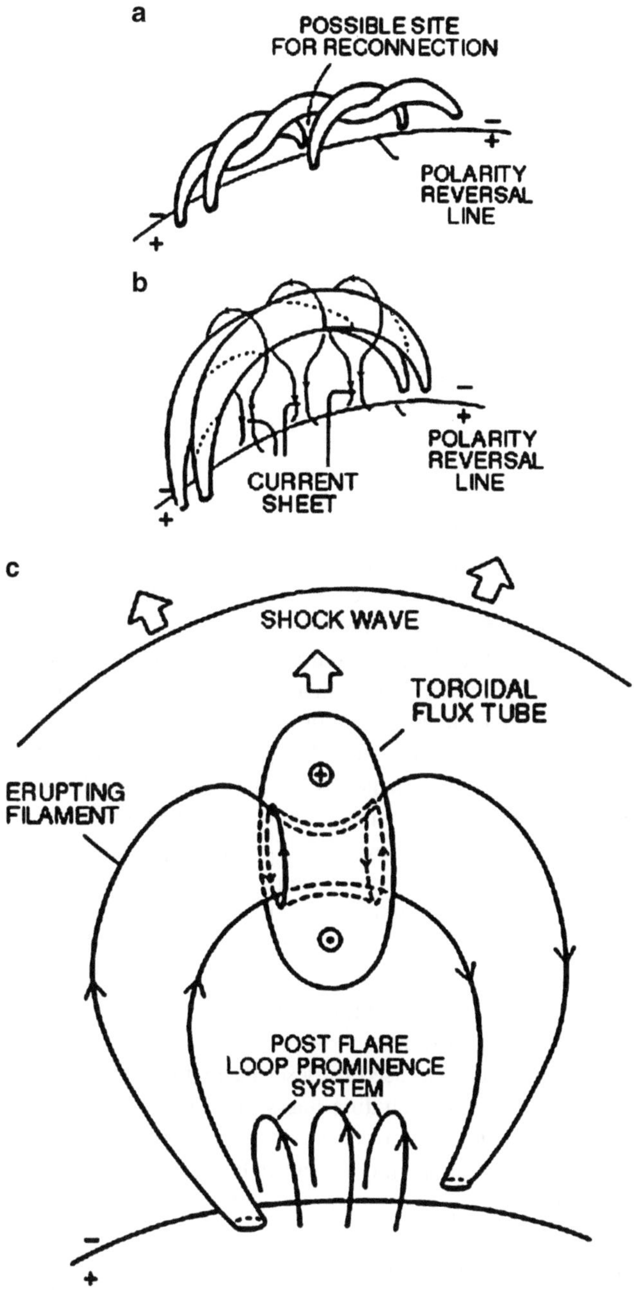

Fig. 8.5 Basic diagram of the Tether Cutting model according to Sturrock [54]. (**a**) The pre-launch magnetic field configuration associated with a prominence, showing where the reconnection may occur triggering the CME onset. (**b**) The launch of the structure after magnetic reconnection and once the structure is no longer connected to the photosphere. (**c**) Final form of the magnetic structure, following the onset. (Reproduced with kind permission of Springer Science and Business Media)

energy limit. That is, even with magnetic reconnection in play there is still an energy gain between the pre-launch and post-launch coronal magnetic fields. It is also difficult to establish a CME onset mechanism with tether cutting.

8.3.2 Breakout

An alternative to the Tether Cutting model, the magnetic breakout model [4] requires magnetic reconnection between a sheared arcade and neighbouring magnetic field structures during the eruption. Breakout gets around the Aly-Sturrock energy limit by assuming the CME takes the form of a closed plasmoid, which originates from an existing magnetic field structure and is "broken off" from the coronal field, with its energy provided via magnetic reconnection from the surrounding field (which remains closed). It also interacts with the strapping field via reconnection and pushes it aside during launch. A diagram of the process is shown in Fig. 8.6.

Figure 8.6a shows the initial flux system for the breakout. There are three neutral lines and four flux systems. The flux systems are (blue) a central arcade across the equator, (green) two arcades associated with neutral lines at mid-latitudes, and (red) a polar flux system overlying the other arcades. A large shear is introduced along the equatorial neutral line, the source of which may be either photospheric flows or magnetic flux emergence. The shear opens the equatorial (blue) flux system without opening the others, thereby avoiding the Aly-Sturrock constraint. Magnetic pressure provided by photospheric shear flow forces the expansion of the equatorial flux system (Fig. 8.6b), which expands as it is provided with additional energy via magnetic reconnection with the surrounding (red and green) flux systems. As it expands the restraining field is moved aside, producing further reconnection and providing more energy to the system. Eventually, a second reconnection process occurs at the base of the shear channel (Fig. 8.6c), and the original flux system is separated from the Sun by a current sheet. This second reconnection may provide the mechanism for the associated solar flare, which is known to lag CME onset in timing. Finally, (Fig. 8.6d) the remaining field relaxes as the disconnected CME flux system continues to move outward.

The Breakout model can be used to describe many of the phenomena known to be associated with CMEs, including flares [5] and the classic three-part CME [43]. It also provides a means to overcome the Aly-Sturrock limit by reducing the energy in the corona without opening the entire field, and allows a means by which energy may be rapidly and increasingly provided to the CME structure as it evolves. Breakout can also be accomplished without a reconnection onset mechanism, as for onset it is not the reconnection that is important but the instability that allows the penetration of the strapping field in 3-D. Rachmeler et al. [50] demonstrated breakout-style eruption in the absence of reconnection via an ideal MHD instability that they called herniation (i.e. a structure emerging from below the photosphere). To date, however, the breakout model has yet to be compared rigorously with any observed CME. This comparison is probably approaching quickly, as 3-D versions of the Breakout model are emerging [42].

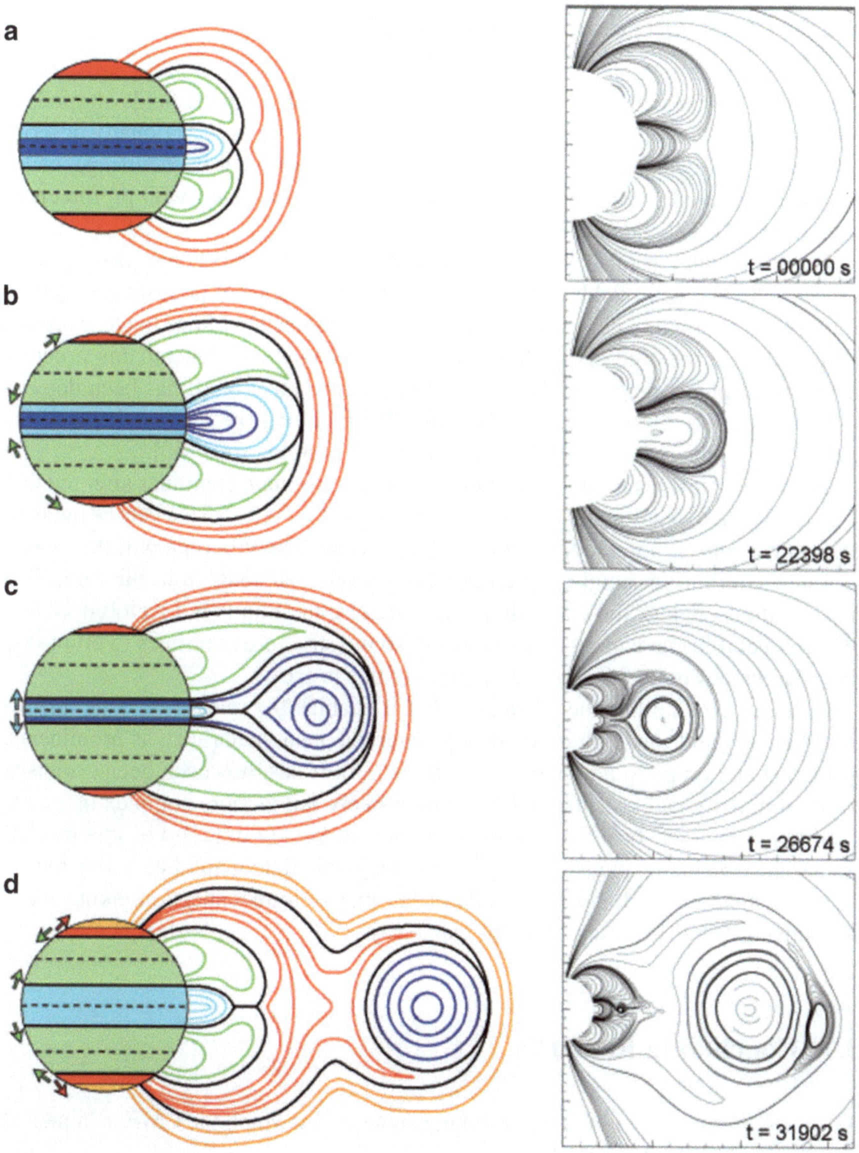

Fig. 8.6 The breakout model (Lynch et al. [42]). (*Left*) Diagrams of the main four stages of field evolution. (**a**) Initial topology, (**b**) Shearing of the field and initiation, (**c**) Flare reconnection starting deep in the shear channel and (**d**) Reconnection allowing the relaxation of the field. The colours for the chosen fields indicate (*blue*) the central arcade straddling the equator, (*green*) two arcades associated with neutral lines at mid-latitudes, and (*red*) a polar flux system (*Right*) Corresponding field lines plots from the MHD simulation of MacNeice et al. [44]. (Reproduced by permission of the AAS)

8.3.3 Flux Cancellation (Catastrophe)

Magnetic Flux Cancellation [45] is the disappearance of magnetic fields of opposite polarity at their separating neutral line. The term is purely observationally descriptive (i.e. it conveys no information about the physics of what is occurring), and it has been observed at all regions of the Sun [38] and associated with prominences [36], flares [27,37] and CMEs [34,53].

Flux Cancellation (also known as the Catastrophe Model) begins with prominences. The magnetic fields within prominences often have a polarity opposite to that of the surrounding photospheric region. This enables field lines to close around the prominence material suspending it above the photosphere. Beneath this structure at the photosphere, magnetic flux is canceled by reconnection. It has been demonstrated using force-free equilibria that cancellation along the neutral line (i.e. the site of highly sheared field) leads to the formation of a flux rope [60].

The theory states that flux cancellation at the photosphere continues after the flux rope has formed, resulting in a loss of equilibrium [17,19,33]. This causes the field lines to become compressed, increasing the pressure. The flux rope will then move to a new equilibrium height, which can be several solar radii from the Sun. This initiates the launch of the CME. In reality, because the Sun is not a closed system this new equilibrium cannot be maintained, so the structure continues to erupt into the solar wind. Figure 8.7 shows a diagram of this process.

Flux cancellation has the advantage that it enables the release of large quantities of energy and can explain physically observed phenomena such as prominence formation and eruption and helmet streamers [35]. It has, however, been criticised for its simplicity and speculated that it would likely fail in more complex magnetic conditions. This has been challenged by some workers (e.g. [2]). For this mechanism to be considered plausible, an accurate magnetic field map of an active region must first be identified and the details of the magnetic reconnection taking place determined. *SDO*/HMI may have this capability.

8.4 Concluding Remarks

As shown in this chapter, the proposed mechanisms by which the CME is launched are many and varied. They do have some similarities. They are all magnetically based, all involve the rapid transfer of great amounts of energy, and all can be used to describe certain types of CMEs. None, however, can adequately describe the launch of all types of CME nor can any one model account for every associated phenomena. For example, some (tether cutting) explain the action of the flare very well while others (kink instability) account for the x-ray sigmoid prior to eruption. Some (breakout) are better at describing the transfer of the very large amounts of energy required for the more massive CMEs, while others (magnetic buoyancy) are better at describing the slower, less energetic CMEs.

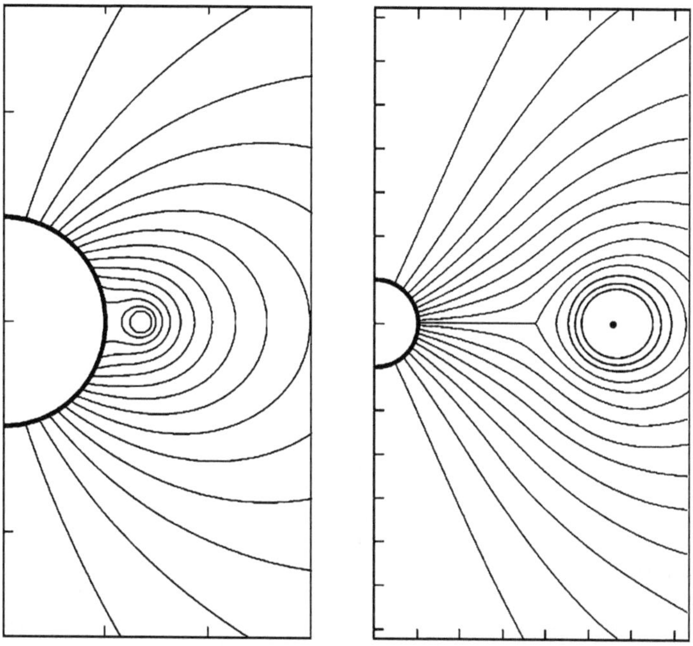

Fig. 8.7 Simplified azimuthally-symmetric flux rope model showing the launch of a structure via flux cancellation [18]. This shows the transition from (*left*) the flux rope close to the Sun to (*right*) after its transition. The line behind the erupting structure is the current sheet forming behind. (Reproduced with kind permission of Springer Science and Business Media)

While it is generally accepted that an instability is probably involved in the onset process, no consensus has been reached on the fundamental mechanism required for that instability. Is magnetic reconnection required, or can the process be achieved without reconnection? If reconnection is involved, where and at what stage in the process does it occur? Can a single mechanism be used to describe all CMEs, or do different CMEs erupt via different mechanisms? And finally, how can we equate the associated phenomena with the launch mechanism? Given the nature of the Sun it is likely that more than one mechanism is responsible for CME onset and early acceleration. Models have shown, for example, that certain processes can take place with or without reconnection, and that the onset mechanism may not be physically related to the acceleration. One may even go as far to say that the instability leading to the CME onset is not relevant, as the Sun will find any that are available. Careful comparison between data and models will help to resolve these critical questions.

References

1. Aly, J.J.: Astrophys. J. **283**, 349–362 (1984).
2. Amari, T., Aly, J.J., Mikic, Z., Linker, J.: Astrophys. J. **671**, L189–L192 (2007).
3. Antiochos, S.K., DeVore, C.R.: In Burch, J.L., Carovillano, R.L., Antiochos, S.K. (eds.), Sun-Earth Plasma Connections, AGU, Washington DC, pp.113–120 (1999).

4. Antiochos, S.K., DeVore, C.R., Klimchuk, J.A.: Astrophys. J. **510**, 485–493 (1999).
5. Aulanier, G., DeLuca, E.E., Antiochos, S.K., McMullan, R.A., Golub, L.: Astrophys. J. **540**, 1126–1142 (2000).
6. Chen, J.: Astrophys. J. **338**, 453–470 (1989).
7. Chen, J.: J. Geophys. Res. **101**, 27499-27520 (1996).
8. Chen, J.: Space Sci Rev. **95**, 165–190 (2001).
9. Chen, J., Krall, J.: J. Geophys. Res. **108**, 1410, doi:10.1029/2003JA009849 (2003).
10. Chen, J., Santoro, R.A., Krall, J., Howard, R.A., Duffin, R., Moses, J.D., Brueckner, G.E., Darnell, J.A., Burkepile, J.T.: Astrophys. J. **533**, 481–500 (2000).
11. Falconer, D.A.: J. Geophys. Res. **106**, 25185–25190 (2001).
12. Fan, Y.: Astrophys. J. **630**, 543–551 (2005).
13. Fan, Y., Gibson, S.E.: Astrophys. J. **589**, L105–L108 (2003).
14. Fan, Y., Gibson, S.E.: Astrophys. J. **609**, 1123–1133 (2004).
15. Fong, B., Low, B.C., Fan, Y.: Astrophys. J. **571**, 987–998 (2002).
16. Forbes, T.G.: J. Geophys. Res. **105**, 23153–23166 (2000).
17. Forbes, T.G., Isenberg, P.A.: Astrophys. J. **373**, 294–307 (1991).
18. Forbes, T.G., Linker, J.A., Chen, J., Cid, C., Kóta, J., Lee, M.A., Mann, G., Mikić, Z., Potgieter, M.S., Schmidt, J.M., Siscoe, G.L., Vainio, R., Antiochos, S.K., Riley, P.: Space Sci. Rev. **123**, 251–302 (2006).
19. Forbes, T.G., Priest, E.R., Isenberg, P.A.: Solar Phys. **150**, 245–266 (1994).
20. Gerrard, C.L., Hood, A.W.: Solar Phys. **214**, 151–169 (2003).
21. Hagyard, M.J., Moore, R.L., Emslie, A.G.: Adv. Space Res. **4**, 71–80 (1984).
22. Hood, A.W., Priest, E.R.: Geophys. Astrophys. Fluid Dyn. **17**, 297–318 (1981).
23. Hoyng, P., Duijveman, A., Machado, M.E., Rust, D.M., Svestka, Z., Boelee, A., de Jager, C., Frost, K.T., Lafleur, H., Simnett, G.M., van Beek, H.F., Woodgate, B.E.: Astrophys. J. **246**, L155 (1981).
24. Karpen, J.T., Antiochos, S.K., Hohensee, M., Klimchuk, J.A., MacNeice, P.J.: Astrophys. J. **553**, L85–L88 (2001).
25. Kippenhalm, R., Schlüter, R.: Z. Astrophys. **43**, 36 (1957).
26. Klimchuk, J.A., Sturrock, P.A.: Astrophys. J. **385**, 344-353 (1992).
27. Kosovichev, A.G., Zharkova, V.V.: Astrophys. J. **550**, L105–L108 (2001).
28. Krall, J., Chen, J., Duffin, R.T., Howard, R.A., Thompson, B.J.: Astrophys J. **562**, 1045–1057 (2001).
29. Krall, J., Chen, J., Santoro, R.: Astrophys. J. **539**, 964–982 (2000).
30. Kuperus, M., Raadu, M.A.: Astron. Astrophys. **31**, 189 (1974).
31. Kliem, B., Titov, V.S., Török, T.: Astron. Astrophys. **413**, L23–L26 (2004).
32. Lin, C.-H., Gallagher, P.T., Raftery, C.L.: Astron. Astrophys. **516**, doi:1051/00046361/200913167 (2010).
33. Lin, J., Forbes, T.G., Isenberg, P.A., Demoulin, P.: Astrophys. J. **504**, 1006–1019 (1998).
34. Lin, J., Raymond, J.C., van Ballegooijen, A.A.: Astrophys. J. **602**, 422–435 (2004).
35. Linker, J.A., Mikić, Lionello, R., Riley, P.: Phys. Plasmas **10**, 1971–1978 (2003).
36. Litvinenko, Y.E., Martin, S.F.: Solar Phys. **190**, 45–58 (1999).
37. Livi, S.H.B., Martin, S.F., Wang, H., Guoxiang, A.: Solar Phys. **121**, 197–214 (1989).
38. Livi, S.H.B., Wang, J., Martin, S.F.: Australian J. Phys. **38**, 855–873 (1985).
39. Low, B.C.: Astrophys. J. **251**, 352–363 (1981).
40. Low, B.C.: Phys. Plasmas **1**, 1684–1690 (1994).
41. Low, B.C.: Solar Phys. **167**, 217–265 (1996).
42. Lynch, B.J., Antiochos, S.K., DeVore, C.R., Luhmann, J.G., Zurbuchen, T.H.: Astrophys. J. **683**, 1192–1206 (2008).
43. Lynch, B.J., MacNeice, P., Antiochos, S.K., Zurbuchen, T., Fisk, L.: Astrophys. J. **617**, 589–599 (2004).
44. MacNeice, P., Antiochos, S.K., Phillips, A., Spicer, D.S., DeVore, C.R., Olson, K.: Astrophys. J. **614**, 1028–1041 (2004).
45. Martin, S.F., Livi, S.H.B., Wang, J.: Australian J. Phys. **38**, 929–959 (1985).
46. Moore, R.L., Hagyard, M.J., Davis, J.M.: Solar Phys. **113**, 347–352 (1987).

47. Moore, R.L., Hagyard, M.J., Davis, J.M., Porter, J.G.: In Uchida, Y., Canfield, R.C., Watanabe, T., Hiei, E. (eds.), Flare Physics of Solar Activity Maximum 22, Springer-Verlag, Berlin, p.324 (1991).
48. Moore, R.L., Labonte, B.J.: In Dryer, M., Tandberg-Hanssen, E. (eds.), Solar and Interplanetary Dynamics, Reidel, Boston, p.207 (1980).
49. Moore, R.L., Roumeliotis, G.: In Svestka, Z., Jackson, B.V., Machado, M.E. (eds.), Eruptive Solar Flares, Springer-Verlag, New York p.69–78 (1992).
50. Rachmeler, L.A., DeForest, C.E., Kankelborg, C.C.: Astrophys. J. **693**, 1431–1436 (2009).
51. Rust, D.M.: J. Geophys. Res. **106**, 25075-25088 (2001).
52. Schuck, P.W.: Astrophys. J. **714**, 68–88 (2010).
53. Somov, B.V., Kosugi, T., Hudson, H.S., Sakao, T., Masuda, S.: Astrophys. J. **579**, 863–873 (2002).
54. Sturrock, P.A.: Solar Phys. **121**, 387–397 (1989).
55. Sturrock, P.A.: Astrophys. J. **380**, 655–659 (1991).
56. Sturrock, P.A., Weber, M., Wheatland, M.S., Wolfson, R.: Astrophys. J. **548**, 492–496 (2001).
57. Titov, V.S., Démoulin, P.: Astron. Astrophys. **351**, 707–720 (1999).
58. Török, T., Kliem, B.: Astron. Astrophys. **406**, 1043–1059 (2003).
59. Török, T., Kliem, B.: Astrophys. J. **630**, L97–L100 (2005).
60. van Ballegooijen, A.A., Martens, P.C.H.: Astrophys. J. **343**, 971–984 (1989).
61. Wolfson, R., Bongani, D.: Astrophys. J. **483**, 961–971 (1997).
62. Wu, S.T., Guo, W.P., Wang, J.F.: Solar Phys. **157**, 325–248 (1995).
63. Zhang, M., Low, B.C.: Astrophys. J. **600**, 1043–1051 (2004).

Chapter 9
CME Evolution

In Chap. 8, we reviewed a number of mechanisms that may be responsible for the launch of the CME. While no consensus had been reached as to which mechanism is responsible (or whether they all may be), they were all based on a number of known facts about the environment surrounding CME onset and on the conditions that must apply for the launch to be physically feasible. For example, the CME was driven by magnetic forces and the energetic state of the corona after launch must be lower than that beforehand. The main reason for why the launch mechanism had not been proven was the lack of observations of the onset region of the CME. Magnetic fields in the corona are very difficult to measure.

Likewise, once the CME is clear of the Sun the mechanism by which it propagates through the heliosphere is not well understood. This is for the same reasons as for the launch mechanism – sparsity of observational evidence and the flexibility of models. As with the launch, we do have physical constraints, physically reasonable assumptions, and some observations to provide us with clues. Much of the observational evidence has traditionally been from in-situ observations and observations near the Sun such as coronagraphs. More recently we have had the advantages of heliospheric imaging, and there is also IPS to provide us with information on the CME density structure en route. To date however, as with the onset, CME evolution must be described using models with the advantages and disadvantages this brings.

This chapter reviews the most common mechanisms for describing the evolution of the CME (ICME) through the heliosphere. We will divide them into three categories:

1. **Aerodynamic Drag** assumes the behaviour of the CME is entirely governed by its interaction with the solar wind and there is no significant intrinsic influence.
2. **Shock Dynamics** assumes the evolution of the CME is governed by the shock front formed ahead of the CME, as this is the only structure that interacts with the interplanetary medium in a significant way.
3. **Separate Ejecta** Treats the ICME is an additional ejecta that is injected in the background solar wind. Some models regard the ICME as a plasma ejecta, others treat it as a magnetic flux rope.

Each of these may be regarded as different aspects of essentially the same MHD problem, but with different bases and assumptions.

T. Howard, *Coronal Mass Ejections: An Introduction*, Astrophysics and Space Science Library 376, DOI 10.1007/978-1-4419-8789-1_9,

9.1 Theoretical Basis

Let us consider what we do know about ICMEs and the physical environment into which the CMEs are launched. Firstly, beyond a few solar radii, the surrounding plasma β exceeds one (i.e. becomes a high-β plasma), meaning that the magnetic field plays less of a significant role than the plasma and surrounding fluid. This means that magnetic processes such as reconnection and magnetic pressure have less of an influence on the evolution of the CME than the hydrodynamics of the plasma. It also means that energy-carrying Alfvén mode waves can more effectively convert to fast mode waves, where they may more efficiently dissipate their energy into the corona. It is also known, however, that ICMEs are typically low-β phenomena (near 1 AU, values as low as 0.1 are common), so we have a typically low-β intrinsic magnetic field moving through a typically high-β environment.

We know from coronagraph observations that shocks often form early in the evolution of the CME and that a well-defined shock with a build-up of compressed plasma (known as a sheath) is often observed ahead of the ICME near 1 AU (Sects. 2.3.1 and 7.3.1). We also know that many ICMEs at 1 AU contain highly structured magnetic flux ropes called magnetic clouds (Sects. 2.3.1.1 and 7.3.2). This means that magnetic structure that is created near the Sun continues to maintain its structure at large distances. We also know about the ionic composition, temperature and bulk plasma speed of CMEs at 1 AU (Sect. 2.3.1), along with the behaviour of associated energetic particles (Sect. 7.2.9), which provides empirical data on the CME there. Finally, with heliospheric imagers (Sect. 2.7.2), IPS (Chap. 6) and radio bursts we can monitor the kinematic evolution and trajectory of ICMEs (Chap. 5). It is not entirely clear as to whether heliospheric imagers and IPS observe the CME itself or the sheath. Recently developments are beginning to provide some clues.

Any working model that describes the evolution of ICMEs through the interplanetary medium must not only appropriately explain all of the observations of ICMEs, but also connect them with the Sun in a physically reasonable way, ideally connecting it with one of the launch mechanisms discussed in Chap. 8. In the following sections we review some popular models for CME evolution, but it should be noted that while some models have emerged with seemingly all of the appropriate physics, to date no evolution model matches the above definition of a working model. We begin with the review of the medium through which the CME must travel – the solar wind.

9.2 The Solar Wind

We present here a very brief discussion of the solar wind with a focus on its interaction with a transiting ICME. A great number of texts on the solar wind can be found in the literature. Some good reviews include Parker [34] for a review of early work, and Meyer-Vernet [29] or any of the proceedings from the Solar Wind Conferences over the years (most recently Solar Wind 12 (http://www.lesia.obspm.fr/SW12/)) for more recent reviews.

The possibility that particles may be continuously streaming from the Sun was developed by Birkeland [4] and Lindemann [24]. Chapman [10] calculated that the corona must extend large distances away from the Sun, but it was Parker who came up with the idea of the solar wind [33]. He recognised that the corona was such an efficient heat conductor that its plasma not only escaped the gravitational effects of the Sun, but it did so supersonically. In the same publication, Parker demonstrated that as the solar wind expands through the heliosphere it drags the solar magnetic field with it, which is anchored to the Sun and so corotates. The result is that the field resembles an Archimedian spiral. This is now known as the Parker spiral and forms the fundamental structure of the interplanetary magnetic field to this day. Figure 9.1 shows a simplified diagram of the solar wind with the interplanetary magnetic field.

As the solar wind is supersonic, it forms shocks with relatively stationary objects in the heliosphere. Planetary magnetospheres, for example, have bow shocks on their sunward side and the structure and behaviour of their magnetospheres (mostly at high altitudes) is strongly influenced by the behaviour of the solar wind. The solar wind is also the reason for the tails observed behind comets, as when they approach the Sun their icy composition begins to vapourise, which is then swept away from the Sun by the solar wind. This is why they always point away from the Sun even when moving an in antisunward direction.

It is this medium through which ICMEs must propagate once they leave the Sun. It is generally believed that when the CME is faster than the surrounding medium, the interplanetary magnetic field (IMF) ahead of the CME is compressed and drapes

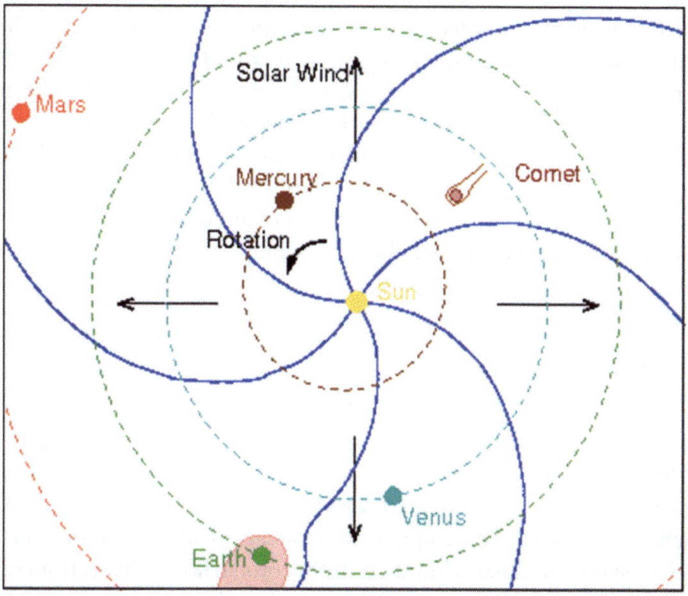

Fig. 9.1 Simplified diagram of the ecliptic plane solar wind (*black*) with the embedded interplanetary magnetic field (*blue*). The Sun, the planets and their orbits are shown

around the CME structure. Solar wind material accumulates in the sheath, but interaction between the flux rope comprising the CME and the surrounding plasma is only possible via magnetic reconnection (Sect. 2.6.1).

9.2.1 *Interplanetary Magnetic Field*

As the IMF originates from the Sun which is known to be highly dynamic, there is no predictable fixed structure for it on the local scale. Locally it is observed to vary greatly (see Figs. 7.13 and 7.17), and different regions of the heliosphere behave differently at different times of the solar cycle. The most significant differences in IMF behaviour is between open field lines and closed field lines. Figure 7.1 shows how near the Sun this may change at solar minimum and maximum and Fig. 7.4 shows how the field may change in the presence of an active region. Coronal holes are known to be sources of open magnetic field lines and these may occur at low latitudes as well as high. These structures extend out into the heliosphere, but other solar features (such as high-latitude coronal holes) make an appearance as well.

Keeping things in their most general sense in order to gain a simple idea of the relationship between ICMEs and their environment, there are known general trends in the field that can be followed. For example, because of the nature of its expansion the strength (magnitude) of the radial component of the field is generally expected to follow an inverse square law with distance. As mentioned in Sect. 6.2.2.1, this is because each dimension varies at a rate proportional to the distance from the Sun R, and magnetic flux is proportional to the cross-sectional area through which the field flows, that is the radial component of B_{IMF} varies with R^{-2}. Typical magnetic fields at 1 AU are of the order of 5 nT, but over 100 nT can be achieved with large magnetic disturbances. Near the Sun this translates to values of the order of 1 Gauss (10^5 nT).

It remains to be seen how strongly small-scale variations in the IMF affect the ICME structure and kinematic evolution. Only recently with measurements from SMEI and *STEREO* have we been able to begin to directly observe the changes that occur in the ICME en route through the solar wind.

9.2.2 *Solar Wind Plasma*

The plasma comprising the solar wind moves supersonically radially outward from the Sun (Fig. 9.1). The same is true of the elements of the magnetic field – while the overall structure is a spiral, each element of the spiral moves radially outward. There are known to be two main components of solar wind plasma: fast and slow flows. Fast flowing solar wind arises from coronal holes, that is from regions on the Sun where the coronal magnetic fields are open. It is not known exactly where the slow solar wind comes from, but it may arise from the opening up of closed coronal field lines. It is often associated with closed field structures. During solar minimum when the magnetic field lines are more simplified, these regions tend to be divided with the slow solar wind around the equator and fast around the poles. During solar maximum this is more complex (Fig. 7.1b).

As with the IMF, local variations in the solar wind plasma may be numerous and complex. However, there are trends in the plasma properties that can be used in a more general sense. The plasma density, for example, decreases with a trend that is approximately with the square of the distance from the Sun. Within 1 AU this trend is more closely approximated by $r^{-2.45}$ [1]. That is

$$\rho = \rho_0 r^{-2.45} \tag{9.1}$$

where ρ is the density and ρ_0 is a constant boundary density value. In order to conserve conservation of mass and momentum across a fixed solid angle, the following must therefore be the case for the general variation of solar wind speed V_{sw} with distance,

$$V_{sw} = V_0 r^{0.45}, \tag{9.2}$$

where V_0 is a boundary speed. This is because across a solid angle the rate of flow of mass is given by the change in sector density σ, where $\sigma = \rho r^2$ [43]. Beyond around 1.0 AU the density varies with approximately R^{-2} and the solar wind speed is approximately constant.

9.3 Aerodynamic Drag

Aerodynamic drag is based on two fundamental assumptions:

1. ICME evolution is entirely governed by plasma and fluid dynamics (i.e. the magnetic field plays no role);
2. ICME dynamics always move to achieve kinematic equilibrium with the solar wind.

Under these assumptions, the theoretical problem of ICME evolution becomes one of momentum transfer between the ICME and its surroundings, and of kinematic equilibrium. Fundamentally, this means that a fast CME will slow down and a slow CME will speed up until its speed matches that of the solar wind. Once there, the ICME structure will simply cruise along at the solar wind speed, carrying accompanying structures along with it. Aerodynamic drag models typically regard the difference in speed between the ICME and the solar wind as the dominating parameter. The tendency for fast CMEs to slow down and slow CMEs to speed up has been observed with coronagraphs [50] and IPS [26].

In this section, we consider two versions of the aerodynamic drag model: one where solar wind material is allowed to accumulate with the ICME and the other where it is not. The fundamental difference between the two is that the mass M is a function of time with one model, where it remains constant with the other.

The former is often referred to as the "Snow Plough" model as snow piling up in front of a snow plough is analogous to the solar wind piling up in front of the ICME shock front. We refer to the latter simply as the "Drag" model. Both models consider the entire mass of the combined ICME structure as observed in the heliosphere, that is they include material in the sheath region as well as the ICME itself.

9.3.1 Snow Plough

The following is from Tappin [43]. The snow plough model is based on the conservation of momentum as the ICME sweeps up the solar wind ahead of it in the sheath which, in turn, is accelerated. This can be represented as a set of two coupled differential equations:

$$\frac{d^2R}{dt^2} = -\frac{dM}{dt}\frac{(v_c - v_{sw})}{M} \tag{9.3}$$

$$\frac{dM}{dt} = \sigma\Omega(v_c - v_{sw}), \tag{9.4}$$

where σ is the solar wind speed sector density (i.e. mass per unit solid angle per unit radial distance $= \rho R^2$), Ω is the heliocentric solid angle, and $M = M(t)$. These can be solved numerically using the basic relationships for density and solar wind speed.

9.3.2 Drag

The paper by Cargill [8] provides a description of the drag model. The drag force is governed by the drag coefficient, C_D which is a dimensionless parameter of order unity for motion in a uniform medium in Cartesian geometry. Assuming the ICME has a cylindrical shape of diameter Δr, he assumed the deceleration was dependent on the difference in speed between the ICME and that of the surrounding solar wind. In units of solid angle Ω and sector density σ [43], this relationship was shown to be

$$\frac{d^2R}{dt^2} = -\gamma C_D(v_c - v_{sw})|v_c - v_{sw}|, \tag{9.5}$$

where

$$\gamma = \frac{\sigma\Omega}{M + \sigma\Omega\frac{\Delta r}{2}}, \tag{9.6}$$

and the mass M in this case is constant. As with the Snow Plough (9.3) and (9.4), these can be solved numerically.

One important parameter intrinsic to both the Snow Plough and Drag models is what Cargill terms the "virtual mass" of the ICME [8, 9]. When one measures the mass of a CME from white light data (coronagraph, heliospheric imager) the images are processed in such a way that only the excess mass is calculated. A difference image, for example, removes the background intensity to enhance the intensity of the CME. However, removing the background also removes a component of the CME mass. This may be restored by estimating the solar wind density across the CME volume and adding it to the excess mass. Cargill [8, 9] defines this as $M_v \sim \rho\tau/2$ where τ is the volume of the ICME.

9.3.3 *Comparison With Data*

A number of studies have been performed to test the validity of the Snow Plough and Drag models. Tappin [43] showed that the differences between the two did not become significant until the ICME was a substantial fraction of an AU from the Sun (around 0.75 AU). Howard et al. [20] studied two events comparing LASCO and SMEI data and found that one of them could be described very well using the models. The other event, as with the one studied by Tappin, required an additional acceleration for the ICME to match the model. Figure 9.2 shows the results from these studies. These and other studies suggest that aerodynamic drag alone can accommodate for the evolution profile for some ICMEs, but not them all.

9.4 Shock-Based

The previous section discussed models that regarded the ICME as a massive body moving through a background solar wind. The dynamics were described in terms of momentum transfer and equilibrium between the two bodies. Here we look at models that regard the ICME as a shock wave moving through the solar wind. In such circumstances the physics of the ICME are governed by those describing shock evolution, and so the ICME is regarded as a perturbation in the surrounding medium. Some of these models consider the magnetohydrodynamics of the medium while others do not. Further information on the theory of ICME-propagated shocks can be found in Hundhausen [21].

While many of the models are originally based on an incorrect premise (that the solar flare creates the CME), dynamics based on shock mechanics may not be an inaccurate description of ICME evolution at large distances from the Sun, or at least the density structure of the ICME as observed by heliospheric imagers and IPS. Recall that the surrounding plasma β is high at these distances and so the dynamics are expected to be driven by the fluid, i.e. the mechanics of the shock (see Sect. 7.3.1) may be appropriate for some ICMEs. A number of these models have been developed [12, 48], but we review here three of the more popular models for ICME shock propagation. Keep in mind that the following models do not regard the CME as having any intrinsic structure, so are not physically accurate.

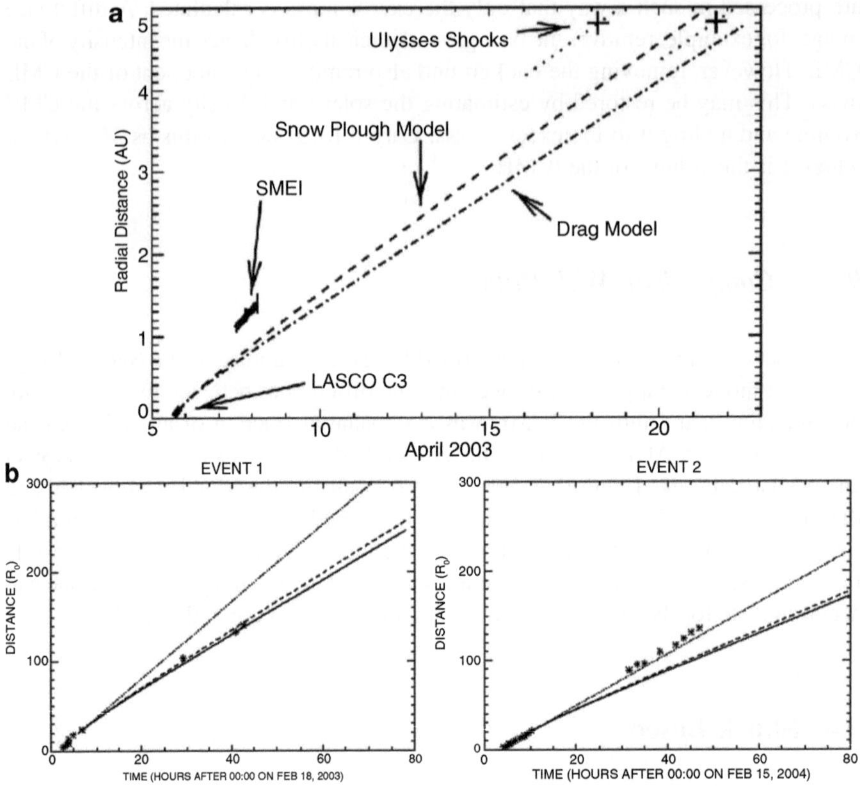

Fig. 9.2 Distance-time plots for three events studied comparing LASCO and SMEI data with the aerodynamic drag model. (**a**) Results from a single event in April 2003 reported by Tappin [43]. Labeled are the LASCO and SMEI height-time datasets, the arrival times of two shocks at *Ulysses* near 5 AU and the results from the Snow Plough and Aerodynamic Drag Models. (Reproduced with kind permission of Springer Science and Business Media). (**b**) Results from two events observed in February 2003 and 2004 (Howard et al. [20]). The crosses are the LASCO datasets, asterisks are SMEI datasets, and the heavy and dashed curves are for the Snow Plough and Aerodynamic Drag models respectively. Note that Event 1 (2003) shows an excellent fit with the drag models, while Event 2 requires an additional acceleration. This event is almost aligned with the constant speed line (dotted), which shows the trajectory of the CME it if were to retain the same speed it had when leaving the LASCO field of view. (Reproduced by permission of the AAS)

9.4.1 STOA and ISPM

The Shock Time Of Arrival (STOA) model [14] uses the theory of blast waves [13] and drives a shock wave based on an eruption from a point source. It assumes the flare is the source of the blast wave and uses the soft x-ray flare duration as one of its boundary conditions. The shock speed is derived from the Type II radio frequency drift (Sect. 7.3.3) via the surrounding plasma frequency. The structure of the shock

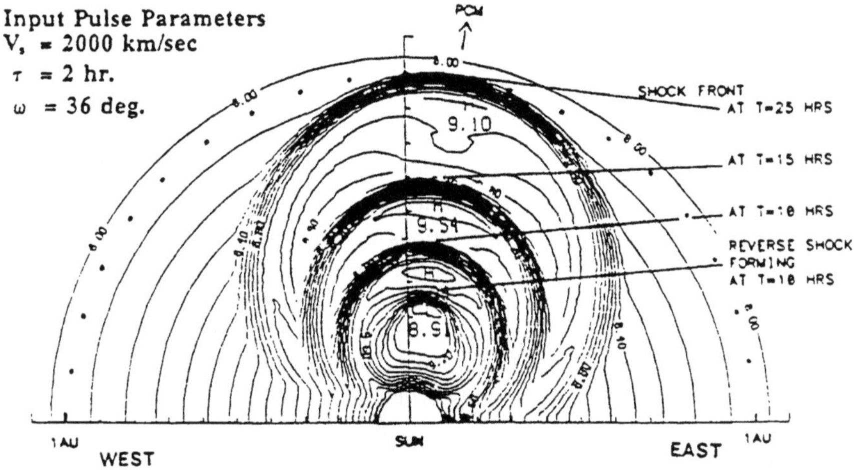

Fig. 9.3 Results from the ISPM model for a single event with a central angle (PCM) offset of 9°E, shown at times of several hours after launch [39]. The reverse shock is shown for the time at 10 h. The contours are in units of solar mass flux. Listed input pulse parameters are: V_s = shock speed, τ = input duration, ω = pulse width. (Reproduced with kind permission of Springer Science and Business Media)

is derived from a cosine function along with the relative variation in speed along the shock front. The background solar wind varies in the radial direction, but not in longitude. Results from this model include Dryer [14], Smart et al. [38] and Smart and Shea [37].

The Interplanetary Shock Propagation Model (ISPM) [39] is derived from the premise that the energy input into the solar wind is the driving parameter. It uses similar boundary conditions to STOA and makes use of the Rankine-Hugoniot equations (Sect. 7.3.1) with an additional equation set for induction [47]. The shock is introduced as a perturbation to the surrounding solar wind lasting of the order of a couple of hours and a realistic shock speed selected. Figure 9.3 shows the results with this model for a single simulated event.

Studies with ISPM include Smith and Dryer [39, 40] and Smith et al. [41]. In a recent space weather survey by Webb et al. [44] ISPM provided the most accurate predicted arrival times when compared with the other surveyed models by a small margin. These results suggest that at least for a subset of ICMEs the shock model can describe ICME evolution at large distances from the Sun in some cases.

9.4.2 HAF

The Hakamada-Akasofu-Fry kinematic solar wind model, version 2 (HAFv2) predicts solar wind conditions using observations at the Sun [18, 19]. It projects fluid parcels outward from the rotating Sun along fixed radials at successive time steps, in an inertial frame. The velocity is radial and the speed distribution on the inner boundary is inhomogeneous. Therefore, parcels move outward with different speeds along fixed radials as the Sun rotates beneath the radial grid. The frozen field condition mandates that, along each radial, the faster parcels do not move through or pass slower parcels. Therefore if the kinematic flow is modified, in that the fluid parcel positions are adjusted to account for the stream–stream interaction as fast parcels (solar wind streams) overtake slower ones. So if the speed gradient along a radial is steep enough, corotating interactions and interplanetary shocks are formed. This is how ICMEs are worked into the background solar wind model – they are disturbances in the background medium driven by solar conditions.

The model uses source surface maps derived from synoptic solar observations [2, 3] to establish the background solar wind conditions. It is also driven by proxies for energy released during solar events, which are obtained from solar flare, white light CME and radio Type II burst information, to model CMEs and interplanetary shock propagation. It produces chronological sequences of the ecliptic-plane IMF and other solar wind parameters.

The initiation of the CME is achieved by modulating the inner boundary velocity field. The solar wind speed on the inner boundary at 2.5 solar radii is increased exponentially to a maximum value and allowed to fall back exponentially to the pre-event value at a slower rate. The initial shape of the CME is determined by this Gaussian distribution in velocity, which is circularly-symmetric on the source surface. As the CME moves outward from the Sun, its shape is modified by its interaction with the background solar wind through which it propagates. Figure 9.4 shows some results for HAFv2 on an event observed in January 2007.

Publications utilising HAFv2 include Fry et al. [16–18], McKenna-Lawlor et al. [28] and Intriligator et al. [22]. As with the STOA and ISPM models, HAFv2 has been utilized as a prediction tool for space weather forecasting. Its "fearless forecasts" [16] are routine in the tools used by NOAA and the Space Weather Prediction Center (SWPC). More recently, HAFv2 has been compared with heliospheric image (SMEI) observations [20, 44, 45], with favourable results.

9.5 Separate Ejecta

In this section two models are discussed which treat the CME as a separate ejecta added to a background solar wind. The first introduces a dense structure with no intrinsic magnetic field. The second considers the ICME as an expanding magnetic flux rope and considers the physics acting upon it from the surrounding medium, but does not produce a complete model of the solar wind. Under certain conditions these two can describe very well the trajectory of ICMEs. The first, the so-called ENLIL model is a major tool used to predict ICME arrival times at the Earth by NOAA/SWPC. The latter is the extension of the Flux Injection model (Sect. 8.2.2) into interplanetary space.

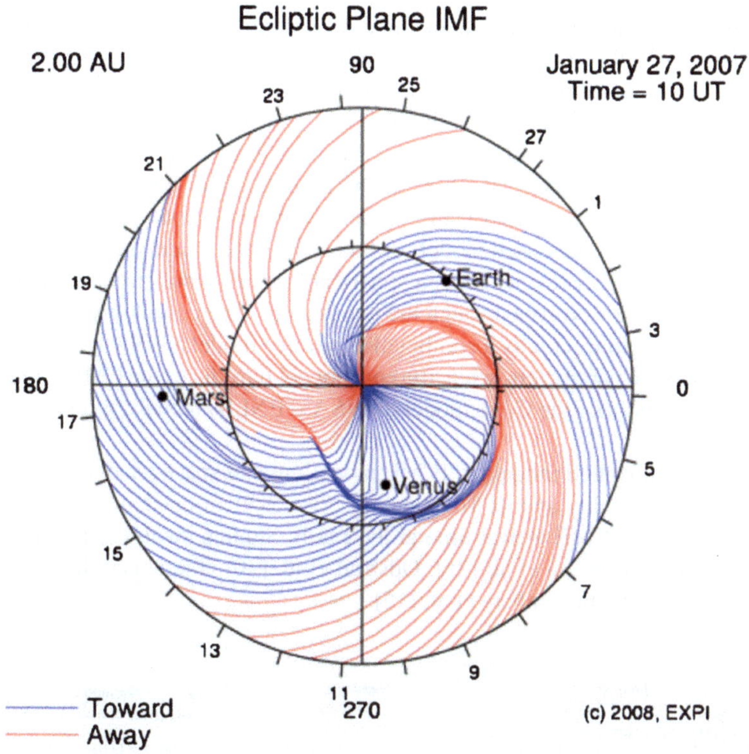

Fig. 9.4 HAFv2 simulation of an ICME observed in January 2007 [45]. The interplanetary magnetic field (IMF) is shown in the ecliptic plane with a radius of 2 AU, with the magnetic field direction shown a red (*southward*) and blue (*northward*). The locations of the Earth, Venus and Mars are shown. The CME is shown as the disturbance toward the bottom right. (Reproduced with kind permission of Springer Science and Business Media)

9.5.1 ENLIL

ENLIL [32] is based on ideal MHD equations that are solved for plasma mass, momentum, energy density, and magnetic field. The inner boundary is at 21.5 solar radii which can interface with a number of time-dependent numerical solar MHD codes [31]. Contemporary versions of ENLIL are commonly driven by the so-called WSA (Wang-Sheeley-Arge) empirical model [3].[1] The WSA/ENLIL combination is commonly used, publicly available from the Community Coordinated Modeling Center (CCMC) web site (http://ccmc.gsfc.nasa.gov).

[1] WSA uses ground-based magnetogram observations to feed a magnetostatic potential field source surface model which extends the coronal magnetic field out to 2.5 solar radii.

As with the interplanetary shock models described in the previous section, ENLIL begins with a model of the solar wind, and the CME can be introduced as an over-pressurised plasma cloud in addition to the background. This is different because the shock models treat the ICME as a disturbance in the interplanetary medium alone. While ENLIL does not assume a shock is always formed, under certain conditions, a shock can form in the model. CME parameters (e.g. location, size, speed) are estimated using models based on coronagraph observations of CMEs. One commonly used model for this is the cone model (see Xie et al. [49] or Sect. 5.3). The CME launched as a spherical homogeneous structure, but it becomes distorted as it evolves through the heliosphere. The commonly-used version of ENLIL is therefore actually a combination of three models: WSA feeds the background solar wind parameters, while the cone model introduces the CME. Figure 9.5 shows the results from the Cone model + WSA – ENLIL model for a single event in January 2007.

ENLIL is a favoured model for ICME modelling and space weather forecasting because of its flexibility and its MHD nature. Also as mentioned above, it performs well in predicting ICME arrival time at the Earth. Studies involving ICMEs using ENLIL include Luhmann et al. [25], Odstrcil et al. [30], Taktakishvilli et al. [42], Webb et al. [45] and Felkenberg et al. [15].

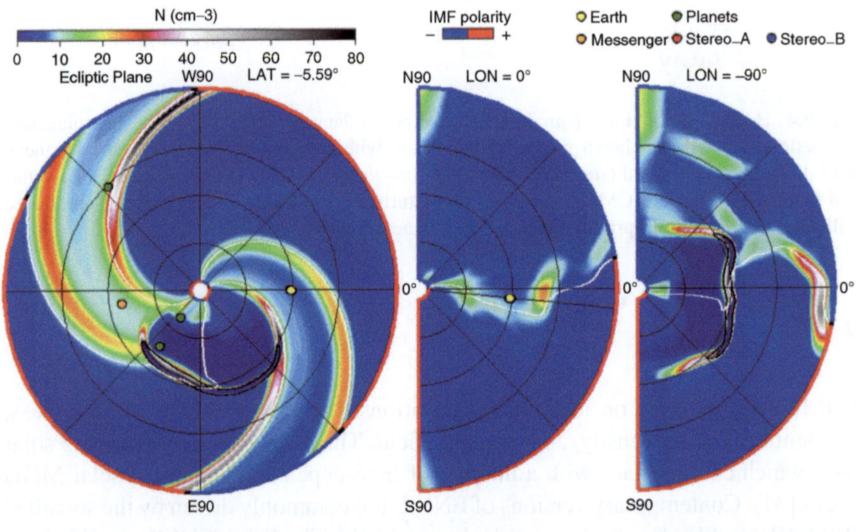

Fig. 9.5 ENLIL simulation of an ICME observed in January 2007 [45]. Shown are solar wind density maps in three different projections: (*left*) Ecliptic plane, (*middle*) meridional along the Sun–Earth line, (*right*) meridional along the east limb relative to the Earth. The location of the Earth is indicated by the yellow circle, the other planets by red circles and the locations of the *Messenger* and *STEREO* spacecraft by the orange, red and blue circles respectively. The ICME is shown as the dark (high density) structure moving along the E90 meridian. The maps extend out to 2.5 AU. (Reproduced with kind permission of Springer Science and Business Media)

9.5.2 Flux Injection

Chen [11] describes the extension of the Flux Injection model into interplanetary space. As described in Sect. 8.2.2 it begins with a toroidal flux rope in a surrounding poloidal field which is subjected to a Lorentz force. Near the Sun it is described by (8.1) but away from the Sun the forces of gravity, coronal and internal pressure and solar magnetic field are no longer significant. So the equation is reduced to

$$F = \frac{\ell I_t^2}{c^2 R} = \left[\ln\left(\frac{8R}{a}\right) - \frac{B_t^2}{2B_p a^2} - 1 + \frac{\zeta}{2} \right] + F_d. \tag{9.7}$$

The boundary conditions from each of the parameters are determined empirically and the model evolved numerically.

Figure 9.6 shows a plot of the Lorentz force (A), drag force (B) and net force (dashed line) for a simulated flux rope evolving for 80 h. Here the force of gravity was assumed by Chen [11] to be

$$F_d = C_D \rho_a m_i a (V_{sw} - V)|V_{sw} - V|, \tag{9.8}$$

where ρ_a is the density inside the loop and m_i is the internal mass. Notice that the net force is reduced from a positive value to zero around 24 h after launch, which is an indicator of the time at which the drag force begins to become the dominant force driving the evolution of the ICME. This indicates that in this case the ICME was under a state of acceleration for 24 h after launch before the drag forces began to take over. Given that the approximate speed of this simulated ICME was 550 km/s this would mean that the ICME was around 70 solar radii (0.3 AU or 20° elongation) away from the Sun before it stopped accelerating and started decelerating down to the solar wind speed. While it has not been confirmed whether the Lorentz force is responsible, a long-duration acceleration has been observed in IPS and heliospheric

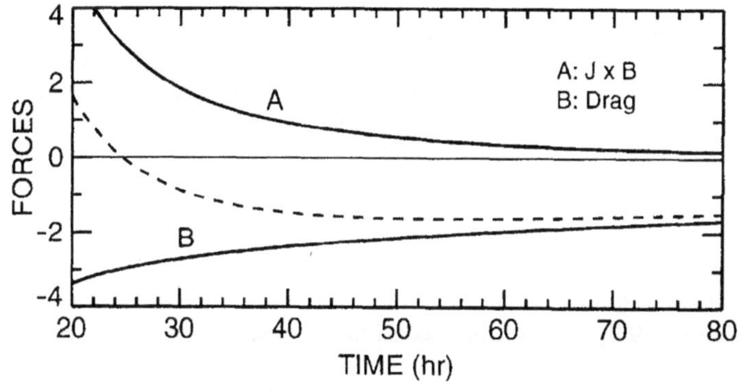

Fig. 9.6 Plot of the Lorentz (**a**), drag (**b**) and net (dashed) forces on a CME using the Flux Injection Model of Chen [11]. The total time duration is for 80 h and the units of force are 10^{12} N

image measurements of some ICMEs at large distances [20, 26, 27, 43]. See for example Fig. 9.2. It is noteworthy that when Howard et al. [20] applied the flux rope model to the two events the second event (Event 2 in Fig. 9.2b) showed a good match of the model with the data.

9.6 Summary

While it is not yet known which evolution model is the most appropriate for describing ICME evolution, like the onset models from Chap. 8 some models describe some types of ICME very well. For example the shock driven model may describe the faster ICMEs well because these are most likely to form shocks in the solar wind, while the flux rope model may best describe those ICMEs that are observed to continue to accelerate well into the solar wind. The most accurate model may be a combination of the models above, say an ENLIL-type CME plasma cloud with a flux rope interior. Research into this area is ongoing as more sophisticated models and datasets continue to be formed.

9.7 The End of the Road – The Fate of CMEs

Section 7.4.1 discusses corotating interaction regions (CIRs) which are regions of high pressure in the solar wind brought about by the interaction between fast and slow streams. Because they originate from solar wind streams they assume a structure similar to that of a Parker spiral. When several CIRs interact in this manner they form a complex structure called a Merged Interaction Region (MIR). MIRs can also form when the reverse shock upstream of a CIR catches up with the forward shock downstream [23] and a variety of other means. They form largely because of the expansion of CIRs and/or ICMEs and because of overall speed differences between different solar wind structures. According to Burlaga [5]:

> Corotating MIRs can be formed in several ways, including (1) the interaction between two distinct streams, (2) the evolution of a compound stream, (3) the interaction between a corotating pressure wave (CIR without a stream) and a corotating stream that overtakes it, and (4) the interaction between two corotating pressure waves. The stream–stream interactions can be of at least three types: (a) a fast corotating stream overtaking a slower corotating or transient flow, (b) two identical corotating streams, "twin streams", and (c) a narrow, fast stream followed by a broad, slower stream (p. 139 [5]).

MIRs are commonplace at large distances from the Sun, beyond around 5 AU where the Parker spirals begin to assume structures resembling circles centred at the Sun (see Fig. 9.7), but they are occasionally observed much closer in at around 1 AU [6].

A good review of MIRs can be found in Burlaga [5]. In this text he demonstrates that CMEs may eventually form MIRs. Sooner or later all of the faster flowing compression regions must catch up with the slower regions ahead of them

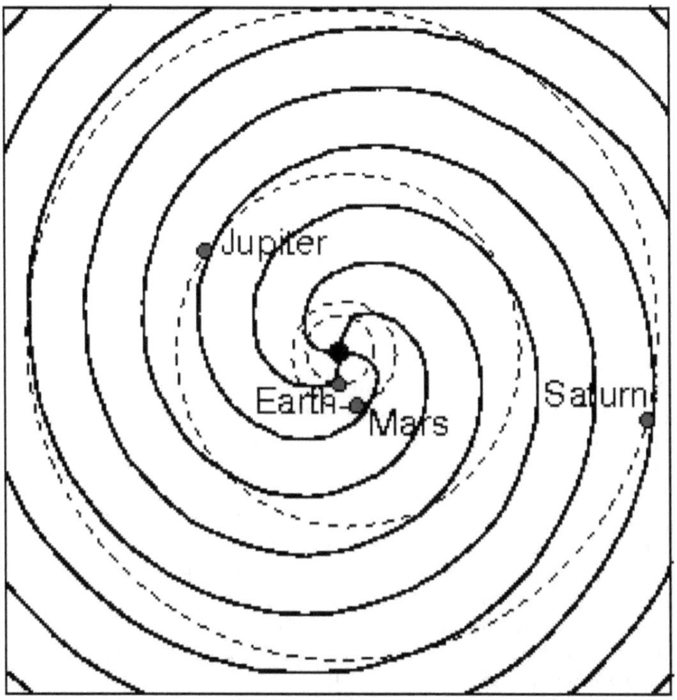

Fig. 9.7 Diagram of the Parker spirals (heavy curves) in the ecliptic plane similar to that shown in Fig. 9.1, but extended out to 10 AU. The orbits of the Earth, Mars Jupiter and Saturn are shown (dashed ellipses), along with their locations at 1 January 2010

to form an MIR layer somewhere in the outer heliosphere. Given that CMEs travel in a generally radial direction and are faster than CIRs they must also interact with this layer when they reach it. One may envision this region as a large dense structure forming a ring around the Sun, slowly moving radially outwards but being replenished at its inner edge. CIRs and CMEs eventually merge with other MIRs, which will catch up with slower MIRs ahead of them and so on, continually merging until they reach this outer layer. The locations at which the MIR formation will vary depending on the local conditions and the recent history of solar activity in that region.

Studies discussing the evolution of CMEs into MIRs include Burlaga et al. [7], Whang et al. [46] and Richardson et al. [35]. Richardson et al., for example, tracked a CME from the Sun out to 58 AU where its resulting structure was observed by *Voyager 2*. It was also observed by *Ulysses* at 5.3 AU and at the Earth in September 1998. They showed that a trailing CME travelling faster than a leading CME will eventually catch up with it forming a compressed plasma region between the two. Thus the MIR was formed in this case by the interaction of two CMEs.

It would seem therefore that the life of a CME does not end per se, but rather its structure is merged deep in the heliosphere with the compressed regions that came before it. Eventually, the outer component of this large MIR may dissipate to form the background turbulence in the outer heliosphere, with the inner component being replenished with new arriving MIRs originating with other CMEs and CIRs. How far from the Sun this occurs is not known, but some studies (e.g. Roelof et al. [36]) have observed evidence of MIRs as far out as the termination shock.

References

1. Allen, C.W.: Astrophysical Quantities, 3rd edn, Athlone Press, London (1976).
2. Arge, C.N., Luhmann, J.G., Odstrĉil, D., Schrijver, C.J., Li, Y.: J. Atmos. Solar Terr. Phys. **66**, 1295–1309 (2004).
3. Arge, C.N., Pizzo, V.J.: J. Geophys. Res. **105**, 10465–10460 (2000).
4. Birkeland, K.: In Videnskapsselskapets Skrifter, I Mat - Naturv. Klasse **1**, Christiania (1916).
5. Burlaga, L.F.: Interplanetary Magnetohydrodynamics, Oxford Univ. Press, New York (1995).
6. Burlaga, L., Berdichevsky, D., Gopalswamy, N., Lepping, R., Zurbuchen, T.: J. Geophys. Res. **108**, 1425, doi:10.1029/2003JA010088 (2003).
7. Burlaga, L.F., Ness, N.F., Richardson, J.D., Lepping, R.P.: Solar Phys. **204**, 399–411 (2001).
8. Cargill, P.J.: Solar Phys. **221**, 135–149 (2004).
9. Cargill, P.J., Chen, J., Spicer, D.S., Zalesak, S.T.: J. Geophys. Res. **101**, 4855–4870 (1996).
10. Chapman, S.: Smithsonian Contrib. Astrophys. **2**, 1 (1957).
11. Chen, J.: J. Geophys. Res. **101**, 27499-27520 (1996).
12. De Young, D.S., Hundhausen, A.J.: J. Geophys. Res. **76**, 2245–2253 (1971).
13. Dryer, M.: Space Sci. Rev. **15**, 403–468 (1974).
14. Dryer, M., Smart, D.F.: Adv. Space Res. **4**, 291–301 (1984).
15. Falkenberg, T.V., Vršnak, B., Taktakishvilli, A., Odstrcil, D., MacNiece, P., Hesse, M.: Space Weather **8**, S06004, doi:10.1029/2009SW000555 (2010).
16. Fry, C.D., Detman, T.R., Dryer, M., Smith, Z., Sun, W., Deehr, C.S., Akasofu, S.-I., Wu, C.-C., McKenna-Lawlor, S.: J. Atmos. Solar-Terr. Phys. **69**, 109–115 (2007).
17. Fry, C.D., Dryer, M., Deehr, C.S., Sun, W., Akasofu, S.-I., Smith, Z.: J. Geophys. Res. **108**, 1070, doi:10.1029/2002JA009474 (2003).
18. Fry, C.D., Sun, W., Deehr, C.S., Dryer, M., Smith, Z., Akasofu, S.-I., Tokumaru, M., Kojima, M.: J. Geophys. Res. **106**, 20985–21002 (2001).
19. Hakamada, K., Akasofu, S.-I.: Space Sci. Rev. **31**, 3–70 (1982).
20. Howard, T.A, Fry, C.D., Johnston, J.C., Webb, D.F.: Astrophys. J. **667**, 610–625 (2007).
21. Hundhausen, A.J.: In Collisionless shocks in the heliosphere: A tutorial review, A87-25326 69–92, Washington, DC, AGU, pp.37–58 (1985).
22. Intriligator, D.S., Sun, W., Dryer, M., Fry, C.D., Deehr, C., Intriligator, J.: J. Geophys. Res. **110**, A09S10, doi:10.1029/2004JA010939 (2005).
23. Lazarus, A.J., Richardson, J.D., Decker, R.B., McDonald, F.B.: Space Sci. Rev. **89**, 53–59 (1999).
24. Lindemann, F.: Philosophical Magazine, Series 6, **38**, 674 (1919).
25. Luhmann, J.G., Soloman, S.C., Linker, J.A., Lyon, J.G., Mikic, Z., Odstrcil, D., Wang, W., Wiltberger, M.: J. Atmos. Solar Terr. Phys. **66**, 1243–1256 (2004).
26. Manoharan, P.K.: Solar Phys. **235**, 345–368 (2006).
27. Manoharan, P.K., Tokumaru, M., Pick, M., Subramanian, P., Ipavich, F.M., Schenk, K., Kaiser, M.L., Lepping, R.P., Vourlidas, A.: Astrophys. J. **559**, 1180–1189 (2001).
28. McKenna-Lawlor, S.M.P., Dryer, M., Smith, Z., Kecskemety, K., Fry, C.D., Sun, W., Deehr, C.S., Berdichevsky, D., Kudela, K., Zastenker, G.: Ann. Geophys. **20**, 917–935 (2002).
29. Meyer-Vernet, N.: In Houghton, J.T., Rycroft, M.J., Dessler, A.J. (eds.), Basics of the Solar Wind, Cambridge Univ. Press, United Kingdom (2007).
30. Odstrcil, D., Pizzo, V.J., Arge, C.N.: J. Geophys. Res. **110**, A02106, doi:10.1029/2004JA 010745 (2005).

31. Odstrcil, D., Pizzo, V.J., Linker, J.A., Riley, P., Lionello, R., Mikic, Z.: J. Atmos. Solar Terr. Phys. **66**, 1311–1320 (2004).
32. Odstrcil, D., Riley, P., Linker, J.A., Lionello, R., Mikic, Z., Pizzo, V.J.: In Wilson, A. (ed.), Solar Variability as an Input to the Earth's Environment, ESA SP-535, Estec, The Netherlands, p.541 (2003).
33. Parker, E.N.: Astrophys. J. **128**, 664–676 (1958).
34. Parker, E.N.: Space Sci. Rev. **4**, 666–708 (1965).
35. Richardson, J.D., Paularena, K.I., Wang, C., Burlaga, L.F.: J. Geophys. Res. **107**, doi:10.1029/2001JA000175 (2002).
36. Roelof, E.C., Decker, R.B., Krimigis, S.M.: Proc. Solar Wind 12 **1216**, 359–362 (2010).
37. Smart, D.F. Shea, M.A.: J. Geophys. Res. **90**, 183–190 (1985).
38. Smart, D.F., Shea, M.A., Barron, W.R., Dryer, M.: In Shea, M.A., Smart, D.F., McKenna-Lawlor, S. (eds.), Proc. STIP Workshop on Solar/Interplanetary Intervals, Book Crafters, Inc., Chelsea, MI, pp. 139 - 156 (1984).
39. Smith, Z., Dryer, M.: Solar Phys. **129**, 387–405 (1990).
40. Smith, Z., Dryer, M.: NOAA Technical Memorandum, ERL/SEL - 89 (1994).
41. Smith, Z., Dryer, M., Ort, E., Murtagh, W.: J. Atmos. Solar-Terr. Phys. **62**, 1265–1274 (2000).
42. Taktakishvilli, A., Kuznetsova, M., MacNeice, P., Hesse, M., Rastätter, L., Pulkkinen, A.: Space Weather **7**, S03004, doi:10.1029/2008SW000448 (2009).
43. Tappin, S.J.: Solar Phys. **233**, 233–248 (2006).
44. Webb, D.F., Howard, T.A., Fry, C.D., Kuchar, T.A., Mizuno, D.R., Johnston, J.C., Jackson, B.V.: Space Weather **7**, S05002, doi:10.1029/2008SW000409 (2009a).
45. Webb, D.F., Howard, T.A., Fry, C.D., Kuchar, T.A., Odstrcil, D., Jackson, B.V., Bisi, M.M., Harrison, R.A., Morrill, J.S., Howard, R.A., Johnston, J.C.: Solar Phys. **256**, 239–269 (2009b).
46. Whang, Y.C., Burlaga, L.F., Ness, N.F., Smith, C.W.: Solar Phys. **204**, 253–263 (2001).
47. Wu, S.T., Dryer, M., Han, S.M.: Solar Phys. **84**, 395–418 (1983).
48. Wu, S.T., Han, S.M., Dryer, M.: Planet. Space Sci. **27**, 255–264 (1979).
49. Xie, H., Ofman, L., Lawrence, G.: J. Geophys. Res. **109**, A03109, doi:10.1029/2003JA010226 (2004).
50. Yashiro, S., Gopalswamy, N., Michalek, G., St. Cyr, O.C., Plukett, S.P., Rich, N.B., Howard, R.A.: J. Geophys. Res. **109**, doi:10.1029/2003JA010282 (2004).

Chapter 10
Interaction With the Earth and Other Planets: Contribution to Space Weather

When CMEs interact with other bodies in the solar system a number of physical processes take place. At the Earth the magnetosphere can be disrupted leading to what is known as a (geo)magnetic storm, but similar effects are known to take place at other planets as well. The behaviour of the magnetosphere and ionosphere are parts of the collective term known as space weather, but the most severe effects of space weather at the Earth are known to be initiated by CMEs. In this chapter we introduce the Earth's magnetosphere and discuss how it is affected by CMEs. We also review the effects of CMEs on other heliospheric bodies such as planets and comets. It is the space weather impact of CMEs that workers are most interested in, and work continues to develop new and improved methods of determining the arrival time, speed and magnetic orientation of CMEs. This is an important part of space weather forecasting and its ultimate goal is to predict these parameters as early and accurately as possible. More in-depth reviews on the Earth's magnetosphere, those of the other planets and space weather can be found in Hargreaves [23], Kivelson and Russell [34], Song et al. [60] and Bothmer and Daglis [7]. A good online reference is the *Oulu Space Physics Textbook* which is now at http://wiki.ouli.fi/display/spacewiki.

10.1 The Earth's Magnetosphere

The outer core of the Earth consists of molten metallic material, whose movement produces convection generating the geomagnetic field. If the Earth was in a vacuum the structure of this field would assume a form similar to that of a bar magnet (i.e. a dipole), and at lower altitudes and latitudes the geomagnetic field closely resembles this. However, as with all bodies in the solar system the Earth lies in the solar wind which is a supersonic magnetised plasma (Sect. 9.2) and a balance is achieved between the two environments. The resulting volume is called the magnetosphere and it extends out to millions of kilometers into space [20, 52].

The solar wind exerts a pressure on the sunward side (dayside) of the geomagnetic field, and because it is supersonic a bow shock forms ahead (Sunward) of the field. This causes a compression of the field there. Magnetic reconnection primarily

T. Howard, *Coronal Mass Ejections: An Introduction*, Astrophysics and Space Science Library 376, DOI 10.1007/978-1-4419-8789-1_10,

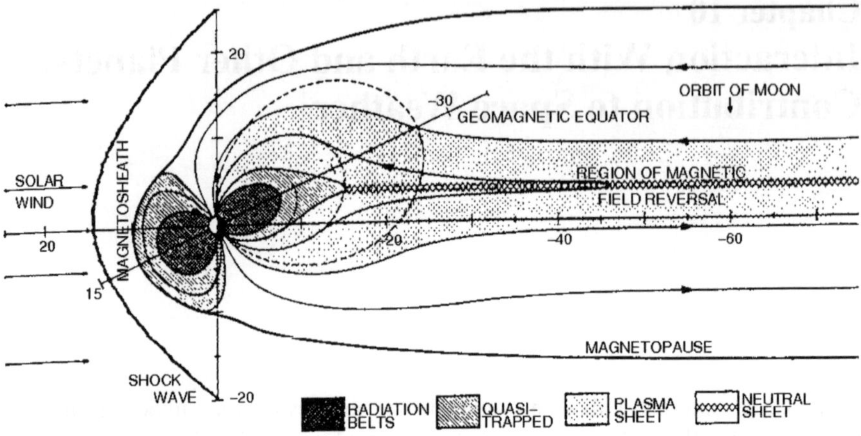

Fig. 10.1 An illustration of the geomagnetic field on the noon–midnight meridian [48] (Copyright Elsevier Academic Press, 1967)

on the dayside at high latitudes opens the field lines to the solar wind, and on the nightside the resulting volume extends out to at least 1,000 Earth radii [12].

Figure 10.1 shows an illustration of the geomagnetic field and some of its plasma regions. In the absence of magnetic reconnection, the geomagnetic field cannot interact with the surrounding IMF, and so ions are trapped in this field and confined to the magnetosphere. These field lines are called closed. On the dayside, reconnection causes the merging of the geomagnetic field lines with the interplanetary field. These field lines are termed open, i.e. with only one end connected with the Earth. Field lines in the high-latitude nightside are also open as a consequence of reconnection. The regions at the base of these open field lines are called the polar caps and a region called the cusp lies in between the last closed field line and the first open one. It is through the cusp region that the solar wind can be funneled into the magnetosphere to the ionosphere below. The aurora is caused by precipitating electrons from the solar wind through this region and along open field lines, energising the atmosphere. A current is also created flowing westward in the Earth's equatorial plane at an altitude of 2–5 Earth radii. This is called the ring current, which is strongly enhanced during the main phase of a geomagnetic storm [10]. Table 10.1 outlines the important features of the magnetic components of the magnetosphere.

Plasma enters the magnetosphere via a number of processes, namely diffusion from the solar wind, particles escaping from the ionosphere and magnetic reconnection. It is important to note that the footpoints of geomagnetic field lines are anchored to their position, so the field lines themselves corotate with the Earth. We can therefore represent the magnetosphere in terms of the plasma regions as well. They are shown in Fig. 10.2 and summarised in Table 10.2. The magnetosphere is therefore a highly dynamic environment that is strongly affected by the behaviour of the surrounding solar wind and interplanetary magnetic field.

Table 10.1 Summary of the important features of the magnetosphere

Feature	Description
Bow shock	A shock formed when the solar wind encounters the geomagnetic field. It occurs on the dayside at a distance of around 15 Earth Radii (R_E). Its thickness is of the order of 50–100 km.
Foreshock	Region immediately sunward of the bow shock populated by particles that leak from behind the bow shock. Particles arriving from the sunward direction are reflected off the bow shock and travel back toward the Sun. These are called back-streaming particles.
Magnetosheath	Turbulent region immediately earthward of the bow shock where the normal component of solar wind particle motions are decelerated to subsonic speeds.
Magnetopause	Outer boundary defined on the dayside by the last closed magnetic field line and threaded elsewhere by field lines opened by reconnection.
Magnetotail	Region on the nightside of the magnetosphere, extending to over 1,000 R_E, containing both open and closed field lines.
Cusp	Funnel-shaped region between the front and rear lobes of the magnetosphere. Through this region solar wind particles gain direct access to the ionosphere.
Auroral oval	Projection to the ground of the region where the auroral brightness maximises.
Polar cap	Projection to the ground of the region containing open field lines.
Ring current	High energy ($\sim 10^{15}$ J) current located at around 2–5 R_E, flowing westward in the equatorial plane. The ring current is enhanced during a geomagnetic storm.

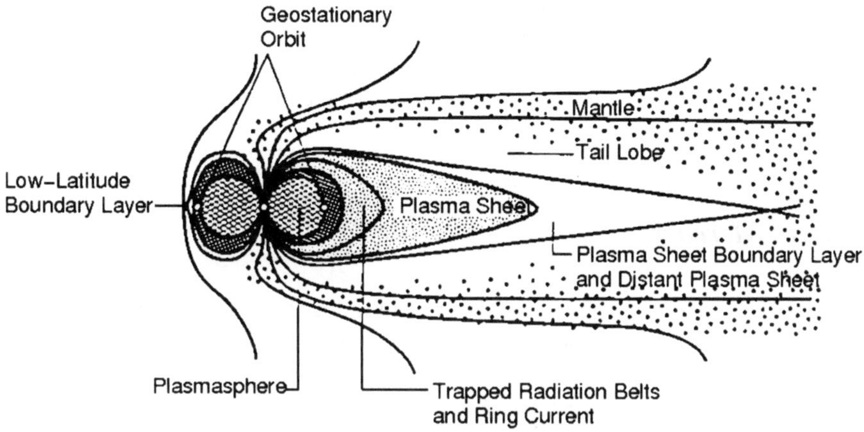

Fig. 10.2 Schematic diagram of plasma regions of the Earth's magnetosphere in the noon–midnight meridian [34]

Table 10.2 Plasma regions within the Earth's magnetosphere

Feature	Description
Plasmasphere	Cold, dense, corotating plasma forming a teardrop-shaped "sphere" around the Earth. Plasma accumulates in a region determined by a balance between the drift plasma, anti-Sunward solar wind pressure, plasma exchange with the ionosphere, and corotation of the plasma with the Earth.
Plasmapause	Boundary between the solar wind driven convection in the outer magnetosphere and corotating plasma in the inner magnetosphere, typically at around $4\,R_E$.
Plasma sheet	Hot accelerated plasma on closed field lines on the nightside.
Tail lobes	Low density, cool plasma regions in the tail on open field lines on the nightside.
Plasmatrough	Cold low-density plasma region just outside the plasmapause.
Boundary layer	Region immediately earthward of the magnetopause populated by a mixture of plasmas of solar wind and magnetospheric/ionospheric origin.
Radiation belts	Two belts containing trapped particles. The inner belt consists mainly of high-energy protons produced when cosmic rays blast particles out of the upper atmosphere. The outer belt has high-energy electrons produced magnetospheric acceleration processes. The particles in these regions can easily move along magnetic field lines and so take the form of the surrounding field. Also called the Van Allen radiation belts.

10.2 The Magnetospheres of Other Planets

As the other planets also lie in the heliosphere, their magnetospheres are also governed by the solar wind and interplanetary magnetic field. The nature of the fields of the planets and their locations relative to the Sun are all different and so their responses to the solar wind and resulting magnetospheric configurations also differ greatly. Here we briefly review the nature of the magnetospheres of the other planets.

10.2.1 Mercury

Mercury has no ionosphere or atmosphere, but it does have a magnetic field which has a similar structure to that of the Earth's. It is a great deal weaker and therefore the magnetosphere is smaller than the Earth's relative to the size of the planet. Nevertheless, the iron core believed to produce this field is still large for a planet of Mercury's size. The field is a distorted dipole field like the Earth but it is met with solar wind of much larger pressure and field strength, and so the effects of space weather here are much more pronounced. NASAs *Messenger* spacecraft (launched August 2004) arrived at Mercury in March 2011, having made three flybys of the planet. A recent review of the Hermian magnetosphere can be found in Anderson et al. [2].

10.2.2 Venus

Venus has no intrinsic magnetic field (or possibly a very weak field), presumably because it has no molten metallic core. Venus therefore does not deflect the solar wind and so its ionosphere is constantly bombarded by it, stripping its atmospheric particles into space. Rather than a magnetopause, Venus has an ionopause, i.e. an interface between the ionosphere and the interplanetary magnetic field. As a result, the ionosphere of Venus contains a great deal of plasma mantle, currents and induced magnetic fields, and is the source of large numbers of VLF and ELF waves. ESA's *Venus Express* (launched in November 2005) is currently in orbit around Venus. Publications regarding the Venetian "magnetosphere" include Bauer et al. [5], Phillips et al. [53] and Luhmann and Russell [43].

10.2.3 Mars

Mars has a very weak magnetic field which may be similar in generation to that of the Earth's moon. It does not seem to generate a magnetic field, and the weak field it does produce appears to arise from metallic bodies (e.g. rocks) which have been magnetised over time by their existence in an external magnetic field.[1] This suggests that Mars may have once generated its own magnetic field. As with Venus, the solar wind often penetrates to the Martian atmosphere, but its effects are not as intense as on Venus, due to its larger distance from the Sun. Unlike with Venus, the Martian magnetic field does provide a small level of deflection of the solar wind. ESA's *Mars Express* (launched in June 2003) and NASA's *Odyssey*, *Reconnaissance Orbiter*, and *Phoenix* (launched October 2001, August 2005 and October 2007 respectively) are currently either in orbit around or on the surface of the planet. Further reading on the Martian magnetosphere may be sought from the *Space Science Reviews* volume on the subject [55].

10.2.4 Jupiter

Jupiter is not only the largest planet of the solar system, it also has the largest magnetic field. If the Jovian magnetosphere could be seen, it would appear from the Earth to be three times as large as the moon. It is considerably larger than a planet of its size and composition should be producing, meaning that there are additional characteristics of the Jovian environment that enhance its field. The Jovian core is

[1] This is analogous to a metallic body being rubbed against a magnet and becoming magnetised itself.

in motion within a liquid metallic shell deep within the planet. It produces mostly a dipole field but has weaker quadrupole and octupole components, which act to enhance the field considerably [31]. A further enhancement is produced by one of its larger moons. Io is a highly active moon with constant eruptions of sulfur from its many volcanos and an ionosphere which is a fairly good electrical conductor. It orbits within the intense radiation belts of Jupiter, and the particles within collide with atoms in Io's atmosphere, spluttering the particles into a cloud of plasma around the planet [59]. The cloud becomes ionised and forms a torus around Jupiter. Both Jupiter and Io's torus are conductors, and so through the magnetic field lines there is a continuous current maintained between the planet and the moon. This changing current system produces an excellent dynamo between Jupiter and Io, and the Jovian magnetosphere is further enhanced.

Other interesting characteristics about the Jovian magnetosphere include:

- The field is oriented in the opposite direction to that of the Earth.
- The Jovian magnetopause is located at around 60 Jovian radii (R_J) and the bow shock is at around 80 R_J.
- The field is large enough to easily engulf the Sun if it could be placed within the field.
- Some of the Jovian moons (Io, Ganymede) may have magnetic fields of their own, carving mini-magnetospheres out of Jupiter's [25, 35, 57].

There are currently no spacecraft in orbit around Jupiter, since *Galileo* descended into the planet's atmosphere in September 2003. NASA/ESA's *Cassini* spacecraft made a flyby in late 2000–early 2001, as did NASA's *New Horizons* in February 2007. Khurana et al. [31] provide a review of the Jovian magnetosphere and a book dedicated to the subject can be found in Dessler [15].

10.2.5 Saturn

The magnetic field of Saturn is produced in a similar fashion to that of the Jovian field, but is only one third as powerful. It is still the second largest of the planets. Saturn provides a problem for some theorists because it has a magnetic axis almost exactly (within 1°) aligned with the rotation axis. This is a problem because there is a well established theorem [13] stating that a planetary dynamo field can never be axially symmetric. Saturn's moon, Titan produces a torus around the planet similarly to Io around Jupiter. This is a hydrogen and nitrogen torus which acts to heat the plasma in Saturn's plasmasphere to an order of 10^6 K. The auroral ovals of Saturn have been observed in visible, infrared and ultraviolet light.

Cassini (launched in October 1997) is currently in orbit about the planet. *New Horizons* also made a flyby in June 2008. Reviews of the magnetosphere of Saturn include Russell and Luhmann [56], and more recently Mauk et al. [44] and Gombosi and Ingersoll [21].

10.2.6 *Uranus*

One concern about Uranus is the fact that its rotational axis is pointing almost completely at the Sun. It was therefore expected that the magnetic axis of Uranus should be close to its rotational axis and the Uranian cusp would be directly exposed to the Sun. This was found not to be the case. *Voyager 2* found that the magnetic axis was steeply inclined to the rotation axis (nearly 60°) causing it to spin like the axis of a top which is about to topple. The Uranian magnetosphere is therefore an extremely dynamic environment and it is expected that the plasma distribution in the surrounding may not be as stable as the other planets. The field is believed to be generated by some mechanism located at relatively shallow regions in the Uranian atmosphere. Publications discussing the Uranian magnetosphere include Ness et al. [49] and Belcher et al. [6].

10.2.7 *Neptune*

Neptune's magnetic field is similar to that of Uranus in both generation mechanism and orientation. It has a magnetic axis which is inclined at $\sim 47^\circ$ from the rotation axis. As the rotational axis of Neptune is around 30° this means that the magnetic axis is almost perpendicular to the Neptune–Sun plane. Triton, Neptune's moon, influences the behaviour of the outer magnetosphere [37]. As with Uranus, Neptune has only been investigated by *Voyager 2* and there are no known expeditions to this planet in the near future. Publications involving Neptune include Ness et al. [50] and Krimigas [36].

10.2.8 *Pluto*

It is unknown as to whether Pluto has a magnetosphere. While the *New Horizons* spacecraft (en route to Pluto) does not have a magnetometer on board it does have an energetic particle detector (SWAP) that will provide information on the solar wind around Pluto and its magnetosphere, if it has one.

Table 10.3 shows a comparison of the location of the bow shock and magnetopause relative to the radius of the planet, along with the difference between the rotational and magnetic axes.

10.3 Magnetic Reconnection

In order to discuss how the magnetosphere is affected by CMEs we must first understand the process of magnetic reconnection. This has been briefly reviewed in Sect. 2.6.1 and mentioned with regard to CME onset and launch mechanisms in

Table 10.3 Comparison of the average locations of the bow shock and magnetopause, and the approximate angular difference between the rotational and magnetic axes (tilt)

	Bow shock	Magnetopause	Tilt angle
Earth	$15\,R_E$	$11\,R_E$	10.8°
Mercury	$1\,R_H$	$0.5\,R_H$	10.0°
Jupiter	$80\,R_J$	$45\,R_J$	9.7°
Saturn	$30\,R_S$	$21\,R_S$	<1.0°
Uranus	$30\,R_U$	$27\,R_U$	59.0°
Neptune	$30\,R_N$	$26\,R_N$	47°

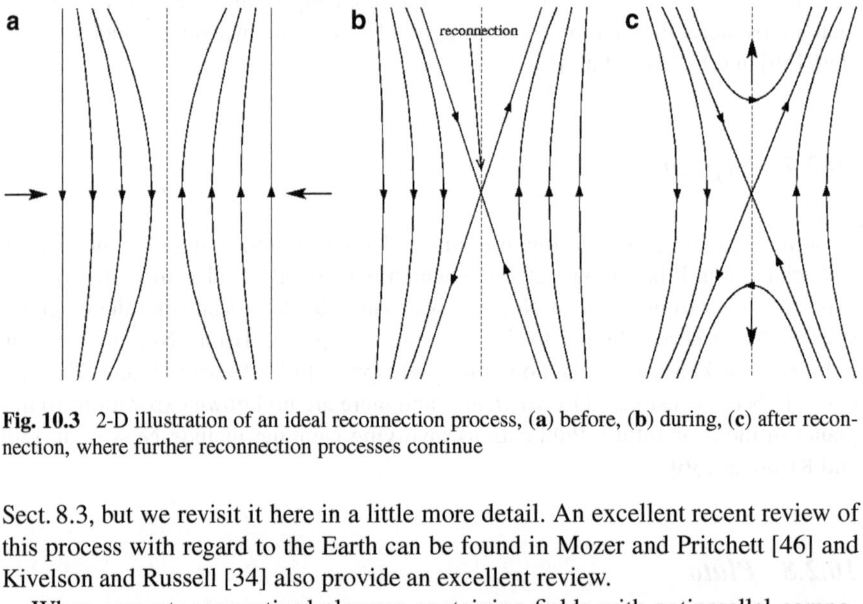

Fig. 10.3 2-D illustration of an ideal reconnection process, (**a**) before, (**b**) during, (**c**) after reconnection, where further reconnection processes continue

Sect. 8.3, but we revisit it here in a little more detail. An excellent recent review of this process with regard to the Earth can be found in Mozer and Pritchett [46] and Kivelson and Russell [34] also provide an excellent review.

When separate magnetised plasmas containing fields with antiparallel components move toward each other, the combined magnetic structure can be reconfigured to a lower energy state. Energy is therefore released during this process. The fields merge and the new configuration moves in a direction perpendicular to the original components. Figure 10.3 shows an illustration of this process.

Reconnection requires the nonphysical result in which the magnetic field points in two different directions at that same point in space. For this to occur there must be a non-zero electric field parallel to the magnetic field present that probably arises from the parallel components of the electron pressure, inertia and/or resistivity. At the point where the field lines meet these properties enable the violation of the following equation

$$\mathbf{B} \times (\nabla \times \mathbf{E}_{||}) = 0, \tag{10.1}$$

where $\mathbf{E}_{||}$ is the parallel electric field. This enables reconnection to occur [46]. Magnetic reconnection can therefore be physically possible in a magnetised plasma if the above circumstances arise.

10.4 Magnetic Storms

Keep in mind that the term *Space Weather* is a general term that encompasses all effects of the planet-space environment and interactions between. For example, small variations in the solar wind have an impact on the Earth's and other planets' magnetospheres, and so are part of space weather. When we discuss magnetic storms we are referring to the most extreme cases of space weather.[2] These are events that significantly disrupt the magnetic field and can cause damage to infrastructure etc. The reader may encounter the term "substorm" during their research on this topic. Substorms are a large disruption to the geomagnetic field (not as large as a storm), but they are caused by a completely different physical mechanism than the storms. Substorms are caused by the release of plasma and magnetic energy that has been stored in the magnetotail [23], storms are caused by an ICME impacting the Earth. It is important to distinguish between the terms space weather, substorms and storms and to understand their differences.

In the previous section we established that magnetic reconnection can occur when two magnetised plasmas and their fields combine under certain conditions. These conditions are met on the dayside of the Earth's magnetosphere. Recall that reconnection requires the fields in the separate plasmas to have anti-parallel components, so for reconnection to occur between an ICME and the Earth's magnetosphere, the north–south components of their fields must be oppositely directed. The geomagnetic field is from south to north (i.e. the north magnetic pole is in the same hemisphere as the south geographic pole), and so an ICME with a strong southward field component will produce the strongest rate of reconnection with the magnetosphere [18]. Figure 10.4 shows a simplified diagram of how this may occur. Note that the green field lines (those connecting the Earth with the Sun and IMF) are in the polar cap region of the magnetosphere.

Magnetic reconnection between the ICME and the geomagnetic field has two major effects on the magnetosphere. Firstly the reconnection process opens the geomagnetic field and connects it with the ICME, allowing energetic particles contained within the ICME to be injected into the magnetosphere. Secondly, the process releases energy, also injected into the magnetosphere. The result is a massive disruption to the magnetospheric system called a (geo)magnetic storm. Over 6 GW of power is typically deposited into the ionosphere during an average storm [23] and for extreme cases, such as the Halloween event, over 100 GW can be injected [16].

Recall also that along with the magnetic structure of the ICME there is also the plasma component, which often produces a shock in the solar wind and exerts pressure on the magnetosphere when it encounters it. The ram pressure is given by

$$P = \rho V^2 \tag{10.2}$$

where ρ and V are the density and speed of the ICME. This pressure reduces magnetic field lines in size on the dayside, expanding the polar cap and causing the cusp

[2] Although large well aimed SEP events can cause extreme space weather events.

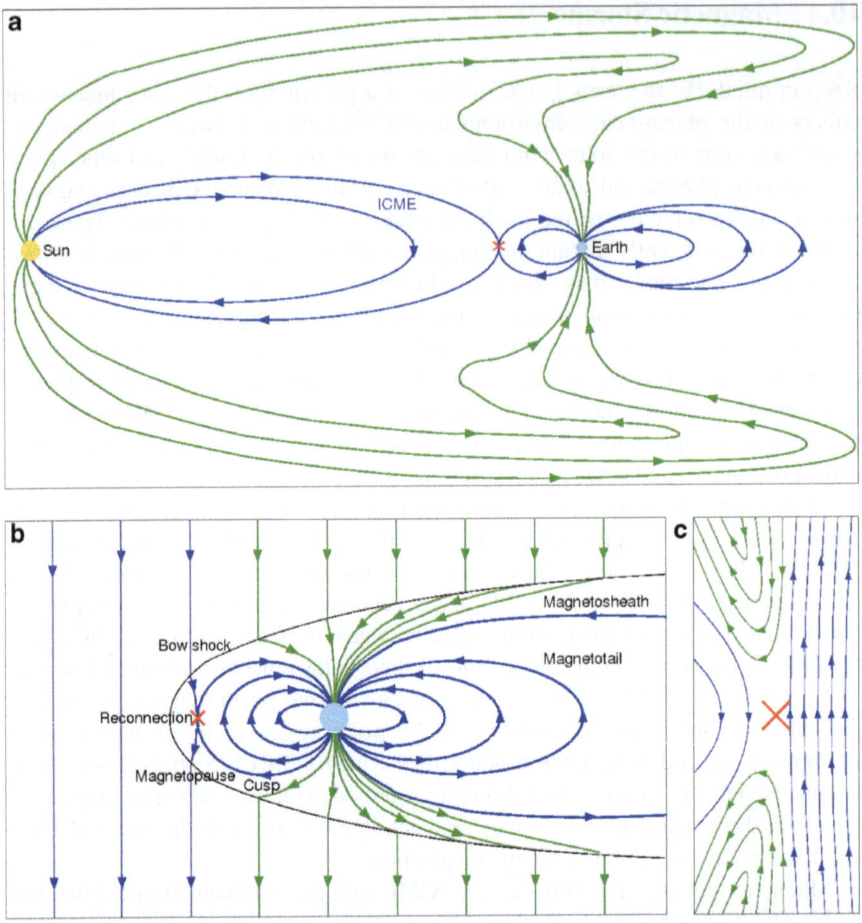

Fig. 10.4 Simplified depiction of magnetic reconnection on the dayside equatorial plane of the Earth's magnetosphere (modified from Mozer and Pritchett [46]). (**a**) Representation of the interplanetary magnetic field and ICME that passes through the Earth and the Sun (*green*) and those that are connected only with the Sun or the Earth (*blue*). (**b**) A close-up view of the Earth's magnetosphere with the green and blue lines shown. Reconnection occurs at the point indicted where the northward-directed geomagnetic field meets the southward-directed interplanetary field. (**c**) Close-up view of the reconnection site on the dayside

to move further toward the equator (see Fig 1.5). This results in the aurora being observed at even lower latitudes and an increase in the area of the ionosphere that is exposed to the solar wind. Also the shock and sheath contain a larger density and are moving faster than the ambient solar wind, which increases the concentration of energetic particles available to the magnetosphere. It is the arrival of an interplanetary shock that produces the so-called sudden (storm) commencement or S(S)C at the Earth.

10.4.1 The Effects of Magnetic Storms

In summary, a geomagnetic storm is the result of two effects: magnetic reconnection and increased ram pressure. The strength of the magnetic storm is therefore governed by two properties of the ICME:

1. The magnitude and duration of the southward component of its magnetic field;
2. Its ram pressure, which is a function of its speed and density.

This has four major effects at the Earth:

1. A large disturbance of the geomagnetic field, causing fluctuations in magnitude, direction and orientation;
2. An increase in the energetic particle population in the magnetosphere and ionosphere;
3. A reconfiguration of magnetic and plasma components of the magnetosphere, e.g. a shrinkage of the plasmasphere;
4. An intensification of the ring current and radiation belts,

which can result in the following:

- Electric charging of spacecraft circuitry, driving currents and short-circuiting electrical components on board [1,4];
- Increased density and temperature in the ionosphere, increasing drag on spacecraft and advancing orbit decay [11];
- Electromagnetic induction in long electrical wires, leading to power station damage and failure [3];
- Electromagnetic interference leading to communications disruption [39,61];
- Increased radiation dosage for high-latitude-flying aircraft passengers, staff and astronauts [19,28].

Specific examples of assets that have most likely been damaged or destroyed by CME-induced magnetic storms include the *Galaxy 4*, *Galaxy 15* and *Equator-S* spacecraft (probably short-circuited by spacecraft charging), *Skylab* (brought down prematurely by increased atmospheric drag), the shutting down of the Hydro-Quebec power grid in Canada in 1989 and damage to a power grid in Sweden in 2003. Spacecraft launches have been delayed and aircraft routes changed as a result of magnetic storms. A review of the effects of magnetic storms can be found a report from a recent workshop on the topic [3].

10.5 CMEs at Other Bodies in the Heliosphere

Given the global nature of CMEs and their extent and influence on the Earth and solar wind, it seems likely that they encounter other bodies in the solar system, with similar major effects as well. Indeed, CMEs have been observed to impact other planets and possibly even comets in the solar system.

10.5.1 Planets

Because of its proximity to the Sun and weakness of its magnetosphere, Mercury suffers extreme effects of space weather. Its weak atmosphere is continually stripped away by the solar wind and photoionisation processes [58]. Models of the Hermian magnetospheric response to the solar wind have been conducted [32, 33], but I have been unable to identify a study involving a CME impacting the planet.

Venus does not have a magnetic field, and so ICMEs act to enhance the already present atmospheric ionisation [30], and the pressure pulse from the interplanetary shock decreases the size of the ionosphere [17], exposing even more of the neutral atmosphere to the solar wind [64]. CMEs have been observed at Venus by *Pioneer Venus Orbiter* [26, 40, 42, 47] and by *Venus Express* [41, 64].

Mars has only a very weak magnetic field, and so like Venus its ionosphere is constantly eroding in the solar wind. This effect is enhanced by the arrival of CMEs, but the effects are not as significant as at Venus. This is because Mars does have a magnetic field (albeit a weak one) and the CME itself is much weaker by the time it reaches Mars. It has been found for stronger CMEs that the magnetic field on the dayside becomes enhanced [14]. Studies of ICMEs impacting Mars include McKenna-Lawlor et al. [45], Crider et al. [14] and Haider et al. [22].

Once CMEs reach Jupiter and the outer planets they are in the region where they are beginning to form MIRs (Sect. 9.7). Here CMEs and CIRs interact to form compressed regions, and so it becomes increasingly difficult to identify individual events. Also, the strength of their magnetic fields are insignificant compared with those of the outer planets, and so their impact does not seem to affect their behaviour to any great extent. CME impacts have been observed to have an impact on Jupiter and Saturn, primarily in the form of an aurora intensity enhancement [54].

10.5.2 Comets

CMEs may also be responsible for some types of comet disconnection events. Disconnection events occur when the tail of the comet appears to be disconnected from its head and moves independently through the solar wind [8, 29, 62]. Comet tails are now known to fluctuate in response to pressure changes in the local solar wind [38]. Three competing theories have been proposed to explain the triggering of cometary disconnection events [62]:

1. A sudden change in the ion production rate [63];
2. A sudden change in the solar wind pressure [24, 27];
3. Magnetic reconnection as the comet crosses the IMF sector boundary [9, 51].

New evidence from SMEI has suggested that in at least one case an ICME may be responsible for a disconnection event. Kuchar et al. [38] identified six disconnections in comets NEAT and LINEAR and for one of them a faint ICME was observed passing the comet tail at around the time of the disconnection. While not entirely conclusive, the evidence strongly suggests that ICMEs interact with comet tails in the heliosphere as well at the Earth and other planets.

References

1. Allen, J.: Space Weather **8**, S06008, doi:10.1029/2010SW000588 (2010).
2. Anderson, B.J., Acuña, M., Korth, H., Slavin, J.A., Uno, H., Johnston, C.L., Purucker, M.E., Soloman, S.C., Raines, J.M., Zurbuchen, T.H., Gloeckler, G., McNutt, R.L., Jr.: Space Sci. Rev. **152**, 307–339 (2010).
3. Baker, D.N., Balstad, R., Bodeau, J.M., Cameron, E., Fennel, J.F., Fisher, G.M., Forbes, K.F., Kintner, P.M., Leffler, L.G., Lewis, W.S., Reagan, J.B., Small, A.A., III, Stansell, T.A., Strachan, L., Jr., Comm. Soc. Econ. Impacts of Severe Space Weather Events, NRC Workshop Report (2009).
4. Baker, D.N., Kanekal, S., Blake, J.B., Klecker, B., Rostoker, G.: EOS Trans. AGU **75**, 401 (1994).
5. Bauer, S.J., Brace, L.H., Hunten, D.M., Intriligator, D.S., Knudsen, W.C., Nagy, A.F., Russell, C.T., Scarf, F.L., Wolfe, J.H.: Space Sci. Rev. **20**, 413–430 (1977).
6. Belcher, J.W., McNutt, R.L., Jr., Richardson, J.D., Selesnick, R.S., Sittler, E.C., Jr., Bagenal, F.: In Uranus, Univ. Arizona Press, Tucson, pp.780–830 (1991).
7. Bothmer, V., Daglis, I.A.: Space Weather: Physics and Effects, Springer, New York (2007).
8. Brandt, J.C., Caputo, F.M., Hoeksema, J.T., Niedner, M.B. Jr., Yi, Y., Snow, M.: Icarus **137**, 69–83 (1999).
9. Brandt, J.C., Snow, M.: Icarus **148**, 52–64 (2000).
10. Burton, R.K., McPherron, R.L., Russell, C.T.: J. Geophys. Res. **80**, 4204–4214 (1975).
11. Compton, W.D., Benson, C.D.: Living and Working in Space: A History of Skylab, NASA SP-4208 (1983).
12. Cowley, S.W.H.: Planet. Space Sci. **39**, 1039–1047 (1991).
13. Cowling, T.G.: Mon. Not. R. Astron. Soc. **94**, 768–782 (1934).
14. Crider, D.H., Espley, J., Brain, D.A., Mitchell, D.L., Connerney, J.E.P., Acuña, M.H.: J. Geophys. Res. **110**, A09S21, doi:10.1029/2004JA10881 (2005).
15. Dessler, A.J. (ed.): Physics of the Jovian Magnetosphere, Cambridge Univ. Press, New York (1983).
16. Dobbin, A.L., Griffin, E.M., Aylward, A.D., Millward, G.H.: Ann. Geophys. **24**, 2403–2412 (2006).
17. Dryer, M., Perez-de-Tejada, H., Taylor, H.A., Jr., Intriligator, D.S., Mihalov, J.D., Rompolt, B.: J. Geophys. Res. **87**, 9035–9044 (1982).
18. Dungey, J.W.: In De Witt, C., Hieblot, J., Lebeau, A. (eds.), Geophysics: The Earth's Environment, Gordon Breach, New York (1963).
19. Dyer, C.: In Sawaya-Lacoste, H. (ed.), Proc. 2nd Solar Cycle Space Weather Euroconference, ESA, Noordwijk, pp.505–512 (2002).
20. Gold, T.: J. Geophys. Res. **64**, 1219–1224 (1959).
21. Gombosi, T.A., Ingersoll, A.P.: Science **327**, 1476–1479 (2010).
22. Haider, S.A., Abdu, M.A., Batista, I.S., Sobral, J.H., Kallio, E., Maguire, W.C., Verigin, M.I.: Geophys. Res. Lett. **36**, L13104, doi:10.1029/2009GL038694 (2009).
23. Hargreaves, J.K.: The Solar-Terrestrial Environment, Cambridge Atm. Space Sci. Ser. **5**, Cambridge (1992).
24. Ip, W.-H., Mendis, D.A.: Astrophys. J. **223**, 671–675 (1978).
25. Isbell, D., Murrill, M.B.: Press Release, NASA, "Galileo Finds Giant Iron Core in Jupiter's Moon Io", Webpage available via Views of the Solar System. http://www.solarviews.com/eng/galpr4.htm. Cited 3 May 1996.
26. Jian, L., Russell, C.T., Luhmann, J.G.: Solar Phys. **239**, 337–392 (2006).
27. Jockers, K.: Astron. Astrophys. Supp. **62**, 791–838 (1985).
28. Jones, J.B.L.: In Daglis, I.A. (ed.), Effects of Space Weather on Technology Infrastructure, Kluwer, The Netherlands, pp.215–234 (2004).
29. Jones, G.H., Brandt, J.C.: Geophys. Res. Lett. **31**, L20805, doi:10.1029/2004GL021166 (2004).
30. Kar, J., Mahajan, K.K., Srilakshmi, M., Kohli, R.: J. Geophys. Res. **91**, 8986–8992 (1986).

31. Khurana, K.K., Kivelson, M.G., Vasyliunas, V.M., Krupp, N., Woch, J., Lagg, A., Mauk, B., Kurth, W.S.: In Bagenal, F., Dowling, T.E., McKinnon, W.B. (eds.), Jupiter: The Planet, Satellites and Magnetosphere, Cambridge Univ. Press, Cambridge (2004).
32. Killen, R.M., Potter, A.E., Reiff, P., Sarantos, M,. Jackson, B.V., Hick, P., Giles, B.: J. Geophys. Res. **106**, 20509–20525 (2001).
33. Killen, R.M., Sarantos, M., Reiff, P.: Adv. Space Res. **33**, 1899–1904 (2004).
34. Kivelson, M.G., Russell, C.T. (eds.): Introduction to Space Physics, Cambridge Univ. Press, Cambridge (1995).
35. Kivelson, M.G., Warnecke, J., Bennett, L., Joy, S., Khurana, K.K., Linker, J.A., Russell, C.T., Walker, R.J., Polanskey, C.: J. Geophys. Res. **103**, 19963–19972 (1998).
36. Krimigis, S.M.: Planet. Rep. **12**, 10–13 (1992)
37. Krimigis, S.M., Armstrong, T.P., Axford, W.I., Bostrom, C.O., Cheng, A.F., Gloeckler, G., Hamilton, D.C., Keath, E.P., Lanzerotti, L.J., Mauk, B.H., Van Allen, J.A.: Science **246**, 1483–1489 (1989).
38. Kuchar, T.A., Buffington, A., Arge, C.N., Hick, P.P., Howard, T.A., Jackson, B.V., Johnston, J.C., Mizuno, D.R., Tappin, S.J., Webb, D.F.: J. Geophys. Res. **113**, A04101, doi:10.1029/2007JA012603 (2008).
39. Lanzerotti, L.J.: In Daglis, I.A. (ed.), Proc. NATO Adv. Study Inst. Space Storms Space Weather Hazards, p.313–341 (2001).
40. Lindsay, G.M., Russell, C.T., Luhmann, J.G., Gazis, P.: J. Geophys. Res. **99**, 11–17 (1994).
41. Luhmann, J.G., Fedorov, A., Barabash, S., Carlsson, E., Futaana, Y., Zhang, T.L., Russell, C.T., Lyon, J.G., Ledvina, S.A., Brain, D.A.: J. Geophys. Res. **113**, E00B04, doi:10.1029/2008JE003092 (2008).
42. Luhmann, J.G., Kasprzak, W.T., Russell, C.T.: J. Geophys. Res. **112**, E04S10, doi:10.1029/2006JE002820 (2007).
43. Luhmann, J.G., Russell, C.T.: In Shirley, J.H., Fainbridge, R.W. (eds.), Encyclopedia Planet. Sci., Chapman & Hall, New York, pp. 905–907 (1997).
44. Mauk, B.H., Hamilton, D.C., Hill, T.W., Hospodarsky, G.B., Johnson, R.E., Paranicas, C., Roussos, E., Russell, C.T., Shemansky, D.E., Sittler, E.C., Jr., Thorne, R.M.: In Dougherty, M.K., Esposito, L.W., Krimigis, S.M. (eds.), Saturn from Cassini-Huygens, Springer, Dordrecht, pp.281–331 (2009).
45. McKenna-Lawlor, S.M.P., Dryer, M., Fry, C.D., Sun, W., Lario, D., Deehr, C.S., Sanahuja, B., Afonin, V.A., Verigin, M.I., Kotova, G.A.: J. Geophys. Res. **110**, A03102, doi:10.1029/2004JA010587 (2005).
46. Mozer, F.S., Pritchett, P.L.: Phys. Today (June edn.), 34–39 (2010).
47. Mulligan, T., Russell, C.T., Luhmann, J.G.: Geophys. Res. Lett. **25**, 2959–2962 (1998).
48. Ness, N.F.: In King, J.W., Newmann, W.S. (eds.), Solar Terrestrial Physics Academic Press, New York (1967).
49. Ness, N.F., Acuña, M.H., Behannon, K.W., Burlaga, L.F., Connerney, J.E.P., Lepping, R.P.: Science **233**, 85–89 (1986).
50. Ness, N.F., Acuña, M.H., Burlaga, L.F., Connerney, J.E.P., Lepping, R.P., Neubauer, F.M.: Science **246**, 1473–1478 (1989).
51. Neidner, M.B. Jr., Brandt, J.C.: Astrophys. J. **223**, 655–670 (1978).
52. Parker, E.N.: J. Geophys. Res. **64**, 1675–1681 (1959).
53. Phillips, J.L., Luhmann, J.G., Russell, C.T.: Adv. Space Res. **5**, 173–176 (1985).
54. Prangé, R., Pallier, L., Hansen, K.C., Howard, R., Vourlidas, A., Courtin, R., Parkinson, C.: Nature **432**, 78–81 (2004).
55. Russell, C.T. (ed.): Space Sci. Rev. **126** (2006).
56. Russell, C.T., Luhmann, J.G.: In Shirley, J.H., Fainbridge, R.W. (eds.), Encyclopedia Planet. Sci., Chapman & Hall, New York, pp.718–719 (1997).
57. Sarson, G.R., Jones, C.A., Zhang, K., Schubert, G.: Science, **276**, 1106–1108 (1997).
58. Schmidt, C., Baumgardner, J., Mendillo, M., Davis, C., Musgrave, I., Proc. EPSC Meeting (2010).
59. Schneider, N.M., Bagenal, F.: In Lopes, R.M.C., Spencer, J.R. (eds.), Io after Galileo: A New View of Jupiter's Volcanic Moon, Springer-Praxis, Chichester, pp.265–286 (2007).

60. Song, P., Singer, H.J., Siscoe, G.L.: Space Weather, AGU, Washington, DC (2001).
61. Thompson, R., McDonald, A.: Communications and Space Weather, Webpage available via IPS (Australia) http://www.ips.gov.au/Educational/1/3/4. Cited 2007.
62. Voelzke, M.R.: Earth, Moon and Planets **90**, 405–411 (2002).
63. Wurm, K., Mammano, A.: Astron. Astrophys. Supp. **18**, 273–286 (1972).
64. Zhang, T.L., Pope, S., Balikhin, M., Russell, C.T., Jian, L.K., Volwerk, M., Delva, M., Baumjohann, W., Wang, C., Cao, J.B., Gedalin, M., Glassmeier, K.-H., Kudela, K.: J. Geophys. Res. **113**, E00B12, doi:10.1029/2008JE003128 (2008).

Chapter 11
Summary and Discussion

The last ten chapters have provided an introduction on the study of coronal mass ejections. We have reviewed general questions related to CMEs (Chap. 1), the history surrounding their discovery, scientific revelations and observation (Chap. 2), the spacecraft that have been used for their observation over the years (Chap. 3), the theory behind how we observe them in white light (Chap. 4) and the effects of their geometry relative to an observer (Chap. 5). We have also discussed their observation and study using astronomical radio sources (Chap. 6) and the phenomena that are known to be associated with CMEs and their likely relationship with them (Chap. 7). Popular models describing their onset and launch (Chap. 8), and evolution (Chap. 9) have also been reviewed. Finally, we have discussed their interaction with the Earth and other planets and their significance for geomagnetic storms and space weather (Chap. 10).

We have revealed that CMEs are an important mechanism for the evolution of the Sun, as they provide a means by which it may remove large quantities of plasma and built-up magnetic flux. We have discussed means by which this may be achieved without increasing the overall energy of the erupting field system and possible ways by which large amounts of magnetic energy may be provided to the accelerating CME. The associated phenomena that are associated with different stages of the CME's evolution have been identified and the means by which we may detect and analyse CME data have been discussed. We have also revealed the means by which the CME evolves through the heliosphere and their eventual fate at large distances from the Sun with the merged interaction regions in the outer heliosphere. Finally, we have discussed the importance of studying CMEs for operational purposes, as their arrival at the Earth and other bodies can cause damaging effects to spacecraft, aircraft and communications and power systems. It is the author's hope that the reader realises the scientific and operational value of studying this important phenomenon.

T. Howard, *Coronal Mass Ejections: An Introduction*, Astrophysics and Space Science Library 376, DOI 10.1007/978-1-4419-8789-1_11,

11.1 The Story of a Life: The Coronal Mass Ejection

To conclude this text, we will summarise the birth, growth, evolution and fate of the CME. It should be noted that the following paragraphs are the author's perspective only derived from his own research and understanding of CMEs. Many workers may disagree with some or all of this perspective.

A CME begins life as a closed magnetic structure, arising from beneath the photosphere as all coronal magnetic structures do, but held in place by an overlying existing closed strapping coronal magnetic field. This strapping field is often much larger than that over an active region, and it is likely that many CMEs have only one of their footpoints at the active region. The author speculates that the absence of an active region at the other footpoint may be due to a greater divergence of field lines at the other end of the CME (see Fig. 7.21).

The pre-launch CME magnetic structure is part of the continually emerging magnetic field which builds up beneath the coronal field over time periods that can last several solar rotations. The plasma that will eventually become part of the CME probably lies at heights of around 1.5–2.0 solar radii. The magnetic field accumulates flux, plasma and helicity (complexity) until a critical stage is reached, the overlying coronal field can no longer confine it, and it explosively erupts away from the Sun. As it erupts, the overlying strapping field is pushed aside to make way for the underlying erupting structure. It is likely that some interaction between the erupting CME and the surrounding field occurs, possibly by magnetic reconnection somewhere (e.g. between the CME and the strapping field), which transfers large amounts of energy allowing a runaway acceleration in these early stages. The eruption of the CME causes a number of associated phenomena including solar flares, lower filament eruptions and EUV and Moreton waves. Filaments, for example, are part of the same magnetic structure being constrained by the overlying coronal field, but lower down in the solar atmosphere. They erupt following the CME.

Once the CME has erupted, a vacancy of plasma and magnetic field remains in the solar corona and a re-orientation of the remaining field and plasma occurs. Some magnetic field and plasma arises from lower in the corona (probably indicated by coronal dimming and post-eruptive arcades) while other parts of the remaining field may sympathetically erupt or reconnect with different fields. Importantly, the overall energy of the remaining system is lower than it was prior to the launch of the CME. The internal magnetic field of the CME continues to provide a means to accelerate it (Lorentz force?) well into the interplanetary medium. Its magnetic structure is initially anchored to the Sun but eventually this structure will break off, producing a separate structure – a bubble or cloud for want of better terminology.

The CME is now a dense magnetic structure travelling through the surrounding solar wind, and may be regarded as an ICME. At these distances the plasma β of the surrounding medium is high, and so the kinematic and structural evolution of the ICME becomes less heavily influenced by the magnetic properties and more influenced by the hydrodynamic properties. The plasma β of the internal field of the CME, however, remains low, so its dynamics remain to some extent magnetically driven. Eventually, at distances around 0.5 AU, the effects of the internal magnetic field will diminish altogether and the ICME cruises to the solar wind speed. That is to say, if by the end of this acceleration process the ICME is travelling faster than the surrounding solar wind then it will decelerate until it reaches the solar wind speed. Likewise if it is slower it will speed up. It will retain its internal magnetic structure until it is much further from the Sun, and if it travels faster than the solar wind sound speed then a forward shock will form ahead of its structure and possibly a reverse shock will form in its wake. The formation of the forward shock enables the build-up of magnetic flux and solar wind material ahead of the ICME, a process known as snow ploughing. The structure of the ICME may be regarded as comprising of three separate parts: shocked material ahead of the ICME itself, followed by a magnetic sheath of turbulent plasma, followed by the magnetic structure of the ICME. It is not uncommon for the magnetic field of the ICME to assume the form of a highly-structured spiral magnetic field called a magnetic cloud. This is probably a result of the initial launch mechanism of the CME at the Sun, perhaps from twisting fields at the source.

The ICME structure continues to expand through the inner heliosphere. When it encounters other bodies in the solar system it interacts with them in different ways. Its shock pulse may disrupt the tails of comets or they may accelerate particles to highly energetic levels, which in turn produce radio bursts. When they encounter a planetary magnetosphere magnetic reconnection between the ICME and the planet's magnetic field may occur. When this happens the ICME may inject some of its particles and energy into the magnetosphere, disrupting its magnetic field and causing a magnetic storm. The pressure pulse from the arriving shock will also compress the planet's magnetosphere. Importantly, the overall structure and behaviour of the ICME remains virtually unchanged as a result of these interactions. Compare, for example, the size of the cross-section of the Earth's magnetosphere with that of an ICME at 1 AU.

The ICME does not propagate in a vacuum and the surrounding solar wind has an event history of earlier ICMEs or corotating interaction regions (CIRs). Also, other CMEs may erupt after the original one, which will typically travel faster because they are travelling through a less dense medium that has been vacated by the earlier CME. Sooner or later, the ICME will either catch up with, or be caught up with, another ICME or corotating interaction region. The plasma between the two structures will become compressed, and they along with the original plasma within the original structures will combine to form a merged interaction region (MIR). The original ICME is still there, but is embedded within the complex MIR structure. The MIR will interact with other MIRs, forming newer and more complex ones, and the initial ICME will look less and less like its original structure. With enough MIR mergers, the ICME structure becomes lost in the complex surrounding interacting structures.

Eventually, the combination of MIRs will form part of a collective MIR sheet, which is being continually replenished with new arriving MIRs in its solar direction, but in the antisolar direction its regions disperse into the outer heliosphere to form part of the background turbulence there. The eventual fate of the initial CME, such as it is at this stage, will be to work its way through the MIR sheath to become part of this background turbulence, and may make it to the termination shock.

11.2 Concluding Remarks

One of the main themes of this book is to emphasise the importance of CMEs in major space weather production at the Earth. Major geomagnetic storms are not produced by solar flares, as is commonly believed, but by CMEs. To emphasise this point further, consider the following analogy:

> Imagine that your laptop is being powered from a power station in your region. Also imagine that the same station provides power to a large factory in your neighbourhood. You want to find out the power output from the station, but cannot measure this directly. You have software that can provide you with detailed information about the power your laptop is using and you can perform some experiments to gain a rough idea of how much power the factory uses. Armed with this information, how much information will the power used by your laptop give you about the output of the station? How about the factory? I put it to the reader that using solar flares to study these energy outputs from the Sun is about as useful as using the power in your laptop to tell you about the power station. Understanding more about the factory usage would be more helpful, as it uses much more power, and therefore a larger proportion of the total output of the station. Hence, it is more beneficial to learn more about the CME (factory) than about the flare (laptop).

As mentioned in Chap. 1, it is my intention that this text be used as an introduction and general point of reference for CMEs, although each of the topics covered has been discussed in great depth in numerous publications specialising in each area. I do not intend this book to be a replacement for these texts, but to serve as a introductory summary and launch pad into these many areas. The reader is encouraged to read the texts recommended throughout this book, and to perform their own research to learn more about CMEs, ICMEs and their related phenomena.

Glossary

Active region A region on the solar surface where the local magnetic field is concentrated. Sunspots and solar flares often occur within active regions.

Alfvén speed The speed of motion of an Alfvén wave, given by $V_A = B_0/\sqrt{\rho\mu}$.

Alfvén waves Magnetic field-guided travelling magnetohydrodynamic wave within a plasma. They are generally transverse in nature.

Alpha (α) Force free parameter for magnetic fields. Also known as the helicity parameter or twist parameter.

Aly-Sturrock Limit Energy limit showing that the completely open coronal magnetic field state is larger than the closed state. This means that it is not energetically favourable for CMEs to simply erupt to an open field state.

Anisotropy The property of being directionally dependent.

Aphelion The furthest point to the focus of an elliptical orbit (point of furthest distance to a body an object is orbiting).

Archimedean spiral The locus of points corresponding to the locations over time of a point moving away from a fixed source with a constant angular speed. Hence every element in an Archimedean spiral is moving radially outwards from the source with a constant speed. Also known as the arithmetic spiral.

Bastille Day Event A famous and heavily studied CME/geomagnetic storm that erupted from the Sun on 14 July 2000.

Bessel Functions Solutions $y(x)$ to Bessel's differential equation: $x^2(\mathrm{d}^2y/\mathrm{d}x^2) + x(\mathrm{d}y/\mathrm{d}x) + (x^2 - \alpha^2)y = 0$. The solutions consist of two orders, for α and $-\alpha$. Also known as cylinder functions.

(Plasma) Beta (β) The ratio of the plasma pressure to the magnetic pressure. So when $\beta < 1$ the magnetic pressure dominates the plasma, and when $\beta > 1$ the plasma pressure dominates.

Broadband Large frequency range.

T. Howard, *Coronal Mass Ejections: An Introduction,* Astrophysics and Space Science Library 376, DOI 10.1007/978-1-4419-8789-1,

Butterfly Diagram Diagram showing the latitudinal location of sunspots throughout the solar cycle. During solar minimum sunspots occur at around 30° of solar latitude and migrate towards the equator as the cycle moves towards maximum. The result, when drawn across the northern and southern hemispheres of the Sun has the appearance of a butterfly.

"Classic" three-part CME CMEs observed in coronagraphs are often described by a three-part structure. That is, a leading edge feature, followed by a low density cavity, followed by a bright filament.

Coherence length Distance over which a wave maintains its coherency.

Coherent scattering Scattering for which emission occurs at the same frequency as the incident radiation. This occurs when the coherence length of the radiation is large compared with the separation of the scattering particles.

Comet Disconnection Event Where a comet's tail is seen to break off from the coma (head of the comet) and travel independently of the comet.

Complex conjugate If a complex number z is defined by $a+bi$ (where $i=\sqrt{-1}$) then the complex conjugate z^* is $a-bi$

Compton scattering Scattering for which the energy of the incident photons are large compared with the rest mass energy of the scattering particles. The result is a decrease in energy of the incident light. Commonly occurring in x-rays and gamma rays.

Cone model A simple model that describes the CME as a spherical shell centred at the Sun.

Convection zone The region of the Sun where the plasma properties are such that energy is more efficiently transmitted through the medium via convection rather than radiation. This begins at around $0.6\,R_\odot$ from the centre of the Sun and continues until near the solar surface, at $1.0\,R_\odot$.

Corona The outer atmosphere of the Sun that evolves into the solar wind. Sunward of the corona lies the chromosphere.

Coronagraph Device used to observe the solar corona. This is achieved by blocking out the bright photosphere of the Sun using an artificial disk called an occulting disk.

Coronal dimming A decrease in intensity in the solar corona often associated with the eruption of a CME. Coronal dimming is typically observed in EUV and x-ray but has also been detected in Hα.

Coronal hole Dark region in the solar corona corresponding to open magnetic field lines. Coronal holes are believed to be responsible for fast flowing solar wind streams.

Coronal mass ejection A large eruption of plasma and magnetic field from the Sun.

Coronal transient The original term for a coronal mass ejection.

Corotating interaction region A region of enhanced density in the solar wind brought about by an interaction between fast and slow solar wind regions, which corotate with the Sun. They are identified by similar white light and in-situ signatures that are used to identify an ICME, and so there is sometimes confusion between the two.

Cosmic rays Energetic particles typically with their origin outside the heliosphere.

Current sheet Electric current that is confined to a surface rather than a volume in space. They typically occur where there is no magnetic field but there are fields surrounding the sheet (i.e. a neutral line).

Cusp region The region in the magnetosphere where the boundary between the dayside closed magnetic field lines and the nightside open magnetic field lines occur. Through the cusp energetic particles from the solar wind are able to directly penetrate the magnetosphere to the ionosphere.

Dayside The hemisphere of the Earth facing the Sun.

Earth radius The radius of the Earth, $\sim$6,360 km.

Eclipse (solar) Where the moon passes between the Sun and the Earth, temporarily blocking out its light. This occurs because the apparent size of the moon from the Earth is the same as the apparent size of the Sun.

Eclipse totality Brief period during a solar eclipse where the moon entirely blocks the photospheric light from the Sun, revealing the faint surrounding corona.

Ecliptic (plane) The plane in the heliosphere containing the orbit of the Earth. Most of the solar planets orbit in planes close to (i.e. within 20° of) the ecliptic.

EIT wave An alternative term for EUV wave.

EUV wave Wave travelling across the solar corona observed with EUV instruments. They are commonly associated with CMEs and CME/flare onset.

Electron cyclotron frequency The frequency with which an electron spirals along a magnetic field.

Elongation The angle between the Sun-observer line and the line from the observer to the point of interest. So, 0° elongation is the direction of the Sun, 90° elongation is the same plane as the observer, and 180° elongation is directly behind the observer relative to the Sun. Elongation can be regarded as an angular measurement of radial distance from the Sun, as images are projections and as such do not contain any depth information.

Entropy A measurement of the amount of disorder in a system.

Erupting magnetic structures An ICME observed by in-situ spacecraft or heliospheric imagers that does not have a coronagraph CME counterpart. It is speculated that this is because the CME does not contain sufficient plasma to be detected by the coronagraph, but its magnetic structure still erupts.

Faraday rotation The rotation of the direction of a plane polarised electromagnetic wave, brought about by its passage through a magnetised ionised medium.

Filament A region on the solar photosphere that appears as a dark line that varies in size and geometry. It is generally believed to be a concentration of relatively cool plasma suspended above the photosphere by magnetic fields. A disappearing filament often occurs when the magnetic structure erupts, carrying the cool plasma with it. Disappearing filaments are often associated with the eruption of a CME. When a filament is observed on the solar limb it is called a prominence.

Flux rope A column containing high levels of magnetic flux. Also known as a flux tube.

Force free A magnetic field in an environment where the plasma pressure is very small compared to the magnetic pressure, thereby allowing the plasma pressure to be ignored.

Geomagnetic field The Earth's magnetic field.

(Geo)magnetic storm A large disturbance in the Earth's magnetosphere and ionosphere. The major storms are typically caused by the arrival of a fast CME with a predominantly southward-directed magnetic field.

Gravitational binding energy Energy binding two massive objects by gravity. One body may escape a more massive body by overcoming the gravitational binding energy.

Halloween Event A famous and heavily studied CME/geomagnetic storm that erupted from the Sun on 28 October 2003.

Halo CME A CME with a large component along the Sun–Earth line, and hence appears on projection in a coronagraph to completely encircle the Sun.

Height-time plot Plot of distance from solar centre against time, typically used to determine CME speeds. Measurements of the height of a CME are usually obtained from the leading edge of the structure as observed by a coronagraph.

Helicity A measurement of the amount of twist in a structure.

Heliocentric Sun-centred.

Heliosphere The region within which the solar wind is contained, roughly a sphere from the Sun out to around 100 AU. For the purposes of this book the inner heliosphere is the region out to a few AU.

Heliospheric current sheet The surface where the polarity of the Sun's magnetic field changes. Generally described as a layer of dense plasma within which a strong current flows.

Helium abundance enhancement (enrichment) An enhancement of helium observed following interplanetary shocks. These were early indicators of ICMEs.

Hermian Related to the planet Mercury.

Herniation As regarding the Sun: where plasma or magnetic flux emerges from beneath the solar photosphere.

Hinode Japanese spacecraft launched in 2006 designed to monitor solar activity with a suite of imagers.

Imager An instrument that captures images (camera).

In-situ Measurements made by instruments in direct contact with a phenomenom, in this case, when an ICME passes through a spacecraft.

Interplanetary coronal mass ejection The interplanetary counterpart of a coronal mass ejection. Typically observed with heliospheric images and in-situ spacecraft and often preceded by an interplanetary shock.

Interplanetary magnetic field The magnetic field that is carried away from the Sun by the solar wind. Every object in the solar system is embedded in both the solar wind and the interplanetary magnetic field. Because the solar wind rotates with the Sun, the interplanetary magnetic field rotates also, resembling an Archimedean spiral (the Parker spiral).

Interplanetary medium General term describing the medium containing the solar wind and interplanetary magnetic field. It is the medium within which the entire solar system exists.

Interplanetary scintillation The distortion of the signal from a distant radio source as a result of a dense structure passing between it and the observer. This is a technique that has been used for ICME detection and tracking.

Interplanetary shock A shock in the interplanetary medium typically brought about by the passage of a supersonic relatively dense structure such as a CME. Forward interplanetary shocks have an in-situ signature of a sudden increase in magnetic field, density and solar wind speed, while reverse shocks have a similar signature, except there is a sudden decrease in magnetic field.

Interplanetary transient The general term for a disturbance in the interplanetary medium. May be used as a more general description for an ICME but does not exclusively describe them.

Ionosphere A relatively thin (relative to the magnetosphere) conducting layer of the Earth's atmosphere, immediately below the magnetosphere.

Isentropic A process that takes place without a change to the entropy of the system.

Isotropic Identical properties in all directions.

Jovian Related to the planet Jupiter.

Kronecker delta Defined as δ_{ij} which equals 1 when $i = j$ and 0 for all other values of i and j.

The L1 Lagrange point The point between the Earth and the Sun where the gravitational effects of the Earth are exactly canceled by the Sun. This is located around 1.5×10^6 km from the Earth, or about 1% of the distance between the Earth and the Sun.

Langmuir waves Oscillations of plasma caused by faster plasma particles catching up with slower ones causing a compression region.

Limb darkening A relative reduction of intensity on the surface of the Sun moving towards the limb, resulting from the curvature of the Sun and the nature of its radiation.

Line of sight The vector from the observer through the point of interest and out to infinity.

Long duration event X-ray enhancements lasting long periods of time (several minutes to several hours).

Magnetic cloud A magnetic flux rope typically associated with ICMEs. It is often preceded by a shock and contains a highly twisted magnetic field structure. The in-situ signature of a magnetic cloud includes: (1) low temperatures, (2) high magnetic pressures, (3) a smoothly rotating magnetic field vector. Magnetic clouds are also of long duration, lasting around 30 h on average.

Magnetic reconnection Where magnetic field lines from different regions connect with each other, such as where the interplanetary magnetic field interacts with the geomagnetic field.

Magnetic shear Region where the magnetic field runs almost parallel to its neutral line, so it is far from potential.

Magnetosphere Region of plasma enclosed by the Earth's magnetic field. It extends to around 10–15 R_E on the sunward side (dayside) and several hundred R_E on the anti-sunward side (nightside).

Magnetotail The nightside of the magnetosphere where the field lines are extended to large antisunward distances by the solar wind and IMF.

Mollweide projection A sky map equal-area projection where the latitude lines are parallel to the equator.

Moreton wave An impulse propagating across the solar photosphere/chromosphere, often associated with CMEs and flares.

Narrowband Small frequency range.

Neutral line Region where the magnetic field is neutral, i.e. its consists of equal quantities of positive and negative flux. These may be regarded as magnetic "source" regions.

Nightside The hemisphere of the Earth facing away from the Sun.

Observer A general term describing an instrument or person looking at or measuring something.

Occulter A disk in a coronagraph that blocks out the light from the Sun's photosphere, to reveal the faint surrounding corona. Also termed the occulting disk.

Orthogonal Vectors that are mutually perpendicular to each other, e.g. in three dimensions an orthogonal set is defined such that each axis is aligned along each dimension (length, breadth, depth).

Orthogonal set Collection of three vectors that are mutually orthogonal to each other, in a right-handed sense.

Parker spiral General structure of the solar wind and interplanetary magnetic field. Field lines and plasma parcels move radially outward from the Sun but undergo corotation. This results in an Archimedean spiral structure.

Partial halo CME A CME with a component along the Sun–Earth line, and hence appears on projection to partially encircle the Sun. A partial halo may be regarded as a CME with an apparent angular width of more than 120°.

Perihelion The closest point to the focus of an elliptical orbit (point of closest approach to a body an object is orbiting).

Permeability The degree of magnetisation of a material in response to a magnetic field. Also termed the magnetic constant.

Permittivity A measurement of how much resistance is encountered when forming an electric field in a material.

Photosphere The top layer of the convection zone of the Sun, where energy is transmitted to the surface via convection rather than radiation. This is the so-called "surface" of the Sun, and the brightest region observed in visible light.

Pitch angle The angle between a particle's velocity vector and the local magnetic field. Low pitch angle indicates that a particle that is field-aligned.

Plasmoid A closed volume of plasma, typically confined by a magnetic field.

Point P approximation A simple approximation for converting elongation to distance, when measuring CMEs. Point P assumes that the part of the CME being observed is spherical and centred at the Sun. This reduces the conversion to simple trigonometry: $p \sim \sin \varepsilon$, where p is the distance from the Sun in AU, and ε is the elongation.

Point spread function The response of an imaging instrument to a point source, or a function describing how a point source is spread across an image.

Polar caps Area mapping to the geomagnetic field lines that are open i.e. are connected to the IMF. Through these field lines energetic particles from the interplanetary medium may enter to lower altitudes of the Earth's atmosphere.

Position angle Angle of a point projected into the sky plane, measured from projected north.

Post-eruptive arcade A region of hot plasma and highly-structured magnetic field from the low corona following the eruption of a CME.

Post-flare loop An alternative term for a post-eruptive arcade.

Projection effects The effects of obtaining a two-dimensional image of a three-dimensional object. The image will represent a projection of the three-dimensional image into the plane of the sky relative to the observer.

Prominence A loop observed on the solar limb suspended above the solar surface by magnetic fields. Erupting prominences are often associated with the eruption of a CME. When a prominence is on the solar disk it is called a filament.

Ring current A current in the equatorial region of the Earth's magnetosphere brought about by the movement of trapped particles. It lies at a distance of 3–5 R_E from the Earth and circulates clockwise around the Earth when viewed from the north.

Running difference A sequence of images where each image has been subtracted away from the previous one. That is, in a running difference image $B_j = A_j - A_{j-1}$.

Sigmoid An S-shaped structure observed in x-rays in the solar corona, often associated with CMEs.

Sky plane The plane of the sky relative to the observer. Images of the Sun and heliosphere are projected into the sky plane.

Snow plough Where an ICME accumulates solar wind material ahead of it which cannot get out of the way. It is possible that much of the material observed by heliospheric imagers may be snow-ploughed material.

Solar cycle The magnetic cycle of the Sun which lasts around 11 years. Throughout the cycle the magnetic complexity in the Sun increases, resulting in a large number of sunspots, solar flares and CMEs (solar maximum), and then the complexity decreases along with activity (solar minimum). At the start of the new cycle the magnetic poles of the Sun are reversed, meaning it takes two cycles to return to the original magnetic orientation.

Solar energetic particles High-energy particles originating from the Sun and observed in the heliosphere. Many are generally believed to be accelerated by solar flares and CMEs.

Solar flare A sudden increase in emission from a localised region of the Sun. Solar flares are generally broadband in nature, and can span the electromagnetic spectrum from below visible light to above x-rays. They are often associated with CMEs, and are known to have effects on the Earth's magnetosphere and ionosphere.

Solar limb The edge of the solar disk.

Solar maximum The maximum phase of the solar cycle: The period in the middle of the solar cycle where activity (e.g. sunspots, flares, CMEs) is maximised.

Solar minimum The minimum phase of the solar cycle: The period at the start and end of the solar cycle where activity (e.g. sunspots, flares, CMEs) is minimised.

Solar Radius The radius of the Sun, $\sim$695,500 km.

SolarSoft A collection of software libraries, databases and system utilities used for data processing and analysis for many instruments in solar physics.

Solar surface The photosphere: The region on the solar disk where the Sun becomes opaque at the top of the convection region. This region is popularly observed with the Hα line.

Solar wind A body of plasma continuously-flowing away from the Sun. It may be regarded as an extension of the corona, and extends to around a 100 AU away from the Sun. It carries a magnetic field with it, known as the interplanetary magnetic field.

Solid angle The angle in three dimensional space subtended by an object at a given point. It is a representation of the apparent size of an object to an observer at that point.

South Atlantic Anomaly A region in the south Atlantic ocean (just off the coast of Brazil), where the van Allen radiation belt makes its closest approach to the Earth's surface. Here, the radiation and energetic particle intensity is greatest compared with surrounding regions.

Space weather A general term embracing many effects, including geomagnetic and magnetospheric activity on the Earth. Large geomagnetic storms are types of severe space weather.

Streamer A bright column of material observed in the solar corona, believed to lie above active regions.

Streamer belt High density, low temperature region around the heliospheric current sheet at solar minimum.

Substorm An enhancement of geomagnetic activity occurring as the result of released particles and energy stored in the magnetotail.

Sudden commencement A sudden increase in geomagnetic activity, usually triggered by the arrival of an interplanetary shock.

Sudden ionospheric disturbance Sudden increase in ionospheric density in the D region of the ionosphere, indicative of an increase in geomagnetic activity.

Sunspot A dark region on the Sun indicative of solar activity there. Sunspots arise from emerging magnetic fields from below the photosphere.

Supercritical twist A highly-twisted flux rope state which may enable the launch of a CME.

Termination shock The region of the heliosphere where the solar wind slows down to sub-sonic speed. This occurs at distances around 80 AU.

Thomson scattering The scattering of electromagnetic radiation from a charged particle, brought about by the acceleration of the particle by incident radiation.

Thomson surface The resulting sphere from obtaining the locus of all points where any line of sight is perpendicular to the solar radial vector. For an observer at the Earth, the Thomson surface is a sphere with a diameter of the Sun–Earth line and the surface crossing both the Earth and Sun.

Triangulation The technique by which the three dimensional location of a point can be determined when observed from multiple locations. The technique involves the application of geometry.

Type I burst Short-duration, narrowband radio bursts occurring during storm periods.

Type II burst Long-duration, varying frequency radio bursts driven by CME shock acceleration.

Type III burst Short-duration, broadband, varying-frequency radio bursts driven by solar flares.

Type IV burst Long-duration radio bursts that often follow Type II bursts.

Type V burst Radio bursts that often follow Type III bursts.

Two ribbon flares A class of solar flare with ribbons of emission spreading apart from each other.

van Allen belts A region of energetic charged particles around the Earth, trapped by the geomagnetic field.

Virtual mass Additional mass of a CME that is removed by image processing such as running difference and background subtraction. This may effectively be regarded the volume of solar wind on which the CME "lies".

Index

MIX
Papier aus verantwortungsvollen Quellen
Paper from responsible sources
FSC® C105338

If you have any concerns about our products,
you can contact us on
ProductSafety@springernature.com

In case Publisher is established outside the EU,
the EU authorized representative is:
Springer Nature Customer Service Center GmbH
Europaplatz 3, 69115 Heidelberg, Germany

Printed by Libri Plureos GmbH
in Hamburg, Germany